MONITORING
STEM CELL RESEARCH

MONITORING
STEM CELL RESEARCH

A Report of
The President's Council on Bioethics

Washington, D.C.
January 2004

www.bioethics.gov

CONTENTS

Letter of Transmittal to
The President of the United States

The President's Council on Bioethics
1801 Pennsylvania Avenue N.W., Suite 700
Washington, D.C. 20006
January 14, 2004

The President
The White House
Washington, D.C.

Dear Mr. President:

I am pleased to present to you *Monitoring Stem Cell Research*, a report of the President's Council on Bioethics. Over the past two years, in keeping with your stated intention, the Council has been monitoring developments in stem cell research, as it proceeds under the implementation of the administration's policy. We have consulted widely, heard presentations, and commissioned review essays (included as appendices in this volume) on all aspects of the topic—scientific, ethical, and legal. Our desire has been both to understand what is going on in the laboratory and to consider for ourselves the various arguments made in the ongoing debates about the ethics of stem cell research and the wisdom of the current policy. Although both the policy and the research are still in their infancy, the Council is now ready to give you and the American people an update on this important area of research.

Because this field and the current policy are so young, this report can be no more than an "update." It summarizes some of the more interesting and significant developments since August 2001, both in the basic science and medical applications of stem cell research and in the related ethical, legal, and policy discussions. It does not attempt to be a definitive or comprehensive study of the whole topic. It contains no proposed guidelines and regulations, nor indeed any specific recommendations for public policy. Rather, it seeks to shed light on where we are now—ethically, legally, scientifically, and medically—in order that you, the Congress, and the nation may be better informed as we all consider where we should go in the future.

The report has four basic aims, three of them the subjects of independent chapters devoted to their themes.

First, we have sought to clarify and explain the current federal policy regarding stem cell research and to make clear the legal, ethical, and prudential foundations on which the policy rests: the desire to promote important biomedical research without endorsing, funding, or creating incentives for the future destruction of human embryos. We have also sought to describe how that policy is being implemented, especially by the National Institutes of Health. Many of these matters have not been well understood or accurately represented in public discussions since August 2001, and we hope that the clarifications introduced in this report will enable future discussions and debates to be better informed.

Second, we have tried to provide an overview of the ethical and policy debates surrounding stem cell research in the past two years. As you already know quite well, these are immensely difficult and challenging matters, with the obligations owed to nascent human life pitted against the obligations to seek knowledge that might someday alleviate much human suffering. Not surprisingly, arguments continue on all aspects of the moral and political debate. We have sought to present the arguments and counter-arguments, faithfully and accurately, so that all may learn what is at stake and where the debate now stands.

Third, we have monitored recent scientific developments in human stem cell research, embryonic and adult, basic and applied. Our goal in the report is to enable (especially non-scientific) readers to appreciate the reasons for the excitement over stem cell research, the complexities of working with stem cells, some early intriguing research and therapeutic findings, and the difficult road that must be traveled before we can reap therapeutic and other benefits from this potentially highly fertile field of research.

The other three specific goals have been informed by a fourth and overarching goal: to convey the moral and social importance of the issue at hand and to demonstrate how people of different backgrounds, ethical beliefs, and policy preferences can reason together about it. We want everyone to understand that biomedical research, being a human activity, must always be regarded as a moral endeavor, to be governed not only by the goals of gaining knowledge and relieving suffering, but also by the obligation to safeguard the inherent freedom and dignity of human life. Throughout the Council's deliberations and in this monitoring report, Council members have tried to acknowledge the strengths and importance of opinions and concerns held by people with whom they disagree. We have aspired to be careful and fair in our approach, precise in our use of language, accurate in presenting data and arguments, and thoughtful in our laying out of the various issues that remain before us. Above all, we want all parties to these debates to understand that their opponents, too, have something vital to defend, not only for themselves but for all of us.

The policy debates over stem cell research that led you to create this Council continue; they, and other debates on related topics, are unlikely to go away any time soon. Our hope is that our work will help to make those debates richer, fairer, and better informed.

Mr. President, allow me to join my Council colleagues and our fine staff in thanking you for this opportunity to offer you and the American people what we hope is a useful and constructive review of where things stand, both in the laboratory and in the public arena, with regard to this promising and ethically challenging area of research.

Sincerely,

Leon R. Kass, M.D.
Chairman

MEMBERS OF
THE PRESIDENT'S COUNCIL ON BIOETHICS

LEON R. KASS, M.D., PH.D., *Chairman.*
> Addie Clark Harding Professor, The College and the Committee on Social Thought, University of Chicago. Hertog Fellow, American Enterprise Institute.

ELIZABETH H. BLACKBURN, PH.D.
> Professor, Department of Biochemistry and Biophysics, University of California, San Francisco.

REBECCA S. DRESSER, J.D., M.S.
> Daniel Noyes Kirby Professor of Law and Professor of Ethics in Medicine, Washington University, St. Louis.

DANIEL W. FOSTER, M.D.
> Donald W. Seldin Distinguished Chair in Internal Medicine, Chairman of the Department of Internal Medicine, University of Texas Southwestern Medical School.

FRANCIS FUKUYAMA, PH.D.
> Bernard Schwartz Professor of International Political Economy, Paul H. Nitze School of Advanced International Studies, Johns Hopkins University.

MICHAEL S. GAZZANIGA, PH.D.
> Dean of the Faculty, David T. McLaughlin Distinguished Professor, Professor of Psychological and Brain Sciences, Dartmouth College.

ROBERT P. GEORGE, J.D., D.PHIL.
> McCormick Professor of Jurisprudence, Director of the James Madison Program in American Ideals and Institutions, Princeton University.

MARY ANN GLENDON, J.D., M. COMP. L.
> Learned Hand Professor of Law, Harvard University.

Council Staff and Consultants

Preface

Monitoring Stem Cell Research is a report of the President's Council on Bioethics, which was created by President George W. Bush on November 28, 2001, by means of Executive Order 13237.

The Council's purpose is to advise the President on bioethical issues related to advances in biomedical science and technology. In connection with its advisory role, the mission of the Council includes the following functions:

- To undertake fundamental inquiry into the human and moral significance of developments in biomedical and behavioral science and technology.

- To explore specific ethical and policy questions related to these developments.

- To provide a forum for a national discussion of bioethical issues.

- To facilitate a greater understanding of bioethical issues.

The President left the Council free to establish its own priorities among the many issues encompassed within its charter, and to determine its own modes of proceeding.

Stem cell research has been of interest to, and associated in the public mind with, this Council since its creation. Taking up the charge given to us by President Bush in his August 9, 2001, speech on stem cell research, the Council has from its beginnings been monitoring developments in this fast-paced and exciting field of research. The first formal discussions of the science and ethics of stem cell research took place at our third meeting, in April of 2002, where presentations were made by two prominent stem cell researchers (John Gearhart of Johns Hopkins University and Catherine Verfaillie of the University of Minnesota) and an ethicist (Gene Outka of Yale University), at a time when we were still mostly engrossed in our discussions of human cloning. Over the course of the following year and a half, even as the Council was preparing the reports *Human Cloning and Human Dignity: An Ethical Inquiry* (July 2002) and *Beyond Therapy: Biotechnology and the Pursuit of Happiness* (October 2003), it was gathering information on stem cell research and devoting increasing portions of its meeting agendas to this topic, which was ultimately discussed at six meetings (specifically, in April and July of 2002; and in June, July, September, and October of 2003).

The Council heard presentations from numerous experts in the relevant scientific, ethical, social, advocacy, and entrepreneurial arenas, and received public comment, oral and written. The Members engaged in serious deliberation throughout the process. All told, fourteen sessions, of ninety minutes each, were devoted to the subject at public meetings. Complete transcripts of all these sessions are available to the public on the Council's website at www.bioethics.gov.

The present monitoring report draws directly upon those sessions and discussions, as well as on written material prepared by Council members, staff, and consultants. As noted in Chapter 1, it is in the spirit of an "update" and contains no recommendations for policy.

We hope the report, with its overview chapters on the law, ethics, and science of stem cell research, and its extensive supporting material located in the appendices, will serve as a source of clear, intelligible, and useful information for both policymakers and the general public regarding the current state of this important research and of the debates that surround it.

In creating this Council, President Bush expressed his desire to see us

consider all of the medical and ethical ramifications of biomedical innovation. . . . This council will keep us apprised of new developments and give our nation a forum to continue to discuss and evaluate these important issues. As we go forward, I hope we will always be guided by both intellect and heart, by both our capabilities and our conscience.

It has been our goal in the present report, as in all of our work, to live up to these high hopes and noble aspirations.

LEON R. KASS, M.D.
Chairman

1

Introduction

This monitoring report has its origins in President George W. Bush's remarks to the nation on August 9, 2001. It was his first major national policy address, and the topic was unusual: federal funding of research on human stem cells.* In the speech, the President announced that after several months of deliberation he had decided to make federal funding available, for the first time, for research involving certain lines of embryo-derived stem cells. At the end of the speech the President declared his intention to

> name a President's Council to monitor stem cell research, to recommend appropriate guidelines and regulations, and to consider all of the medical and ethical ramifications of biomedical innovation. . . . This council will keep us apprised of new developments and give our nation a forum to continue to discuss and evaluate these important issues.[1]

In keeping with the President's intention, the Council has been monitoring developments in stem cell research, as it proceeds under the implementation of the administration's policy. Our desire has been both to understand what is going on in the laboratory and to consider for ourselves the various arguments made in the ongoing debates about the ethics of

---

* Throughout this report, excluding appendices, all references to embryos, cells, or other biological materials are assumed to be of human origin unless otherwise stated.

1

stem cell research and the wisdom of the current policy. Although both the policy and the research are still in their infancy, the Council is now ready to give the President and the public an update on this important and dynamic area of research.

This report is very much an "update." It summarizes some of the more interesting and significant recent developments, both in the basic science and medical applications of stem cell research and in the related ethical, legal, and policy discussions. It does not attempt to be a definitive or comprehensive study of the whole topic. It contains no proposed guidelines and regulations, nor indeed any specific recommendations for policy change. Rather, it seeks to shed light on where we are now—ethically, legally, scientifically, and medically—in order that the President, the Congress, and the nation may be better informed as they consider where we should go in the future.

I. WHAT ARE STEM CELLS, AND WHY IS THERE CONTENTION ABOUT THEM?

The term "stem cells" refers to a diverse group of remarkable multipotent cells. Themselves relatively undifferentiated and unspecialized, they can and do give rise to the differentiated and specialized cells of the body (for example, liver cells, kidney cells, brain cells). All specialized cells arise originally from stem cells, and ultimately from a small number of embryonic cells that appear during the first few days of development.* As befits their being and functioning as progenitor cells, all stem cells share two characteristic properties: (1) the capacity for unlimited or prolonged *self-renewal* (that is, the capability to maintain a pool of similarly undifferentiated stem cells), and (2) the potential to produce *differentiated* descendant cell types. As stem cells within a developing human embryo differentiate in vivo, their capacity to diversify generally

---

* These cells are grouped together as the "inner cell mass" of the embryo, at the blastocyst stage of its development. Readers should consult the Glossary for definitions of technical terms and Appendix A for an illustrated guide to the embryonic developments referred to in this report.

becomes more limited and their ability to generate many differentiated cell types generally becomes more restricted.

Stem cells first arise during embryonic development and exist at all developmental stages and in many systems of the body throughout life. The best described to date are the blood-forming (hematopoietic) stem cells of the bone marrow, the progeny of which differentiate (throughout life) into the various types of red, white, and other cells of the blood. It appears that some stem cells travel through the circulatory system, from their tissue of origin, to take up residence in other locations within the body, from which they may be isolated. Other stem cells may be obtained at birth, from blood contained in the newborn's umbilical cord. Once isolated and cultured outside the body, stem cells are available for scientific investigation. Unlike more differentiated cells, stem cells can be propagated in vitro for many generations—perhaps an unlimited number—of cell-doublings.

Stem cells are of interest for two major reasons, the one scientific, the other medical. First, stem cells provide a wonderful tool for the study of cellular and developmental processes, both normal and abnormal. With them, scientists hope to be able to figure out the molecular mechanisms of differentiation through which cells become specialized and organized into tissues and organs. They hope to understand how these mechanisms work when they work well, and what goes wrong when they work badly. Second, stem cells and their derivatives may prove a valuable source of transplantable cells and tissues for repair and regeneration. If these healing powers could be harnessed, the medical benefits for humankind would be immense, perhaps ushering in an era of truly regenerative medicine. No wonder that scientists around the world are actively pursuing research with stem cells.

Why, then, is there public contention about stem cell research? Not because anyone questions the goals of such research, but primarily because there are, for many people, ethical issues connected to the means of obtaining some of the cells. The main source of contention arises because some especially useful stem cells can be derived from early-stage human embryos, which must be destroyed in the process of obtaining the cells. Arguments about the ethics of using human embryos in research are not new. They date back to the mid-1970s, beginning not long after in vitro fertilization (IVF) was first successfully accomplished with human egg and sperm in 1969. A decade later, after IVF had entered clinical practice for the treatment of infertility, arguments continued regarding the fate and possible uses of the so-called "spare embryos," embryos produced in excess of reproductive needs and subsequently frozen and stored in the assisted-reproduction clinics. Although research using these embryos has never been illegal in the United States (except in a few states), the federal government has never funded it, and since 1995 Congress has enacted annual legislation prohibiting the federal government from using taxpayer dollars to support any research in which human embryos are harmed or destroyed.

Although the arguments about embryo research had been going on for twenty-five years, they took on new urgency in 1998, when the current stem cell controversy began. It was precipitated by the separate publication, by two teams of American researchers, of methods for culturing cell lines derived, respectively, from: (1) cells taken from the inner cell mass of very early embryos, and (2) the gonadal ridges of aborted fetuses.[2,3] (In this report, we shall generally refer to the cell lines derived from these sources as, respectively, *embryonic stem cells* [or "ES cells"] and *embryonic germ cells* [or "EG cells"].) This work, conducted in university laboratories in collaboration with and with financial support from Geron Corporation, prompted great excitement and has already led to much interesting research, here and abroad. It has also sparked a moral and political debate about federal support for such research: Is it morally permissible to withhold support from research that holds such human promise? Is it morally permissible to pursue or publicly support (even beneficial) research that depends on the exploitation and destruction of nascent human life?

Persons interested in the debate should note at the outset that ES and EG cells are not themselves embryos; they are not whole organisms, nor can they be made (directly) to become whole organisms. Moreover, once a given line of ES or EG cells has been derived and grown in laboratory culture, no further embryos (or fetuses) need be used or destroyed in order to work with cells from that line. But it is not clear whether these lines can persist indefinitely, and only very few lines, representing only a few genetic backgrounds, have been made. Thus there is continuing scientific interest in developing new embryonic stem cell lines, and the existence of large numbers of stored cryopreserved embryos in assisted-reproduction clinics provides a potential source for such additional derivations. Complicating the debate has been the study of another group of stem cells, commonly called "adult stem cells," derived not from embryos but from the many different tissues in the bodies of adults or children—sources exempt from the moral debate about obtaining ES and EG cells. For this reason, we often hear arguments about the relative scientific merits and therapeutic potential of embryonic and adult stem cells, arguments in which the moral positions of the competing advocates might sometimes influence their assessments of the scientific facts. Further complicating the situation are the large commercial interests already invested in stem cell research and the competition this creates in research and development not only in the United States but throughout the world. The seemingly small decision about the funding of stem cell research may have very large implications.

II. BROADER ETHICAL ISSUES

While most of the public controversy has focused on the issue of embryo use and destruction, other ethical and policy issues have also attracted attention.[*] Although entangled with the issue of embryos, the question of the significance and use of federal funds is itself a contested issue: Should moral considerations be used to decide what sort of research may or may not be funded? What is the symbolic and moral-political significance of providing national approval, in the form of active support, for practices that many Americans regard as abhorrent or objectionable? Conversely, what is the symbolic and moral-political significance of refusing to support

---

[*] Introduced here, these issues and the discussions they have produced will be reviewed in Chapter 3.

potentially life-saving scientific investigations that many Americans regard as morally obligatory?

Even for those who favor embryo research, there are questions about its proper limits and the means of establishing and enforcing those limits through meaningful regulation. Under the present arrangement, with the federal government only recently in the picture, what is done with human embryos, especially in the private sector, is entirely unregulated (save in those states that have enacted special statutes dealing with embryo or stem cell research). Is this a desirable arrangement? Can some other system be devised, one that protects the human goods we care about but that does not do more harm than good? What are those human goods? What boundaries can and should we try to establish, and how?

Although well-established therapies based on transplantation of stem cell-derived tissues are still largely in the future, concern has already been expressed (as it has been about other aspects of health care in the United States) about access to any realized benefits and about research priorities: Will these benefits be equitably available, regardless of ability to pay? How should the emergence of the new field of stem cell research alter the allocation of our limited resources for biomedical research? How, in a morally and politically controverted area of research, should the balance be struck between public and private sources of support? As with any emerging discovery, how can we distinguish between genuine promise and "hype," and between the more urgent and the less urgent medical needs calling out for assistance?[4]

There are also sensitive issues regarding premature claims of cures for diseases that are not scientifically substantiated and the potential exploitation of sick people and their families. Some advocates of stem cell research have made bold claims about the number of people who will be helped should the research go forward, hoping to generate sympathy for increased research funding among legislators and the public. A few advocates have gone so far as to blame (in advance) opponents of embryonic stem cell research for those who will die unless the research goes forward today. At the same time, other scientists have cautioned that the pace of progress will be very slow, and that no cures can be guaranteed in advance.

Which of these claims and counterclaims is closer to the truth cannot be known ahead of time. Only once the proper scientific studies are conducted will we discover the potential therapeutic value of stem cells from any source. How, then, in the meantime should we discuss these matters, offering encouragement but without misleading or exploiting the fears and hopes of the desperately ill?

Finally, questions are raised by some about the social significance of accepting the use of nascent human life as a resource for scientific investigation and the search for cures. Such questions have been raised even by people who do not regard an early human embryo as fully "one of us," and who are concerned not so much about the fate of individual embryos as they are about the character and sensibilities of a society that comes to normalize such practices.[5] What would our society be like if it came to treat as acceptable or normal the exploitation of what hitherto were regarded as the seeds of the next generation? Conversely, exactly analogous questions are raised by some about the social significance of *refusing* to use these 150-to-200-cell early human embryos as a resource for responsible scientific investigation and the search for cures. What would a society be like if it refused, for moral scruples about (merely) nascent life, to encourage every thoughtful and scientifically sound effort to heal disease and relieve the suffering of fully developed human beings among us?[6]

It is against the background of such moral-political discussion and argument that the Council has taken up its work of monitoring recent developments in stem cell research. We are duly impressed with the difficulty of the subject and the high stakes involved. All the more reason to enable the debate to proceed on the basis of the best knowledge available, both about science and medicine and about ethics, law, and policy. Before proceeding to the results of our monitoring, we complete this introduction with some additional comments on the different types of stem cells, a few terminological observations and clarifications, and an overview of the report as a whole.

III. TYPES OF STEM CELLS: AN INTRODUCTION

Although we shall report later (in Chapter 4) on recent developments in basic and clinical research using various types of stem cells, we think the following introduction to the "cast of characters" would be useful at the start.[*]

A. Embryonic Stem (ES) Cells

As noted above, ES cells are derived from the inner cell mass of embryos at the blastocyst stage, roughly five to nine days after fertilization—after the zygote has divided enough times to result in about 200 cells, but before it has undergone gastrulation and differentiation into the three primary germ layers (see Appendix A).[†] The inner cell mass is the part of the blastocyst-stage embryo whose cells normally go on to become the body of the new individual. The outer cells of the blastocyst-stage embryo (the trophoblast cells) normally (that is, in vivo) go on to become the fetal contribution to the placenta and other structures that connect the developing individual to the mother's bloodstream and that otherwise support the embryo's further development. Collecting the cells of the inner cell mass results in the destruction of the developing organism.

---

[*] The remarks about embryo-derived cells presented in the next two sections apply to human embryonic stem cells, as opposed to, say, mouse embryonic stem cells (which will be referred to in several places because they have provided the basis for much of what we now know about embryonic stem cells).

[†] In this report, we will not call the cells contained in the inner-cell-mass "stem cells," so long as they remain inside the intact embryo. We reserve the term "stem cells" for those cells that are successfully cultured outside the embryo, following artful derivation, and that demonstrate the characteristic capacities of "stemness": a capacity for self-renewal and a capacity for differentiation. Inner-cell-mass cells may or may not be identical to ES cells, though in an intact embryo the inner-cell-mass cells are still part of a nascent organic whole. Indeed, it is important to remember that the developmental fate of all cells inside the body is in part a function of their location within the larger whole and of the influences of the local embryonic environments to which they are subject.

The embryos from which human stem cells can be derived are available (so far) only from in vitro fertilization (IVF): they have been conceived by a combination of egg and sperm, occurring outside the body.*

B. Embryonic Germ (EG) Cells

EG cells are stem cells that are isolated from the gonadal ridge of a developing fetus. These are the cells that ultimately give rise to sperm cells or egg cells, depending on the sex of the fetus. The EG cells are collected from the bodies of five-to-nine-week-old fetuses that have been donated after induced abortions.† In federally funded research, collection of the EG cells is governed by existing federal regulations for fetal-tissue donation, designed (among other things) to ensure the separation of the decision to terminate pregnancy from the decision to donate the fetal tissue for research.[7]

Cell lines established from either of these two sources (ES and EG cells, from embryos and fetal gonads, respectively) have demonstrated two important properties: great ability to multiply and form stable lines that can be characterized, and great flexibility and plasticity. Their progeny can differentiate in vitro into cells with characteristics of those normally derived from all three embryonic germ layers (ectoderm, endoderm, and mesoderm), which layers (in vivo) give rise in turn to all the different types of cells in the body. Because they are so flexible, it also seems likely that they could be used to produce cell preparations that could then be transplanted (assuming that the recipient's immune response could be managed) to repopulate a part of the body such as the pancreas or spinal cord that has lost function due to disease or injury. As with stem cells derived from the various tissues of the adult body,

---

* As of this writing, experiments in asexual methods of conceiving a human embryo, such as parthenogenesis or cloning, have not, to our knowledge, been successful beyond the very early stages of development. Embryos, fertilized in vivo, could also be procured for use in research by flushing them from the womb, but this procedure, though technically feasible, has a very low yield and is rarely done.

† Abortion is legal throughout the United States, pursuant to a series of federal Supreme Court decisions, the most important cases being *Roe v. Wade* (1973) and *Casey v. Planned Parenthood* (1992).

ES cells and EG cells seem to hold out hope for an era of regenerative medicine.

C. Adult (or Non-embryonic) Stem Cells

Adult stem cells are more differentiated than ES or EG cells, but not yet fully differentiated. Like stem cells of embryonic origin, they can give rise to lineages of cells that are more specialized than themselves. The term "adult" is a bit of a misnomer ("non-embryonic" would be more accurate): these cells are found in various tissues in children as well as adults (and in fetuses as well), and they have been isolated from umbilical cord blood at the time of delivery. Despite its inaccuracy regarding the *origin* of the cells, the term "adult" helpfully emphasizes that the cells have been partially differentiated. Although they can give rise to various cell types, these non-embryonic stem cells are generally all within the same broad type of tissue (for example, muscle stem cells, adipose stem cells, neural stem cells). For this reason, it had long been thought that they are less flexible than those derived from embryos or fetal gonads. Yet this presumption has been disputed in recent years by those who think that certain forms of adult stem cells may be equally or nearly as plastic as non-adult stem cells. Indeed, possible exceptions to the generalization that adult stem cells give rise only to cell types found within their own broad type of tissue have recently been reported (though most of these cells may well be shorter-lived than ES cells, and, if so, potentially less useful in therapy). This finding has ignited a debate about the relative merits of embryonic stem cells and adult stem cells: which is more valuable, both for research and (especially) for clinical treatment?*

Research involving adult stem cells raises few difficult ethical concerns, beyond the usual need to secure free and

---

* We shall review recent work with both kinds of cells in Chapter 4. Anticipating the implications of that discussion, we may safely say that not enough is known to answer this empirical question. Work with both kinds of cells seems promising. Some people argue that research with non-embryonic stem cells, being a morally unproblematic path, should be given priority. Most researchers, meanwhile, support the advancement of work with both kinds of cells simultaneously, to explore their potential.

fully informed consent from donors and recipients, a favorable benefit-to-risk ratio for all participants in attempts at therapy and protection of privacy. Adult stem cells are less controversial than embryonic ones, as we have noted, because the former can be collected without lasting harm to the donor.

D. Cord Blood Stem Cells

Though clearly a type of non-embryonic stem cell, cord blood stem cells deserve some special mention. Blood found in the umbilical cord can be collected at birth and hematopoietic stem cells (and other progenitor cells) isolated from it. It has been proposed that individually banked cord blood cells may, at some later time, offer a good match for a patient needing stem cell-based treatments, whether the individual cord-blood-donor himself or a close relative, and in unrelated recipients may require a less exact genetic match than adult bone marrow.[*][†]

---

[*] Several companies in the United States have sprung up to offer commercial storage services for cord blood in case the child or a closely genetically-matched sibling should later need the stem cells contained in the cord blood for medical use. It is unclear whether individual banking of cord blood will turn out to be valuable. It may turn out that, for the vast majority of people, the cells are never needed, or that, when therapy is needed, the stored cells are found to be unsuitable or incapable of meeting the need in the time required. At the same time, Congress has recently allocated funds to create a national non-commercial cord blood bank potentially available to all patients. The authors of the legislation argue that a national bank would have cord blood of many different types, increasing the odds that a patient would find a match.

[†] The possibility of therapeutic use of cord blood stem cells has raised a serious question unrelated to the ethics of stem cell research: whether parents of a sick child may morally conceive another child, of genetic make-up appropriate for providing compatible cord blood cells, primarily to treat the first child. In generating the second child, a prospective parent or parents might screen preimplantation embryos for genetic suitability to provide the cells (both compatible blood type and freedom from the genetic disease affecting the older sibling). These and other ethical questions surrounding preimplantation genetic diagnosis (PGD) go beyond our present subject and will not be considered further in this report.

IV. TERMINOLOGY

In considering complicated or contested public questions, language matters—even more than it ordinarily does. Clear thinking depends on clear ideas, and clear ideas can be conveyed only through clear and precise speech. And fairness in ethical evaluation and judgment depends on fair framing of the ethical questions, which in turn requires fair and accurate description of the relevant facts of the case at hand. Such considerations are highly pertinent to our topic and to the arguments it generates.

Confounding the discussions of stem cell research, there are, to begin with, difficult technical concepts, referring to complicated biological entities and phenomena, that can cause confusion among all but the experts. Some of these concepts we will clarify in Chapter 4, and others are defined in the Glossary and, in some cases, illustrated in Appendix A on early embryonic development. But the more important terminological issues are those used to formulate the ethical and policy issues about which people so vigorously disagree. We pause to comment on three of them: "the embryo" (or "the human embryo"), "spare embryos," and "the moral status of the embryo."

Strictly speaking, there is no such *thing* as "*the* embryo," if by this is meant a distinctive being (or *kind* of being) that deserves a common, reified name—like "dog" or "elephant." Rather, the term properly intends a certain *stage of development* of an organism of a distinctive kind. Indeed, the very term comes from a Greek root meaning "to grow": an embryo is, by its name and mode of being, an immature and growing organism in an early phase of its development.* The advent of in vitro fertilization, in which living human embryos

---

* In classical embryology, "embryo" is the name given—somewhat arbitrarily—to the developing human organism from the time of fertilization until roughly eight weeks, the time that the first calcification of bone occurs. After that, the developing human organism is called a "fetus," equally a reified name for a *dynamic* entity, an entity-in-the-process-of-becoming-more-fully-the-kind-of-organism-it-already-is.

from their first moments are encountered as independent entities outside the body of a mother, before human eyes and in human hands, may also have contributed to this tendency to reify "*the* embryo" in its early stages (though such reification has likely always played a role in embryology). The ex vivo existence of nascent human life is genuinely puzzling and may invite terminology that can be distorting.[*]

If the term "*the* embryo" risks conveying the false notion that embryos are distinct kinds of beings or things, the term "spare embryo" risks making a difficult moral question seem easier than it is. The term is frequently used to describe those embryos, produced (each with reproductive intent, but in excess of what is needed) in assisted-reproduction clinics, that are not transferred to a woman in attempts to initiate a pregnancy. No longer needed to produce a child, they are usually frozen and stored for possible later use, should the first efforts fail. But the "spareness" of a "spare embryo" is not a property of a particular embryo itself; it bespeaks rather our attitude toward it, now that it may no longer be needed to serve the purpose for which it was initially brought into being. Calling something "spare," or only "extra," invites the thought that nothing much is lost should it disappear, because one already has more than enough: one has "embryos to spare." It also abstracts from the distinct genetic individuality of each embryo and invites the view that embryos are, like commercial products, simply interchangeable—an outlook that may affect the further judgment of any embryo's moral standing. To be sure, most of these unused embryos will die or be destroyed. To be sure, if these unused embryos are otherwise destined for destruction, a case can be made—and debated—that their unavoidable loss should be redeemed by putting them to use beforehand. But the moral question regarding their possible

---

[*] The Council is well aware of the fact that the debate about abortion shadows all these discussions about "embryos." Yet in all of our work to date, on cloning and stem cell research, we have called attention to the fact that we face a rather different moral situation when we are dealing with embryos in the laboratory, in the absence of concerns for a pregnant woman's life and future. Accordingly, we explore the ethical issues of embryo research by addressing what we know (and how we know) about ex vivo human embryos, separate from any issues that enter when the interests of pregnant women are engaged.

use and destruction should not be decided—here, as elsewhere—on terminological grounds, in this case, by the naming of the embryo "*spare.*" Rather it should be decided on the basis of a direct moral appraisal of the rights and goods involved: on the basis of what we owe to suffering humanity and the obligations we have to seek the means of its relief; and on the basis of the nature of human embryos, what we owe them as proper respect and regard, and whether and why such respect or regard may be overridden.* For many people, the moral question depends, in other words, on what some bioethicists call—and we ourselves will sometimes call—"the *moral status* of the embryo." If embryos lacked all "moral status," there would be little moral argument about their use and destruction.

Yet the notion "moral status" is problematic, even though it is easy to understand why it has come into fashion. For many people, the central ethical question regarding embryonic stem cell research is whether an embryonic organism from which cells may be removed to develop ES cells is fully "one of us," deserving the same kind of respect and protection as a newborn baby, child, or adult. What they want to know is the *moral* standing of these organisms—entities that owe their existence, their extra-uterine situation, and their "spare-ness" to deliberate human agency—at such early stages of development. As we shall see, some people try to find structural or functional markers—for example, the familiar human form or the presence or absence of sensation—to decide the moral worth of a human embryo. Others use an argument from continuity of development to rebut any attempt to find a morally significant boundary anywhere along the continuum of growth and change. But, to judge from countless efforts to provide a biologically based criterion for ascribing full human worth, it seems certain that we shall never find an answer to our moral question in biology *alone*, even as the answers we give must take into account the truths of embryology. At least until now, philosophical attempts to draw moral inferences

---

* Some Members of this Council (including Alfonso Gómez-Lobo and Robert George) hold that the moral question should be decided on the basis of the prior consideration of the rightness or wrongness of intentionally destroying human beings for the sake of further goals, and then on whether or not human embryos are human beings in the relevant sense.

from the biological facts have not yielded conclusions that all find necessary or sound.

Under these circumstances, some people believe that we have no choice but to stipulate or ascribe some degree of moral "status" to the entity, based either on how it strikes us and the limited range of what we are able to know about it, or on what we wish to do with it: we confer upon it some moral status *in regard to us*, much as we confer one or another class of immigration status upon people.[8] For this very reason, others object to the term, fearing that it enables us to beg the question of the intrinsic moral worth or dignity of the entity *itself*, seen in its own terms and without regard to us. Different Members of this Council hold different views of this terminological and ontological matter, but we all recognize the moral freight carried by attempts to speak about and ascribe "moral *status*" to human embryos in their earliest stage of development.[*] We encourage readers to be self-conscious about this and similar terms, even as we proceed ourselves to make use of them.

V. ABOUT THE REPORT

Monitoring stem cell research can be a bit like watching Niagara Falls. Not only do scientific reports pour forth daily, as they do in many other areas of research, but a kind of mist rises up for the torrent of news flashes and editorials, making it difficult to separate knowledge from opinion and hope from hype. The underlying biology—whether viewed at the level of the gene, cell, tissue, organ, or organism—is dauntingly complex, as is all cell biology. At any of these levels, in this new and dynamic field, it is frequently difficult for even the most knowledgeable scientist to be truly certain of "what really causes what." For example, how exactly do certain kinds of stem cells have their apparently beneficial effects on heart

---

[*] It is, of course, possible to hold the view that the earliest human embryos have *no* moral status or worth, because they are so small and undifferentiated or because they lack the ordinary human shape and form or the specifically human capacities for sensation or consciousness or the capacity to develop on their own ex vivo. Some of these arguments are reviewed in Chapter 3. Here it suffices to observe that at least one Member of this Council (Michael Gazzaniga) holds this view.

disease when the cells are extracted from a cardiac patient's bone marrow or muscle, expanded in culture, and injected into the patient's heart? Or what is responsible for the positive effects on a Parkinson Disease patient when cells from his own brain are similarly extracted, treated, and re-injected? We do not yet really know precisely what stem cell-based preparations do when put into the body.

At the same time, all discussion in this area suffers from a persistent background tension. The stakes are high, or seem so, to many of the discussants, and there is much politicking involved. As noted earlier, opponents of embryo research try to tout the virtues of adult stem cells, because they regard their use as a morally permissible alternative. Proponents, for their part, often find it tempting to disparage or downplay all adult stem cell studies and to emphasize instead what they believe to be the superior potential of embryonic stem cells for successful future therapeutic use. Navigating between these tendencies in search of the full truth can be daunting, and few people are altogether immune to the partial but seductive calls from the scientific or moral side they prefer.

Yet without denying our individual differences on the ethical and policy questions at issue, the Council has sought in this monitoring report to present a fair-minded and thorough overview, both of the ethical and policy debates and of the scientific and medical results to date. To aid us in our task of monitoring, we have commissioned six review articles and heard several oral presentations on the state of research, covering studies using embryonic and studies using adult stem cells. We have commissioned a review article and heard a presentation on the problem of immune rejection, a potential major stumbling block to effective cell transplantation therapies.

We have read papers, commissioned writings, heard presentations, and debated among ourselves about the various ethical and philosophical issues involved, from "the moral status of the embryo," to the existence of a moral imperative to do research, to the meaning of federal funding of morally controversial activities. We have read and heard public testimony from both supporters and opponents of the current policy on federal funding of ES cell research.

We have considered arguments—presented by scientists and patient-advocacy groups, and shared by some Members of the Council—that the current policy is impeding potentially life-saving research, for example, by offering researchers too few useful ES cell lines to work with, by causing a chilling effect on the whole field, or by allowing the field to be dominated by private companies, less given (than are publicly-funded academic scientists) to publishing and sharing the results of their research. We have considered arguments—presented by various critics and opponents of embryonic stem cell research, and shared by some Members of the Council—that the current policy has opened the path toward the possibility of "embryo farming" or that it risks weakening our respect for nascent life and our willingness to protect the weakest lives among us. We have heard from ethicists and scientific researchers, representatives of biopharmaceutical companies and disease research foundations, and senior government officials from such agencies as the National Institutes of Health and the Food and Drug Administration. We benefited from working papers prepared by the Council's staff and from existing reports on stem cell research, and in particular reports by the National Bioethics Advisory Commission (1999) and the National Academies (2001).[9] Holding our own personal views in abeyance, we have tried in the three chapters that follow to synthesize accurately and fairly what we have heard and learned: about current law and policy, about the state of the ethical debate, and about the current state of scientific research.

Chapter 2, "Current Federal Law and Policy," describes and explains the current federal policy regarding stem cell research. It locates that policy in relation to previous law and policy touching this area of research and tries to make clear the ethical, legal, and prudential foundations on which the policy rests. It then describes the implementation of the policy and other relevant considerations. Our goal in that chapter is to describe and understand the present policy situation, in its legal, political, scientific, and ethical colorations, and to present accurately the various features of the current federal policy, many of which are not generally well understood.

Chapter 3, "Recent Developments in the Ethical and Policy Debates," provides an overview of the ethical and policy debates surrounding stem cell research in the past two years. Special attention is, of course, given to arguments about what may (or may not) be done with human embryos, and why. But those arguments are also reviewed in relation to larger debates about the other ethical and policy issues mentioned earlier. Our goal in that chapter is to present the arguments and counter-arguments, faithfully and accurately, rather than finally to assess their validity.

Finally, in Chapter 4, "Recent Developments in Stem Cell Research and Therapy," we offer an overview of some recent developments in the isolation and characterization of various kinds of stem cell preparations and a partial account of some significant research and clinical initiatives. In addition, by means of a selected case study, we consider how stem cell-based therapies might some day work to cure devastating human diseases, as well as the obstacles that need to be overcome before that dream can become a reality. Our goal in that chapter, as supplemented by several detailed commissioned review articles contained in the appendices, is to enable (especially non-scientific) readers to appreciate the reasons for the excitement over stem cell research, the complexities of working with these materials, some early intriguing research and therapeutic findings, and the difficult road that must be traveled before we can reap therapeutic and other benefits from this potentially highly fertile field of research.

After these three substantive chapters—on policy, ethics, and science—we offer a Glossary and a series of appendices, beginning (in Appendix A) with a brief primer on early human embryonic development. That primer aspires to provide the basic facts and concepts that any thoughtful and public-spirited person needs to know about human development and especially about (early) human embryos if he or she is to participate intelligently in the ethical and political deliberations that are certain to continue in our society for some time. There follow the texts of President Bush's August 9, 2001, stem cell speech and the NIH guidelines (for both the Clinton and Bush administrations) regarding the funding of embryonic stem cell research. Completing the appendices are the texts of all the papers that the Council commissioned, as revised by their authors in light of subsequent developments or comments received. These papers appear in the authors' own words, unedited by the Council.

In all that we offer in this monitoring report, we have aspired to be careful and fair in our approach, precise in our use of language, accurate in presenting data and arguments, and thoughtful in our laying out of the various issues that remain before us. It is up to our readers to judge whether or not we have succeeded. The policy debates over stem cell research that led to the creation of this Council continue; they, and other debates on related topics, are unlikely to go away any time soon. Our hope is that our work will help to make those debates richer, fairer, and better informed.

ENDNOTES

[1] "Remarks by the President on Stem Cell Research," Crawford, Texas, August 9, 2001. Text made available by the White House Press Office, August 9, 2001. (Also available in full at Appendix B of this report.)

[2] Thomson, J., et al., "Embryonic stem cell lines derived from human blastocysts," *Science* 282: 1145-1147 (1998).

[3] Shamblott, M., et al., "Derivation of pluripotent stem cells from cultured human primordial germ cells," *Proceedings of the National Academy of Science* 95: 13726-13731 (1998).

[4] See, among others, Dresser, R., "Embryonic Stem Cells: Expanding the Analysis," *American Journal of Bioethics* 2(1): 40-41 (2003); and the personal statements of Council Members Rebecca Dresser and William May, appended to the Council's July 2002 report *Human Cloning and Human Dignity: An Ethical Inquiry*.

[5] See, among others, Kass, L., "The Meaning of Life – In the Laboratory," *The Public Interest*, Winter 2002; Cohen, E., "Of Embryos and Empire," *The New Atlantis* 2: 3-16 (2003); and "The Moral Case against Cloning-for-Biomedical-Research" presented by some Members of the Council in the Council's July 2002 report *Human Cloning and Human Dignity: An Ethical Inquiry*, Chapter 6.

[6] See, for instance, "The Moral Case for Cloning-for-Biomedical-Research" presented by some Members of the Council in the Council's July 2002 report *Human Cloning and Human Dignity: An Ethical Inquiry*, Chapter 6; and the personal statements of Council Members Elizabeth Blackburn, Daniel Foster, Michael Gazzaniga, and Janet Rowley appended to that report.

[7] 45 C.F.R. § 46.204(h-i).

[8] See, for instance, Green, R., *The Human Embryo Research Debates*, New York: Oxford University Press (2001). Also see the discussion of the Council in its October 17, 2003, meeting, particularly the comments of Council Member Alfonso Gómez-Lobo. A transcript of that session is available on the Council's website at www.bioethics.gov.

[9] National Bioethics Advisory Commission (NBAC), *Ethical Issues in Human Stem Cell Research*, Bethesda, MD: Government Printing Office (1999); National Research Council/Institute of Medicine (NRC/IOM), *Stem Cells and the Future of Regenerative Medicine*, Washington, D.C.: National Academy Press (2001).

2

Current Federal Law and Policy

Any overview of the state of human stem cell research under the current federal funding policy must begin with a thorough understanding of that policy. This is not as simple as it may sound. From the moment of its first announcement, on August 9, 2001, the policy has been misunderstood (and at times misrepresented) by some among both its detractors and its advocates. Its moral foundation, its political context, its practical implications, and the most basic facts regarding the policy's implementation have all been subjects of heated dispute and profound confusion. Whether one agrees with the policy or not, it is important to understand it as it was propounded, accurately and in its own terms, in the light also of the historical and political contexts in which it was put forward.

This chapter attempts to place the policy in its proper context; to articulate its moral, legal, and political underpinnings (as put forward by its authors and advocates); to offer an overview of its implementation thus far; and to begin to describe its ramifications for researchers and for medicine. By articulating the policy in its own terms, we intend neither to endorse it nor to find fault with it.* Indeed, in the next chapter

---

* Some Members of the Council oppose the current policy, some Members support it. Yet the descriptive account that we offer here aspires to be seen as accurate and fair, regardless of where one personally stands on the issue. Nearly all Members of this Council recognize, as we said in our report *Human Cloning and Human Dignity*, that "all parties to this debate

we present an overview of arguments on all sides of the question. Here we mean only to clarify, as far as we are able, the original meaning and purpose of the policy, so as to be better able to monitor its impact.

I. A BRIEF HISTORY OF THE EMBRYO RESEARCH FUNDING DEBATE

The federal government makes significant public resources available to biomedical researchers each year—over $20 billion in fiscal year 2003 alone—in the form of research grants offered largely through the National Institutes of Health (NIH). This level of public expenditure reflects the great esteem in which Americans hold the biomedical enterprise and the value we place on the development of treatments and cures for those who are suffering. But such support is not offered indiscriminately. Researchers who accept federal funds must abide by ethically based rules and regulations governing, among other things, the use of human subjects in research. And some policymakers and citizens have always insisted that taxpayer dollars not be put toward specific sorts of research that violate the moral convictions and sensibilities of some portion of the American public. This has meant that controversies surrounding the morality of some forms of scientific research have at times given rise to disputes over federal funding policy. Among the most prominent examples has been the three-decade-long public and political debate about whether taxpayer funds should be used to support

---

have something vital to defend, something vital not only to themselves but also to their opponents in the debate, and indeed to all human beings. No human being and no human society can afford to be callous to the needs of suffering humanity, cavalier regarding the treatment of nascent human life, or indifferent to the social effects of adopting in these matters one course of action rather than another." (*Human Cloning and Human Dignity,* p. 121.) Thus, whatever we think of the current funding policy, we recognize that this is a genuine ethical dilemma and that reasonable people of good will may come to different conclusions about where the best ethical or policy position lies. We therefore also believe that not only results but also reasons matter, and that it behooves us to understand the principled or prudential reasons for the current policy (as well as for any alternative policy that might be offered to replace it).

research that involves creating or destroying human embryos or making use of destroyed embryos and fetuses—practices that touch directly on the much-disputed questions of the moral status and proper treatment of nascent human life.

In the immediate aftermath of the Supreme Court's 1973 *Roe v. Wade* decision legalizing abortion nationwide, some Americans, including some Members of Congress, became concerned about the potential use of aborted fetuses (or embryos) in scientific research. In response to these concerns, the Department of Health, Education and Welfare (DHEW, the precursor to today's Department of Health and Human Services) initiated a moratorium on any potential DHEW sponsorship or funding of research using human fetuses or living embryos. In 1974, Congress codified the policy in law, initiating what it termed a temporary moratorium on federal funding for clinical research using "a living human fetus, before or after the induced abortion of such fetus, unless such research is done for the purpose of assuring the survival of such fetus."[1] Concurrently with that moratorium (and also addressing concerns not directly related to embryo and fetal research), Congress established a National Commission for the Protection of Human Subjects of Biomedical and Behavioral Research. Among its other tasks, Congress explicitly assigned the Commission responsibility for offering guidelines for human fetal and embryo research, so that standards for funding might be established and the blanket moratorium might be lifted. The statutory moratorium was lifted once the Commission issued its report in 1975.[2]

In that report, the Commission called for the establishment of a national Ethics Advisory Board within DHEW to propose standards and research protocols for potential federal funding of research using human embryos and to consider particular applications for funding. In doing so, the Commission looked ahead to the possible uses of in vitro embryos, since the first successful in vitro fertilization (IVF) of human egg by human sperm had been accomplished in 1969.[*] The Department

---

[*] In its discussion of "fetal" research, the commission defined the fetus as the product of conception from the time of implantation onward, which therefore included what we generally think of (and define in this report) as embryos in utero. Its separate consideration of embryo research was therefore directed at in vitro embryos.

adopted the recommendation in 1975, established an Ethics Advisory Board, and put in place regulations requiring that the Board provide advice about the ethical acceptability of IVF research proposals. The Board first took up the issue of research on in vitro embryos in full in the late 1970s and issued its report in 1979.[3]

By that time, human IVF techniques had been developed (first in Britain) to the point of producing a live-born child (born in 1978). These techniques, and their implications for human embryo research, raised unique prospects and concerns that were distinct from some of those involved in human fetal research. As a consequence, starting in the late 1970s, funding of embryo research and funding of fetal research came to be treated as mostly distinct and separate issues. The Ethics Advisory Board concluded that research involving embryos and IVF techniques was "ethically defensible but still legitimately controverted." Provided that research did not take place on embryos beyond fourteen days of development and that all gamete donors were married couples, the Board argued, such work was "acceptable from an ethical standpoint," but the Board decided that it "should not advise the Department on the level of Federal support, if any," such work should receive.[4]

This left the decision in the hands of the DHEW, which decided at that stage not to offer funding for human embryo studies. The Ethics Advisory Board's charter expired in 1980, and no renewal or replacement was put forward, creating a peculiar situation in which the regulations requiring the Ethics Advisory Board to review proposals for funding remained in effect, but the Board no longer existed to consider such requests. Funding was therefore rendered impossible in practice. Because the Ethics Advisory Board was never replaced, a de facto ban on funding remained in place through the 1980s.

In 1993, Congress enacted the NIH Revitalization Act, a provision of which rescinded the requirement that research protocols be approved by the non-existent Ethics Advisory Board.[5] This change opened the way in principle to the possibility of NIH funding for human embryo research using IVF embryos. The following year, the NIH convened a Human

Embryo Research Panel to consider the issues surrounding such research and to propose guidelines for potential funding applications. The panel recommended that some areas of human embryo research be deemed eligible for federal funding within a framework of recognized ethical safeguards. It further concluded that the creation of human embryos with the explicit intention of using them only for research purposes should be supported under some circumstances.[6] President Clinton overruled the panel on the latter point, ordering that embryo creation for research not be funded, but he accepted the panel's other recommendations and permitted the NIH to consider applications for funding of research using embryos left over from IVF procedures.[7]

Congress, however, did not endorse this course of action. In 1995, before any funding proposal had ever been approved by the NIH, Congress attached language to the 1996 Departments of Labor, Health and Human Services, and Education, and Related Agencies Appropriations Act (the budget bill that funds DHHS and the NIH) prohibiting the use of any federal funds for research that destroys or seriously endangers human embryos, or creates them for research purposes.

This provision, known as the "Dickey Amendment" (after its original author, former Representative Jay Dickey of Arkansas), has been attached to the Health and Human Services appropriations bill each year since 1996. Everything about the subsequent debate over federal funding of embryonic stem cell research must be understood in the context of this legal restriction. The provision reads as follows:

None of the funds made available in this Act may be used for—

(1) the creation of a human embryo or embryos for research purposes; *or*

(2) research in which a human embryo or embryos are destroyed, discarded, or knowingly subjected to risk of injury or death greater than that allowed for research on fetuses in utero under 45 CFR 46.204 and 46.207, and

subsection 498(b) of the Public Health Service Act (42 U.S.C. 289g(b)).*

(b) For purposes of this section, the term 'human embryo or embryos' includes any organism, not protected as a human subject under 45 CFR 46 as of the date of the enactment of the governing appropriations act, that is derived by fertilization, parthenogenesis, cloning, or any other means from one or more human gametes or human diploid cells.[8]

This law effectively prohibits the use of federal funds to support any research that destroys human embryos or puts them at serious risk of destruction. It does not, however, prohibit the conduct of such research using private funding. Thus, it addresses itself not to what may or may not be lawfully done, but only to what may or may not be supported by taxpayer dollars. At the federal level, research that involves the destruction of embryos is neither prohibited nor supported and encouraged.

The Dickey Amendment was originally enacted before the isolation of human embryonic stem cells, first reported in 1998 by researchers at the University of Wisconsin, whose work was supported only by private funds (largely from the Geron Corporation and the University of Wisconsin Alumni Research Foundation). The discovery of these cells and their unique and potentially quite promising properties aroused great excitement both within and beyond the scientific community. It led some people to question the policy of withholding federal funds from human embryo research. Most Members of Congress, however, did not change their position, and the Dickey Amendment has been reenacted every year since. For many of its supporters, the amendment expresses their ethical conviction that nascent human life ought to be protected against exploitation and destruction for scientific research, however promising that research might be, and that at the very least such destruction should not be supported or encouraged by taxpayer dollars.

---

* These legal citations refer to the federal regulations and federal statute relating to research on living human fetuses.

On its face, the Dickey Amendment would seem to close the question of federal funding of human embryonic stem cell research, since obtaining stem cells for such research relies upon the destruction of human embryos. But in 1999, the General Counsel of the Department of Health and Human Services argued that the wording of the law might permit an interpretation under which human embryonic stem cell research could be funded. If embryos were first destroyed by researchers supported by private funding, then subsequent research employing the derived embryonic stem cells, now propagated in tissue culture, might be considered eligible for federal funding. Although such research would presuppose and follow the prior destruction of human embryos, it would not itself involve that destruction. Thus, the Department's lawyers suggested, the legal requirement not to fund research "in which" embryos were destroyed would still technically be obeyed.[9]

This has generally been taken to be a legally valid interpretation of the specific language of the statute, and indeed the subsequent policies of both the Clinton and Bush administrations have relied upon it in different ways. But some critics of the 1999 legal opinion argued that, though it might stay within the letter of the law, the proposed approach would contradict both the spirit of the law and the principle that underlies it.[10] It would use public funds to encourage and reward the destruction of human embryos by promising funding for research that immediately follows and results from that destruction—thereby offering a financial incentive to engage in such destruction in the future. By so doing, these critics argued, it would at least implicitly state, in the name of the American people, that research that destroys human embryos ought to be encouraged in the cause of medical advance. Supporters of the Clinton administration's proposed approach, however, argued that promoting such research—especially given its therapeutic potential—was indeed an appropriate government function, and that the policy proposed by DHHS was neither illegal nor improper, given the text of the statute and provided that the routine standards of research ethics (including informed consent and a prohibition on financial inducements) were met.[11]

The Clinton administration adopted this course of action and drew up specific guidelines to enact it.[*] But the guidelines, completed just before the end of the Clinton administration, never had a chance to be put into practice, and no funding was ever provided. Upon entering office in 2001, the Bush administration decided to take another look at the options regarding human embryonic stem cell research policy and therefore put the new regulations on hold, pending review.

In conducting its review, the Bush administration stated that it sought a way to allow some potentially valuable research to proceed while upholding the spirit (and not just the letter) of the Dickey Amendment, a spirit that the President himself has advocated.[12] The expressed hope was that the government, while continuing to withhold taxpayer support or encouragement for the destruction of human embryos, might find a way to draw some moral good from stem cell lines that had already been produced through such destruction—given that this deed, even if immoral, could not now be undone. This is the ethical-legal logic of the present stem cell funding policy: it seeks those benefits of embryonic stem cell research that might be attainable without encouraging or contributing to any future destruction of human embryos.

II. THE PRESENT POLICY

The current policy on federal government funding of human embryonic stem cell research, then, must be understood in terms of the constraints of the Dickey Amendment and in terms of the logic of the moral and political aims that underlie that amendment.

At the time of the policy's announcement, a number of embryonic stem cell lines had already been derived and were in various stages of growth and characterization. The embryos from which they were derived had therefore already been destroyed and could no longer develop further. As President Bush put it, "the life and death decision had already been made."[13]

---

[*] These regulations, as published in the Federal Register, are provided in Appendix D.

The administration's policy made it possible to use taxpayer funding for research conducted on those preexisting lines, but it refused in advance to support research on any lines created after the date of the announcement. In addition, to be eligible for funding, those preexisting lines would have had to have been derived from excess embryos created solely for reproductive purposes, made available with the informed consent of the donors, and without any financial inducements to the donors—standard research-ethics conditions that had been attached to the previous administration's short-lived funding guidelines, as well as to earlier attempts to formulate rules for federal funding of human embryo research. The policy denies federal funding not only for research conducted on stem cell lines derived from embryos destroyed after August 9, 2001 (or that fail to meet the above criteria), but also (as the proposed Clinton-era policy would have) for the creation of any human embryos for research purposes and for the cloning of human embryos for any purpose.[*]

The moral, legal, and political grounds of this policy have been hotly contested from the moment of its announcement. Debates have continued regarding its aims, its character, its implementation, and its underlying principles, as well as the significance of federal funding in this area of research. For example, many scientists, physicians, and patient advocacy groups contend that the policy is too restrictive and thwarts the growth of a crucial area of research. On the other side, some opponents of embryo research believe the policy is too liberal and legitimates and rewards (after the fact) the destruction of nascent human life. Some ethicists argue that there is a moral imperative to remove all restrictions upon potentially life-saving research; other ethicists argue that there is a moral imperative to protect the lives of human beings in their earliest and most vulnerable stages. These and similar arguments are reviewed in the next chapter. But before one can enter into these debates, it is essential first to understand the relevant elements of the policy itself as clearly and distinctly as possible.

---

[*] The official NIH statement of this policy is provided in Appendix C.

III. MORAL FOUNDATION OF THE POLICY

In articulating its proposed funding policy in 1999 and 2000, the Clinton administration expressed a firm determination that funded research could use only those human embryos that had been left over from IVF procedures aimed at reproduction and that had been donated in accordance with the standards of informed consent and in circumstances free of financial inducements. Provided that these crucial conditions were met, the administration argued that the potential benefits of stem cell research were so great that publicly funded research should go forward. In August of 2000, reflecting on the guidelines put forward by his administration, President Clinton remarked,

> Human embryo research [as approved for funding by the NIH guidelines] deals only with those embryos that were, in effect, collected for in-vitro fertilization that never will be used for that. So I think that the protections are there; the most rigorous scientific standards have been met. But if you just—just in the last couple of weeks we've had story after story after story of the potential of stem cell research to deal with these health challenges. And I think we cannot walk away from the potential to save lives and improve lives, to help people literally to get up and walk, to do all kinds of things we could never have imagined, as long as we meet rigorous ethical standards.[14]

Given the promise of embryonic stem cell research, the existence of many embryos frozen in IVF clinics and unlikely ever to be transferred and brought to term, and the willingness of some IVF patients to donate such embryos for research, the Clinton administration reasoned that research using cell lines derived from these embryos could ethically be supported by federal funds. That position implies, of course, that the destruction of embryos is not inherently or necessarily unethical, or so disconcerting as to be denied any federal support. The Clinton-era NIH Embryo Research Panel put succinctly one form of this view in stating that "the preimplantation human embryo warrants serious moral consideration as a developing form of human life, but it does not have the same moral status as infants and children."[15] If there is sufficient promise or reason to support research, the claim of a human embryo to "serious moral consideration" (or, as others, including some of us, have put it, to "special respect"[16]) could be outweighed by other moral aims or principles.

This (at least implicit) understanding of the moral status of human embryos might be seen to have put the Clinton administration at odds with the principle animating the operative law on this subject (the Dickey Amendment). But given its responsibility to carry out the laws as they are enacted, the administration sought a way to advance research within the limitations set by the statute. Its approach to the funding of embryonic stem cell research, therefore, seems to have sought an answer to this question: *How can embryonic stem cell research, conducted in accordance with standards of informed consent and free donation, be maximally aided within the limits of the law?* The NIH guidelines published in 2000 represent the answer the Clinton administration found: funding research on present and future embryonic stem cell lines, so long as the embryo destruction itself is done with private funds.

The Bush administration appears to have been motivated by a somewhat different question, arising from what seems to be a different view of the morality of research that destroys human embryos. President Bush put the matter this way, in discussing his newly announced policy in August of 2001:

> Stem cell research is still at an early, uncertain stage, but the hope it offers is amazing: infinitely adaptable human cells to replace damaged or defective tissue and treat a wide variety of diseases. Yet the ethics of medicine are not infinitely adaptable. There is at least one bright line: We do not end some lives for the medical benefit of others. For me, this is a matter of conviction: a belief that

life, including early life, is biologically human, genetically distinct and valuable.[17]*

While expressing a desire to advance medical research, this argument describes a line that such research should not cross, and therefore past which funding should not be offered. That line, in this context, is the destruction of a human embryo for research purposes. The Bush administration thus appears to share the view that underlies both the word and spirit of the Dickey Amendment. In its approach to the stem cell issue it has sought to answer a question that differs, subtly but significantly, from that formulated by the previous administration: *How can embryonic stem cell research, conducted in accordance with basic research ethics, be maximally aided within the bounds of the principle that nascent human life should not be destroyed for research?*

In seeking to answer that question, the Bush administration (like the Clinton administration) had to take account of the existing situation and—as always in such instances—to mix prudential demands and opportunities with an effort at principled judgment. Given the existence of some human embryonic stem cell lines, derived from human embryos that had already been destroyed, the administration determined that it might not simply have to choose between funding research that relies on the ongoing destruction of embryos (and therefore tacitly supporting and encouraging such destruction by paying for the work that immediately follows it) and funding no human embryonic stem cell research at all. The decision regarding the funding of research on already-

---

* Using similar language, but speaking even more unambiguously, President Bush reiterated his ethical view of the destruction of human embryos for medical research in a speech on human cloning legislation, saying, "I believe all human cloning is wrong, and both forms of cloning ought to be banned, for the following reasons. First, anything other than a total ban on human cloning would be unethical. Research cloning would contradict the most fundamental principle of medical ethics, that no human life should be exploited or extinguished for the benefit of another. Yet a law permitting research cloning, while forbidding the birth of a cloned child, would require the destruction of nascent human life." ("Remarks by the President on Human Cloning Legislation," as made available by the White House Press Office, April 10, 2002.)

derived human embryonic stem cells came down to this question: *Can the government support some human embryonic stem cell research without encouraging future embryo destruction?*

The present funding policy is therefore *not* an attempt to answer the question of how the government might best advance embryonic stem cell research while conforming to the law on the subject. Rather, it is an attempt to answer the question of how the government might avoid encouraging the (presumptively) unethical act of embryo destruction and still advance the worthy cause of medical research. Whether or not one agrees with the premises defining the question, and whether or not one accepts the logic of the answer, any assessment of the policy must recognize this starting point.

From the very beginning, the policy has been described—even by many of its supporters and defenders—as occupying a kind of middle-ground position in the debate over the morality of embryo research. It has been termed a "Solomonic compromise." But while it may be a prudential compromise on the question of funding, it has been argued that the policy—as articulated by its authors—does not seem to be intended as a compromise on the question of the moral status of human embryos or the moral standing of the act of embryo destruction. In this sense, it appears to be not a political "splitting of the difference" but an effort at a principled solution.[18]

To some extent, the effort reflects a traditional approach in moral philosophy to an ancient and vexing question: Can one benefit from the results of (what one believes to be) a past immoral act without becoming complicit in that act?* The moralists' approach suggests that one may make use of such benefits if (and only if) three crucial conditions are met: (1) *Non-cooperation*: one does not cooperate or actively involve oneself in the commission of the act; (2) *Non-abetting*: one does nothing to abet or encourage the repetition of the act, for

---

* Readers should note that in reporting on this approach, as applicable to President Bush's stem cell decision, the Council is not itself declaring its own views on whether the past act of embryo destruction was "immoral." (Some of us think it was, some of us think it wasn't.) We are rather describing what we understand to be the moral logic of the decision as put forward.

instance by providing incentives or rewards to those who would perform it in the future; and (3) *Reaffirmation of the principle*: in accepting the benefit, one re-enunciates and reaffirms the principle violated by the original deed in question.

As a plan for redeeming some good from embryo destruction that has already taken place, while not encouraging embryo destruction in the future, the administration's policy appears at least to seek to address each of these three conditions: (1) No federal funds have been or, by this policy, would be used in the destruction of human embryos for research. (2) By restricting research funding exclusively to embryonic stem cell lines derived before the policy went into effect, the policy deliberately refuses to offer present or future financial or other incentives to anyone who might subsequently destroy additional embryos for research; this is the moral logic behind a central feature of the policy, the cut-off date for funding eligibility (though some argue that by failing to call for an end to privately funded research the policy does not altogether avoid complicity). And (3) the President, in his speech of August 9, 2001, and since (as in the passage quoted above and elsewhere), has reaffirmed the moral principle that underlies his policy and the law on the subject: that nascent human life should not be destroyed for research, even if good might come of it. The policy as a whole draws attention to that principle by drawing a sharp line beyond which funding will not be made available.

Of course, since these terms from the parlance of moral philosophy were not those explicitly employed by the policy's authors, they can go only so far in helping us to understand the policy's foundation. As in any public policy decision, prudence is here mixed with principle, in the hope that the two might reinforce (rather than undermine) each other, and a variety of moral aims are brought together. The desire to afford some aid to a potentially promising field of research moderates what might otherwise have been an at least symbolically stauncher stance against embryo destruction: no public funding whatsoever, even for work on stem cell lines obtained from embryos destroyed in the past. Moreover, the desire to show regard for established principles and standards of ethical research leads to an insistence that, to be approved, stem cell lines must have been drawn from embryos produced for reproduction and obtained with consent and without financial inducements. In these ways, the policy gives some due to competing moral and prudential demands. But the policy's central feature—the announcement date separating eligible from ineligible stem cell lines—holds firm to the principle that *public funds* should not be used to encourage or support the destruction of embryos *in the future*.

It is perhaps worth pointing out that one's attitude regarding the best federal funding policy is not simply determined by one's view regarding the moral standing of human embryos, and that even persons who hold the same view of the moral standing of human embryos may not all agree on the best policy. For example, support for the current policy does not *necessarily* require a belief that human embryos are persons with full moral standing; and conversely, those who believe that human embryos are persons do not necessarily support the policy. One might believe, for instance, that an embryo is a mystery, not clearly "one of us" but unambiguously a life-in-process, and thus conclude that we should err on the side of restraint (non-destruction) when moral certainty is impossible. Or, one might believe that embryos are not simply persons but are nonetheless either worthy of protection from harm or at least worthy of more respect than ordinary human tissues or animals, and that it would be wrong to begin a massive public project of embryo research that offends the deeply-held beliefs of many citizens. Meanwhile, an individual who believes that human embryos have the same moral standing as children or adults may be deeply unsatisfied with the present policy, since merely denying federal encouragement for future embryo destruction while taking no action to prevent privately-funded stem cell research that destroys embryos may be an insufficient response to the ongoing destruction of nascent human life.

For some of its supporters, the policy goes as far as it seems possible to go within the bounds of the spirit and aims of the law—that the government should not encourage or support the destruction of nascent human life for research. Yet at the same time, it goes farther than the federal government has gone before in the direction of actually funding research involving human embryos, since no public funds had ever before been spent on such research. To go further—say, by funding research on the currently ineligible lines derived after August 9, 2001—would not extend the logic of the policy or of the law, but rather contradict them both: it would be a difference not of degree but of principle. By implying that research using embryos destroyed in the future might one day be supported with public funding, such a policy shift would at least implicitly encourage the very act (embryo destruction) that the current policy aims not to encourage. Of course, such a change might well be in order, but the case for it must address itself to the moral argument and its principles, and not only to the state of research and its progress or promise.

Rather than focus on this principled aspect of the policy, the public debate has tended to concentrate on the precise balance of benefits and harms resulting from the combination of the administration's policy and the state of the relevant science. It has focused on whether there are "enough" cell lines or on whether the science is advancing as quickly as it could. And it has proceeded as though the administration's aim was simply to maximize progress in embryonic stem cell research without transgressing the limits of the letter of the law.

Had the decision been based on that aim alone, then claims or evidence of slowed progress alone might, in themselves, constitute an effective argument against it on its own terms (on the ground that the law technically permits federal funding of research on cells derived from embryos whose destruction was underwritten by private funding). But if one accepts the premise that the decision was grounded also in a discernible (albeit highly controversial) moral aim, one cannot show that the policy is wrong merely by pointing to the potential benefits of stem cell research or the potential harm to science caused by restrictions in federal funding. The present policy aims to support stem cell research while insisting that federal funds not be used to support or encourage the future destruction of human embryos. To argue with that policy on its own terms, therefore, one would need to argue with its view of the significance of that aim. Concretely, this means arguing with its ethical position regarding the destruction of nascent human life and with its ethical-political position regarding the significance of government funding of a contested activity.

This latter point—regarding the meaning of government funding—is much neglected in the current debates and deserves further clarification. That will require delving into the important distinction between government permission (that is, an absence of prohibitions) of an activity and government support for an activity. This ethical-political distinction lies at the heart of the stem cell debate.

IV. THE SIGNIFICANCE OF FEDERAL FUNDING

The national debate over human embryonic stem cell research often raises the most fundamental questions about the moral status of human embryos and the legitimacy of research that destroys such embryos. Yet, looking over this debate, it is easy to forget that the question at issue is not whether research using embryos should be allowed, but rather whether it should be financed with the federal taxpayer's dollars.

The difference between *prohibiting* embryo research and *refraining from funding* it has often been blurred by both sides to the debate. Ignored in the battles over embryo research itself, the ethical-political question regarding funding is rarely taken up in full.

That question arises because modern governments do more than legislate and enforce prohibitions and limits. In the age of the welfare state, the government, besides being an enforcer of laws and a keeper of order, is also a major provider of resources. Political questions today, therefore, reach beyond what ought and ought not be allowed. They include questions of what ought and ought not be encouraged, supported, and made possible by taxpayer funding. The decision to fund an activity is more than an offer of resources. It is also a declaration of official national support and endorsement, a positive assertion that the activity in question is deemed by the nation as a whole, through its government, to be good and worthy. When something is done with public funding, it is done, so to speak, in the name of the country, with its blessing and encouragement.

To offer such encouragement and support is therefore no small matter. The federal government is not required to provide such material support, even for activities protected by the Constitution, let alone for those permitted but not guaranteed.[19] The affording of most federal funding is entirely optional, and the choice to make such an offer is therefore laden with moral and political meaning, well beyond its material importance. In the age of government funding, the political system is sometimes called upon to decide not only the lowest standards of conduct, but also the highest standards of legitimacy and importance. When the nation decides an activity is worth its public money, it declares that the activity is valued, desired, and favored.

The United States has long held the scientific enterprise in such high regard. Since the middle of the twentieth century, the federal government, with the strong support of the American people, has funded scientific research to the tune of many hundreds of billions of dollars. The American taxpayer is by far the greatest benefactor of science in the world, and the American public greatly values the contributions of science to human knowledge, human health, and human happiness. And we Americans have overwhelmingly been boosters of medical science and medical progress, deeming them worthy of support for moral as well as material reasons.

But this enthusiasm for medical science is not without its limits. As already noted, we attach restrictions to federally funded research, for instance to protect human subjects. In fact at times we even use funding to *place* restrictions on research that might otherwise not be constrained. Indeed, federal funding sometimes serves as a means by which *private* research can be subjected to critical standards, since institutions that receive federal funds are often inclined (and given strong administrative incentives) to abide by the prescribed ethical standards throughout all of their activities, not only those directly receiving public dollars. Some supporters of funding therefore argue that extending public money to research is the most effective means of making certain that nearly all researchers, public and private, adhere to basic standards of ethics and safety. Public funding also requires researchers to make their work available to the public and for critical review by their peers, and it may encourage some degree of responsibility not necessarily encouraged by commercial endeavors.*

In addition to conditions attached to government funding of research, law sometimes erects specific limits on certain practices that might be medically beneficial. For example, we put limits on some practices that might offer life-saving benefits, such as the buying and selling of organs for transplantation, currently prohibited under the National Organ Transplant Act. Also, as in the present case, many Americans and their congressional representatives have moral reasons for opposing certain lines of research or clinical practice, for example those that involve the exploitation and destruction of human fetuses and embryos.

The two sides of the embryo research debate tend to differ sharply on the fundamental moral significance of the activity in question. One side believes that what is involved is morally abhorrent in the extreme, while the other believes embryo research is noble or even morally obligatory and worthy of praise and support. It would be very difficult for the government to find a middle ground between these two positions, since the two sides differ not only on what should or should not be done, but also on the moral premises from which the activity should be approached.

To this point, the federal government has pursued a policy whereby it does not explicitly prohibit embryo research but also does not officially condone it, encourage it, or support it with public funds (though state governments have often taken more active roles in both directions, as detailed in Appendix E). This approach, again, combines prudential demands with moral concerns. It has allowed the political system to avoid banning embryo research against the wishes of those who believe it serves an important purpose, while not compelling those citizens who oppose it to fund it with their tax money. This approach is also based, at least in part, on the conviction that debates over the federal budget are not the place to take

---

* Indeed, some even argue that the terms and conditions set for federal funding of research could be defined in such a way as not only to subject private research to general standards but also to help influence the eventual distribution of the products of that research to all those in need, or to serve other goods deemed publicly worthy.

up the anguished question of the moral status of human embryos.

But the position is not only a compromise between those who would have the government bless and those who would have the government curse this activity. It is also a statement of a certain principle: namely, that public sanction makes a serious difference and ought not to be conferred lightly. While embryo destruction may be something that some Americans support and engage in, it is not something that America *as a nation* has officially supported or engaged in.*

Of course, if the funding issue were merely a proxy for the larger dispute over the moral status of human embryos, then the present arrangement might appeal only to those who would protect human embryos, and it would succeed only as long as they were able to enact it. The argument might end there, with a vote-count on the question of the moral status or standing of human embryos. But some proponents of the present law suggest that the particulars and contours of the embryo research debate offer an additional rationale for that arrangement. Here again, it is important to remember that the issue in question is public funding, not permissibility. Opponents of embryo research have in most cases acquiesced (likely owing to various prudential and moral factors) in narrowing the debate at the federal level to the question of funding. They do not argue for a wholesale prohibition of embryo research by national legislation, even though many of them see such work as an abomination and even a form of homicide. In return, proponents of the Dickey Amendment argue that it would be appropriate for supporters of research to agree to do without federal funding in this particular field.

On the other hand, it might reasonably be argued that part of living under majority rule is living with the consequences of sometimes being in the minority. Were the Congress to overturn the current policy of withholding public funds from the destruction of embryos, opponents of funding for embryo research would not be alone in being compelled to pay for activities they abhor. We all see our government do things, in our name, with which we disagree. Some of these might even

---

* The repeated reenactment of the Dickey Amendment by the Congress may be taken as evidence of some support for this assertion.

involve life and death questions of principle, for instance in waging wars that some citizens deeply oppose. The existence of strong moral opposition to some policy is not in itself a decisive argument against proceeding with that policy.

These concerns give the question of funding its own crucial ethical significance, even apart from the more fundamental question of the legitimacy and propriety of the act being funded. This matter of funding broadly understood, together with the moral and prudential aims apparently motivating the administration's policy, as well as the legal context created by the Dickey Amendment, are the essential prerequisites for thinking about the underlying logic of the current policy. The combination of these elements gives form not only to the specific rules set forth in the administration's funding policy, but also to the implementation of that policy, to which subject we now turn.

V. IMPLEMENTATION OF THE PRESENT POLICY

The complex and critical task of implementing the funding policy falls largely to the National Institutes of Health, which administers most federal funding of biomedical research. As noted, the administration's policy attempts to advance stem cell research within the bounds already laid out regarding further destruction of human embryos. Thus, while the funding criteria of the policy set the bounds, the NIH, in its ongoing work, is expected to advance the goal of maximally effective funding and support within those bounds.

To this end, the NIH has worked to "jump-start" this field of research through a series of coordinated activities.[20] To plan and oversee these activities, the NIH has established a Stem Cell Task Force charged with determining the best uses for public funds in the field and with putting in place the resources required to make effective use of those funds.

The most basic material resources in question are the human embryonic stem cell lines themselves. In August 2001, President Bush announced that "more than sixty genetically diverse stem cell lines" (or stem cell preparations) already existed, and so would be eligible for funding under his policy.[21]

The NIH now believes the actual number to be somewhat higher, so that seventy-eight lines (or preparations) are known to be eligible for funding.* The lines are held by universities, companies, and other entities throughout the world. According to the National Institutes of Health's latest report (September 2003), the following organizations have developed stem cell derivations eligible for federal funding (that is, derived prior to August 9, 2001, under the approved conditions):

Name	Number of Derivations
BresaGen, Inc., Athens, Georgia	4
CyThera, Inc., San Diego, California	9
ES Cell International, Melbourne, Australia	6
Geron Corporation, Menlo Park, California	7
Göteborg University, Göteborg, Sweden	19
Karolinska Institute, Stockholm, Sweden	6
Maria Biotech Co. Ltd. – Maria Infertility Hospital Medical Institute, Seoul, Korea	3
MizMedi Hospital – Seoul National University, Seoul, Korea	1
National Centre for Biological Sciences/ Tata Institute of Fundamental Research, Bangalore, India	3
Pochon CHA University, Seoul, Korea	2
Reliance Life Sciences, Mumbai, India	7
Technion University, Haifa, Israel	4
University of California, San Francisco, California	2
Wisconsin Alumni Research Foundation, Madison, Wisconsin	5

Although all of these lines (or preparations) are deemed *eligible* for funding according to the criteria of the administration's policy, not all are presently *available* for use by researchers (nor is it clear that *all* of them will ever be

---

* These numbers took almost everyone by surprise. Prior to the President's announcement, the best estimates of the number of human embryonic stem cell lines then existing worldwide ranged between 10 and 20. But eligibility is not the same thing as availability, as we will discuss.

available for widespread use). Indeed, a point critical to understanding the current situation is that as of the autumn of 2003 only *twelve* lines are available for use,[22] while most of the other lines are not yet adequately characterized or developed (some exist only as frozen stocks) and so have at least not yet become available.* The process of establishing a human embryonic stem cell line, turning the originally extracted cells into stable cultured populations suitable for distribution to researchers, involves an often lengthy process of growth, characterization, quality control and assurance, development, and distribution. In addition, the process of making lines available to federally funded researchers involves negotiating a contractual agreement (a "materials transfer agreement") with the companies or institutions owning the cell lines, establishing guidelines for payment, intellectual property rights over resulting techniques or treatments, and other essential legal assurances between the provider and the recipient.

The entire process—scientific and legal—has tended to take at least a year for each cell line. Thus, determining which of the 78 eligible lines are in sufficiently good condition, characterizing and developing those lines, and establishing the arrangements necessary to make them available has been a demanding task. By September of 2003, slightly over two years after the enactment of the funding policy, twelve of the eligible lines had become available to federally funded researchers.* The NIH has made available "infrastructure award" funds (totaling just over $6 million to date) to a number of the institutions that possess eligible cell-lines, to enable them to more quickly and effectively develop more lines to distribution quality. As a result, while the number of available lines (only one in the summer of 2002 but risen to twelve in the autumn of 2003)* is expected to continue to grow with time, it is unclear how many of the 78 lines will finally prove accessible and useful. According to the NIH, as of the autumn of 2003, the own-

---

* By the time of final publication of this document, in January 2004, the number of available lines had risen to 15. This number is likely to rise further, and readers are advised to keep abreast of the current number and availability of embryonic stem cell lines eligible for funding at the NIH Stem Cell Registry website, stemcells.nih.gov.

ers of the available lines have distributed over 300 shipments of lines to researchers. No information is presently available on the number of individual researchers or institutions that have received lines.[23]

Successful implementation of the current funding policy depends not only on the availability of eligible lines, but also on adequate allocation of financial resources to develop and make use of those lines and to advance the field in general. The funding policy, though it limits the targets of funding to the eligible lines, does not directly delimit or restrict the *amount of money* and other resources that the NIH may invest in human embryonic stem cell research. The amount invested, a decision left to NIH and the Congressional appropriations process, is largely a function of the number of qualified applicants for funding and of the NIH's own priorities and funding decisions. Of course, if more lines were eligible for funding, it is quite possible that more funding would be allocated, but the *amount* that *can* be allocated to work on existing lines is not limited by the funding criteria. In fiscal year 2002, the NIH devoted approximately $10.7 million to human embryonic stem cell research. Based upon a September 2003 estimate, it will have spent approximately $17 million in fiscal year 2003. It is expected that further increases will follow as the field and the number of grant applications grow.

As of November 2003, NIH funds have been allocated to support the following new and continuing awards for human embryonic stem cell research: nine infrastructure awards to assist stem cell providers to expand, test, and perform quality assurance, and improve distribution of cell lines that comply with the administration's funding criteria (aimed at making more of the eligible lines available); 28 investigator-initiated awards for specific projects; 88 administrative supplements (awarded to scientists already receiving funds for work on other sorts of stem cells, either non-embryonic or non-human, to enable them to begin to work with eligible human embryonic stem cell lines); two pilot and feasibility awards; three awards to support exploratory human embryonic stem cell centers; one institutional development award; four post-doctoral training fellowships; one career enhancement award; and six awards to fund stem cell training (including short-term courses) to provide hands-on training to enable researchers to learn the skills and techniques of culturing human embryonic stem cells.

The latter task, of training new researchers, the NIH regards as one of its principal challenges in advancing the field, and, along with available lines and available financial resources, as a key measure of how the field is progressing. As NIH Director Elias A. Zerhouni put it in his presentation before this Council,

> I don't think the limiting factor is the cell lines. I really don't. I really think the limiting factor is human capital and trained human capital that can quickly evaluate a wide range of research avenues in stem cells.[24]

The NIH has therefore devoted funding to the training of investigators and the cultivation of career development pathways, including short-term courses in stem cell culture techniques and (long-term) career enhancement awards in the field. Some critics have contended, however, that the two issues (funding restrictions and the scarcity of personnel) are likely connected, and that limits on the cell lines eligible for funding and the surrounding political controversy cause some potential researchers to stay away from the field, contributing to a shortage of investigators.[25]

These federal resources, then, have been directed toward the advancement of human embryonic stem cell research within the bounds of the determination to refrain from supporting or funding new destruction of human embryos. Scientists may receive federal funding—at any level determined appropriate by the NIH—for any sort of meritorious research, using as many of the approved lines as they are eventually able to use. They can, of course, also receive federal funding for using or deriving new animal embryonic stem cell lines, to assess the potential of these cells for treatment of animal models of human disease (though of course animal models provide only limited information because they are not in many cases exactly extrapolatable to the specific situations that hold in human disease and development, and so cannot replace human cell sources).

Researchers can, in addition, use federal funds for work involving human embryonic germ cells, obtained from aborted fetuses. They can carry out research projects using embryonic germ cell lines already derived, following review and approval of specific institutional assurances, informed consent documents, scientific protocol abstracts, and Institutional Review Board approvals by the NIH's Human Pluripotent Stem Cell Review Group (HPSCRG). They can also receive federal funds for the derivation and study of *new* embryonic germ cell lines following the same HPSCRG review and approval process. In addition, of course, they can develop animal embryonic germ cell lines to assess the potential of these cells through animal models.

Also, researchers can receive federal funds for work conducted on human adult (non-embryonic) stem cells. There are no restrictions regarding what American scientists can do with regard to adult stem cells using taxpayer funds, other than those requiring them to honor the usual human subject protections and clinical research requirements (if they are to be transplanted into human patients). The NIH has devoted substantial resources to the study of human adult stem cells, allocating over $170 million to the field in fiscal year 2002, and approximately $181.5 million in fiscal year 2003 (approximately ten times the amount devoted to human embryonic stem cell work).

Finally, researchers remain free to pursue work (including the derivation of new lines of embryonic stem cells) in the private sector, without government funding. Indeed, as discussed above, embryonic stem cells were first isolated and developed in the private sector, or in university laboratories using private sector funds, and no work in the field was publicly funded at all until 2001. Under present law, work supported by private funds can proceed without restriction. Under rules promulgated in the spring of 2002, such work does not need to be conducted in a separate laboratory, but a clear separation of the funds used to support this work from any federally funded work of the laboratory is required. Of course, because of the highly interlocking and complex nature of the various aspects of operating a laboratory, such separation can still prove extremely difficult to manage. It is not clear precisely how much privately funded work using human embryonic stem cells has been undertaken in the past few years, but some general figures are available. The most recent and thorough survey available, based on figures from 2002,[26] suggests that approximately 10 companies in the United States were actively engaged in embryonic stem cell work, employing several hundred researchers and, cumulatively over the past several years, spending over $70 million in the field, which is well over twice what the NIH has so far spent.[27] Those involved in privately-funded research in the field, however, generally do not see private funding as a substitute for federal funds, but would much prefer that the field had the opportunity to benefit from both. They also argue that restrictions on federal funds, and the controversy surrounding the subject, act to dissuade potential investors from entering the field, and thereby have a "chilling effect" on private as well as publicly funded research.[28]

Moreover, just as federal policy can affect privately conducted research, so too a number of states have enacted policies affecting stem cell research, ranging from all-out prohibitions of such research to official statements of support and positive encouragement.* The status of such research, and the conditions to which it is subject, can vary dramatically from state to state, independent of federal funding policy.

VI. CONCLUSION

The administration's policy on the funding of embryonic stem cell research rests on several moral and ethical-legal principles, set upon the reality of existing law:

1. *The law*: The Dickey Amendment, which the President is required to enforce.

2. *The principle underlying the law*: The conviction, voiced by the administration, a majority of the Congress, and some portion of the public, that federal taxpayer dollars should not be used to encourage the exploitation or

---

* State policies regarding embryo research are detailed in Appendix E.

destruction of nascent human life, even if scientific and
medical benefits might come from such acts.

3. *The principle underlying the desire to offer funding*: That
    efforts to heal the sick and the injured are of great
    national importance and should be vigorously
    supported, provided that they respect important moral
    boundaries.

4. *The significance of federal funding*: That federal funding
    constitutes a meaningful positive statement of national
    approval and encouragement, which should be
    awarded only with care, particularly in cases where
    the activity in question arouses significant public moral
    opposition.

The significance of the policy is best understood in light of
these key elements. Its soundness is most reasonably measured against them and against the policy's implementation
by the National Institutes of Health.

Though the prudential and principled considerations raised
in this chapter governed the formulation of the policy, or at
least defined its articulation by its advocates and authors,
these are not the only terms by which federal funding policy
might be conceived or measured. In the next chapter we
present an overview of the ethical and policy debates that
have raged for the past two years around both the wisdom of
the present policy and the fundamental issues at stake in human embryonic stem cell research.

ENDNOTES

[1] National Research Act, Pub. L. No. 93-348, § 213, 88 Stat. 342 (passed by the 93rd Congress as H.R. 7724, July 12, 1974).

[2] National Commission for the Protection of Human Subjects of Biomedical and Behavioral Research, *Research on the Fetus: Report and Recommendations* (Washington, D.C.) 1975. Reprinted at 40 Fed. Reg. 33,526 (1975).

[3] "HEW Support of Human In Vitro Fertilization and Embryo Transfer: Report of the Ethics Advisory Board," 44 Fed. Reg. 35,033 (June 18, 1979) at 35,055-35,058.

[4] *Ibid.*

[5] National Institutes of Health Revitalization Act of 1993, Pub. L. No. 103-43, § 121(c), 107 Stat. 122 (1993), repealing 45 C.F.R. § 46.204(d).

[6] National Institutes of Health, *Report of the Human Embryo Research Panel*, Bethesda, MD: NIH (1994).

[7] "Statement by the President," as made available by the White House Press Office, December 2, 1994.

[8] The text of the Dickey Amendment can be found in each year's Labor/HHS Appropriations Bill. The original version, introduced by Representative Jay Dickey, is in § 128 of Balanced Budget Downpayment Act, I, Pub. L. No. 104-99, 110 Stat. 26 (1996). For subsequent fiscal years, the rider is found in Title V, General Provisions, of the Labor, HHS and Education Appropriations Acts in the following public laws: FY 1997, Pub. L. No. 104-208; FY 1998, Pub. L. No. 105-78; FY 1999, Pub. L. No. 105-277; FY 2000, Pub. L. No. 106-113; FY 2001, Pub. L. No. 106-554; and FY 2002, Pub. L. No. 107-116. The most current version (identical in substance to the rest) is in Consolidated Appropriations Resolution, 2003, Pub. L. No. 108-7, 117 Stat. 11 (2003).

[9] "Rendering legal opinion regarding federal funding for research involving human pluripotent stem cells," Memo from Harriet S. Rabb, General Counsel of the Department of Health and Human Services to Harold Varmus, Director of the National Institutes of Health, January 15, 1999. (Available through the National Archives.)

[10] This case was made, for instance, in a letter authored by Rep. Jay Dickey and signed by seventy other Members of Congress to DHHS Secretary Donna Shalala, February 11, 1999.

[11] DHHS Secretary Shalala argued this point in a letter responding to the Congressional letter of opposition (see note 10, above), February 23, 1999.

[12] President Bush has made a number of statements articulating the position that nascent human life (including at the early embryonic stage) is deserving of protection and ought not be violated. See especially: "Stem Cell Science and the Preservation of Life," *The New York Times*, August 12, 2001, p. D13; "Remarks by the President on Human Cloning Legislation," as made available by the White House Press Office, April 10, 2002; "Remarks by the President at the Dedication of the Pope John Paul II Cultural Center," as made available by the White House Press Office, March 22,

2001; and "President Speaks at 30th Annual March for Life on the Mall," as made available by the White House Press Office, January 22, 2003.

[13] "Remarks by the President on Stem Cell Research," as made available by the White House Press Office, August 9, 2001.

[14] "Remarks by the President upon Departure for New Jersey," as made available by the White House Press Office, August 23, 2000.

[15] National Institutes of Health, *Report of the Human Embryo Research Panel*, Bethesda, MD: NIH, 1994.

[16] See, for instance, the "Moral Case for Cloning-for-Biomedical-Research" presented by some Members of the Council in the Council's July 2002 report *Human Cloning and Human Dignity: An Ethical Inquiry*, chapter 6.

[17] Bush, G.W., "Stem Cell Science and the Preservation of Life," *The New York Times*, August 12, 2001, p. D13.

[18] See, for instance, Council discussion at its September 3, 2003, meeting. A transcript of that session is available on the Council's website at www.bioethics.gov

[19] This question has been addressed by the Supreme Court on a number of occasions, in which the Court found that even activities protected as rights under the Constitution are not thereby inherently worthy of financial support from the federal government. See, for instance, *Maher v. Roe* 432 U.S. 464 (1977); *Harris v. McRae* 448 U.S. 297 (1980); and *Rust v. Sullivan* 500 U.S. 173 (1991). Also see Berkowitz, P. "The Meaning of Federal Funding," a paper commissioned by the Council and included in Appendix F of this report.

[20] The information provided in this section relies primarily on a presentation delivered before the Council by NIH Director Elias Zerhouni on September 4, 2003, and on data otherwise made available by the National Institutes of Health. The full transcript of Director Zerhouni's presentation may be found on the Council's website at www.bioethics.gov.

[21] "Remarks by the President on Stem Cell Research," as made available by the White House Press Office, August 9, 2001.

[22] As of the autumn of 2003, the following providers have eligible lines available for distribution: BresaGen (2 available lines), ES Cell International, Australia (5 available lines), MizMedi Hospital, South Korea (1 available line), Technion University, Israel (2 available lines), University of California at San Francisco (1 available line), Wisconsin Alumni Research Foundation (1 available line). A complete list of available and eligible lines, updated as more lines become available, can be found at the NIH Stem Cell Registry website at stemcells.nih.gov.

[23] This information has been made available to the Council by the National Institutes of Health.

[24] Quoted from a presentation before the Council by NIH Director Elias Zerhouni, September 4, 2003. The full transcript of Director Zerhouni's presentation may be found on the Council's website at www.bioethics.gov.

[25] See, for instance, the presentation of Thomas Okarma, President and CEO of Geron Corporation, before the Council on September 4, 2003. The full transcript of Okarma's presentation may be found on the Council's website at www.bioethics.gov.

[26] Lysaght, M.J., and Hazlehurst, A.L., "Private Sector Development of Stem Cell Technology and Therapeutic Cloning," *Tissue Engineering* 9(3): 555-561 (2003).

[27] The dollar amount spent specifically on embryonic stem cell research in the private sector is not apparent from Lysaght and Hazlehurst's survey. The $70 million figure is drawn from a presentation before the Council by Thomas Okarma, President and CEO of Geron Corporation, the oldest and largest of the private companies involved in embryonic stem cell research. Okarma told the Council, speaking only of Geron, "We have spent over $70 million on this technology, most of it since 1999 after the cells were derived. That's a number against which the NIH disbursements pale by both absolute and relative terms." The full transcript of Okarma's presentation may be found on the Council's website at www.bioethics.gov.

[28] *Ibid.*, and see also (for instance) Mitchell, S., "U.S. stem cell policy deters investors," *The Washington Times*, November 2, 2002 (original source: UPI).

3

Recent Developments in
the Ethical and Policy Debates

The announcement of the Bush administration's human embryonic stem cell research funding policy in the summer of 2001 certainly did not end the debates surrounding the issue. The policy offered a particular target to which participants in the debate could react, but the basic questions involved in assessing how the federal government should approach embryonic stem cell research remained just as relevant, and just as controversial, as they had been before. In this chapter, we offer an overview of that still continuing debate as it has developed in the past two years. Without attempting to provide anything like a full account of different positions and arguments, we hope, rather, to point to the major items under argument—to the issues that any interested citizen might wish to ponder. First, we will outline the general form of the moral argument as it has developed over time. Second, we will discuss specific questions and critiques regarding the current policy, as those have emerged in public discussion. Third, we will review the various positions on the moral standing of human embryos, seeking again to outline the chief fault lines in that continuing debate. And finally, we will highlight some critical ethical concerns that do not arise directly out of the debate over the moral standing of human embryos but that may be no less important to the larger question confronting the country.

This array of subjects includes both some that are clearly ethical questions and some that may fall closer to questions of policy. We take up both together because we believe both are essential to a full understanding of the issues involved and because even the policy debates—as understood and engaged in by participants on all sides—are clearly directed to a set of underlying moral issues and informed by ethical beliefs and opinions. Critiques (and defenses) of the present policy, and proposals of alternatives, are almost without exception based on moral grounds.

This chapter will, of course, revisit several of the themes raised in the previous chapter. There, our aim was to present some basic facts regarding the current policy, its context, and its execution. Here, we take up arguments on all sides of the issue—including those that dispute the understanding of the policy put forward by its authors and those that raise other issues or alternative courses of action.

While we will raise these arguments and counterarguments, examining problems, questions, and concerns, our purpose is not finally to assess the validity of the competing claims or to arrive at a conclusion, but—in line with the Council's charge to monitor developments in this area—to present them more or less as they have appeared in the public debates of the past several years. This way of proceeding suffers from one especially prominent drawback: it tends to present all arguments as equal in importance or prevalence. We will seek to avoid this whenever possible, by offering some sense of which arguments have been most crucial for the public debate.

I. THE NATURE OF THE MORAL ARGUMENT

Before entering upon a detailed review of the debates, we take a moment to consider the *character* of the moral reasoning and argumentation we will confront. Different participants in the stem cell debates tend to hold different views not only regarding individual substantive judgments, but also regarding the kind of moral question, in the most general terms, we are facing in deciding about stem cell research policy. At first glance, people seem to be disagreeing about whether a balancing of competing interests and goods is called for, or whether some one overriding moral duty ought to shape our judgment, though of course that dichotomy is not in every case quite so stark.

In casual conversation, and sometimes in more theoretical reasoning, moral questions are often analyzed in the language of "weighing" and "balancing," or else in terms of overriding concerns, inviolable principles, or fundamental "rights." These two sets of terms and metaphors point to two distinct ways of making difficult moral judgments. In some circumstances there need be no irresolvable conflict between the two approaches. Those who seek to respect fundamental rights and adhere to inviolable principles need not ignore the complexities of moral situations or the consequences of our actions when they form their judgments. Likewise, those who seek to weigh or balance competing goods may think of those goods as involving not only benefits to be realized but also principles to be upheld. There may also be circumstances, however, in which those differences that do exist between these two approaches constitute a fork in the road, constraining our decision. Thus, to take an example from outside the domain of bioethics, the principle that civilians should be safeguarded from direct, intended attack in time of war may be understood to trump all other competing goods (without denying the importance of those goods), or it may be understood as one good to be balanced against others. Which fork in the road of moral reasoning one takes at such a point will have a decisive influence on the character of the arguments employed and the conclusions reached.

This has often also been the case in the public debate over federal funding of human embryonic stem cell research. Many observers have agreed that federal policy must seek a certain balance between two competing goods or interests: the progress of medical research and therapy, and respect for nascent human life. Indeed, President Bush himself framed the issue in these terms, saying a few weeks before his decision was announced that his policy would "need to balance value and respect for life with the promise of science and the hope of saving life."[1] But not all participants in the debate have had the same idea of what such balance should entail and, therefore, how weight should be assigned to the competing demands.

For some, the degree of medical promise should profoundly affect the degree of government support for embryonic stem cell research. For them, the critical element (though of course not the exclusive one) in establishing policy must be scientific data about what might be achieved with human embryonic stem cells.[2] The greater the promise of the research, the more support it should receive and the more it should outweigh reasons offered for opposing the techniques involved.[3] On this view, the concern given greatest prominence and weight in reaching a judgment ought to be the very great good of medical progress toward the relief of suffering. Most of the arguments in favor of increased taxpayer funding of embryonic stem cell research have begun from this premise—explicitly or implicitly—and have proceeded by laying out the possible medical benefits of the research or the possible harms (to patients or to American science) of withholding support. We shall review a number of these lines of argument in more detail below. But it is worth noting at the start that most of them tacitly assume that the policy decision at hand ought to be based on a reasoned balancing of crucial concerns—all of which matter but none of which simply overrides the others.

Others, however, have suggested that at least a substantial portion of the opposition to the research rests upon the belief that human embryos should not be violated and therefore—if this claim is valid—that the threat to their life and worth cannot be justified by the promise of research.[4] It follows, on this view, that the federal government should do nothing to encourage or support the future destruction of human embryos, regardless of the promise of research. What remains then to be considered is the extent to which the government might advance the additional aim of progress in medical research *within* the bounds of the principle.[5] In this case, the moral reasoning is understood to be decisively affected by an unbreachable boundary, and only the extent of some particular provisions of the policy are left to be settled by a weighing and balancing of other priorities. Proponents of the various forms of this position generally argue that the claim of human embryos to our protection presents us with a fundamental duty, to be overridden, if at all, only in extreme circumstances, rather than with just one good to be balanced off against others.[6]

In presenting the matter this way, adherents of this view consciously appeal to the ethics governing research with human subjects, which obliges those engaged in efforts to advance knowledge and seek cures to keep from trespassing upon human safety, freedom, and dignity.[7] The decades-old and nearly universal adherence of researchers to rules protecting human subjects, these commentators suggest, demonstrates that the needs of research are not always treated as paramount and that the scientific community itself joins the general public in recognizing instances in which research—however important—must be limited for ethical reasons.[8] Researchers do not weigh the interests of human subjects against the importance of their work; rather they respect a principled boundary—that human subjects are not to be harmed or put at risk without their informed consent—the importance of which trumps even the most promising experiment. For some defenders of human embryos, the prospect of embryo research raises precisely these concerns; accordingly, they argue that this issue too should be decided on the basis of a moral rule, not by a shiftable tally on a balance sheet of benefits.[9]

But this assertion about the proper form of moral argument depends on the truth of the claim that human embryos are indeed human subjects of research. Therefore, one's position in the debate about the basic character of the moral issue may depend, in many cases, on one's understanding of the moral standing of human embryos. As we shall see, the question of the moral standing of embryos is by no means the only relevant question. But in the actual public debate, as it has developed, this question seems to have been most central and prominent and probably most responsible for shaping the different basic approaches pursued. It is this question that very often informs the differing views regarding which aims or goods are more weighty or which should not be compromised at all.

We turn next to arguments regarding that very issue: Which moral aims or which concerns should be given priority in shaping government policy toward human embryonic stem cell research?

II. THE MORAL AIMS OF POLICY

A significant part of the public debate surrounding the policy governing funding of embryonic stem cell research has involved differing views of just which purposes or ideals should most directly guide policymakers in this arena. The current funding policy—while it appears to strike some balance between protecting human embryos and advancing biomedical research—seems to take as its overriding concern the insistence that the federal government not encourage or support the deliberate violation or destruction of human embryos. But a number of commentators in recent years—and, of course, particularly those who ascribe lesser moral status to human embryos—have proposed alternative principles to govern policy.

A. The Importance of Relieving Suffering

Many observers argue that the proper governing principle should be the duty to relieve the pain and suffering of others— the purpose that ultimately motivates the work of biomedical science. This aim is broadly, perhaps universally, shared. Indeed, the current policy, as outlined by its advocates, while it seeks to protect nascent human life, explicitly seeks to advance medical research as far as its authors believe is morally permissible. Some commentators argue, however, that the administration has chosen the wrong one of these aims as the governing principle of its approach.

The cause of curing disease has a human face, the face of a loved one or neighbor, bent under the suffering of an incompletely understood or treated disease. As a result, the aspiration to know is linked to a desire to relieve. How, wonder many commentators, could anyone think of withholding support for research that might yield therapies for devastating diseases and conditions such as spinal cord injury, diabetes, and Parkinson disease?[10] Surely, they argue, the pain and suffering of those in need should outweigh concerns for human embryos frozen in a laboratory.[11] Indeed, for many this is an especially critical issue in light of the likely ultimate fate of these embryos. Will they be frozen indefinitely? If not, then will they be thawed out and discarded and destroyed, or used to potentially benefit mankind?

One form of this argument, heard increasingly over the past few years, begins with data about disease prevalence and suggests that anyone who obstructs public funding of research that might someday help patients suffering from such diseases must bear some of the responsibility for their future suffering and (perhaps) death.[12] Some opponents have countered that any assertion that makes relief of human suffering the highest moral principle to this extent might logically impugn any and all deflection of resources into less ultimate concerns such as recreation, beautification, or social ceremony. Others respond more directly that an unwillingness to violate one's moral principles in order to help relieve the sick does not make one responsible for their sickness.[13]

In most cases, however, the arguments for grounding federal funding policy in the importance of biomedical research do not blame opponents for the suffering of the sick. Rather, they focus on the promise of bringing relief to those who most need it. They point to the immense benefits already delivered to us by modern medicine and argue that the federal government should advance this cause in whatever ways it reasonably can. Many advocates of this view agree that nascent human life should receive respectful treatment, but they argue that the claims of our duties to human embryos—whether in general or in particular circumstances, like those stored in fertility clinic freezers—cannot simply trump the claims of promising medical research and our duties to suffering humanity. The obligation to aid the sick, they contend, and the fact that the research in question might relieve the pain and terrible suffering of countless patients and their families, should lead us to do all that can reasonably be done to find treatments and cures, and to offer help. This does not mean simply ignoring the significance of human embryos, or taking lightly the decision to destroy them in research, but, they suggest, it should mean taking seriously the moral calling to help the suffering and deciding how to proceed based on more than one sort of obligation.[14]

In response, one commentator has argued in broad terms against the underlying assumption that the demands of biomedical research should somehow be seen as "imperative." Such work, he contends, should not be seen as inherently and always obligatory, and claims to support it may be overridden even by a level of respect for nascent human life that does *not* suppose that embryos possess full human moral standing. He suggests that it is not at all obvious that individuals or the government have a definite responsibility to support such research or that such a responsibility would override other moral duties.[15]

Others point out that the duty to find cures for disease cannot be an unqualified or absolute imperative. Pointing to the present rules governing the treatment of human subjects in research, they argue that the case for embryonic stem cell research cannot rest on an alleged and overriding imperative to pursue that research. These rules prohibit certain sorts of procedures on human subjects of research, and the same, they suggest, should be required in stem cell research. Yet those who argue that the importance of medical research and treatment should override the aim of protecting human embryos presumably would not propose to override protections for human subjects in research on children or adults. Rather, they approach the present matter differently because they do not consider human embryos the developmental, anthropological, or moral equivalent of children or adults, or worthy of the same protection. The difference, therefore, has to do not so much with a dispute over the imperative importance of research, but rather (at least to a significant extent) with the status of nascent human life, which again turns out to be the fundamental point at issue.[16]

Nevertheless, the moral claims of medical research and treatment are extremely powerful in the debate over human embryonic stem cell policy, and they are acknowledged as profoundly important even by those who do not finally take them to be decisive. Most arguments in opposition to the present funding policy and in support of expanded embryonic-stem cell research are grounded in these claims.

B. Freedom to Conduct Research

Other opponents of the present policy argue not from the value of medical benefits but on the basis of freedom to conduct research, which they believe is the principle by which federal policy ought to be governed. They regard government restraints on scientific research as inherently offensive and generally unjustifiable.[17] The cherished ideals of freedom of thought, freedom of conscience, and—specifically in this context—freedom of inquiry, trump concerns over the moral status of human embryos, they contend.[18]

The most common claim advanced for protecting research as a basic right (employed, among others, by the American Bar Association) involves some form of an appeal to the First Amendment's protection of free speech, interpreted through the years by the Supreme Court to encompass free expression and perhaps freedom of inquiry and thought.[19] Some argue that research is a form of expression, particularly when it is politically or socially controversial, and when restraints upon it are imposed for moral reasons.[20] "One could make the case that research is expressive activity and that the search for knowledge is intrinsically within the First Amendment's protection for freedom of thought," says one ethicist.[21]

Others, however, contend that this claim, never tested in the courts, seems far-fetched. Most currently controversial biological research involves experimental manipulation of living matter, rather than theoretical exploration or mere observation of natural objects. It is therefore as much action as expression, as much creation as inquiry. It is difficult to see, they argue, how such activity (as opposed to the reporting of the results of such activity) could be classified as a form of expression. "Scientists may have the right to pursue knowledge in any way they want cognitively, intellectually," argues one observer, "but when it comes to concrete action in the lab, that becomes conduct and the First Amendment protection for that is far, far weaker."[22]

Moreover, argues at least one commentator, even if one did stipulate that research activity is to be protected from governmental proscription or restriction, it is far from clear that failure to provide federal funding would constitute such a restriction.[23] Indeed, legislative and judicial precedents suggest it would not. The government routinely refrains from funding activities it otherwise permits or even guards as constitutionally protected. A line of Supreme Court decisions stretching from 1977 to 1991, dealing with abortion and government funding, established the principle that the Constitution does not require the government to fund even those activities that the Constitution protects.[24] Because the only issue in the present debate is one of federal funding, the protected status of scientific activity seems not to be a determining factor.[25]

Finally, some critics of the case for a paramount right to research point again to the fact that scientific research—conducted both with private and (especially) with government funding—is already subject to certain restrictions, particularly with regard to protecting human subjects. The proposition that embryo research should not be subject to the same restrictions hinges on an argument about the standing of human embryos, rather than about the unrestrictable standing of research as such.[26] Once more, an important part of the question turns on the status of extra-uterine human embryos.

C. The Moral Standing of Human Embryos

However they approach the matter, then, many people engaged in the debate over federal funding policy find they must consider the fundamental question of the moral standing of human embryos: What *are* early human embryos, and how should we regard them morally? Approaches to the question of federal funding of embryonic stem cell research that propose some other guiding principle—relief of suffering, freedom of research—seem almost by necessity to assume that human embryos do not possess the same human moral standing as persons already born.[27] Conversely, if human embryos ought rightly to be treated as inviolable—as some have argued—then questions of balancing other goods or giving priority to other principles are largely rendered moot. Thus, to many observers, some of the central questions in this arena would appear to be those that surround human embryos: How ought we to think about and act toward human embryos? Should all ex vivo human embryos be treated the same, or are some, because of their circumstances, origins, or prospects, to be treated differently from others? Or are embryos of sufficiently little moral significance that we should simply decide the funding question solely on the basis of the merits and promise of the proposed biomedical research?

We shall, in due course, review the recent arguments on these critical questions. But before doing so, we examine a series of other more specific objections to the present policy. These arguments recognize that the current policy rests on the conviction that federal funds should not support or encourage the violation or destruction of human embryos. Rather than disputing this premise, a number of commentators—including both supporters and critics—have assessed the policy on its own grounds and judged it against its own claims and terms. As a result, several general categories of criticism of the policy itself have emerged.

III. THE CHARACTER OF THE POLICY

In addition to debating which aims ought to guide federal policy, many observers in the past two years have also assessed the *particulars* of the present funding policy, considering them in scientific, political, and moral terms. Critics have generally found fault with the policy through one or more of three general lines of argument: that it is arbitrary, that it is unsustainable, and that it is inconsistent. Defenders of the policy, meanwhile, have usually sought to counter these critics on these terms and to rebut these assertions and criticisms.

A. Arbitrary

One quite common line of argument criticizes the present funding policy as essentially arbitrary, because it relies on what is deemed a capricious cutoff date. Cells derived from embryos destroyed on August 9, 2001 are eligible for federal funding, but those obtained from embryos destroyed the next day are not. The only difference is the date of embryo destruction, argue some critics, and what moral difference could that possibly make? If the policy of funding human embryonic stem cell research serves a genuine good, these commentators suggest, would it not be equally good regardless of when the cell lines were derived? Would it not, therefore, make sense to permit funding for *all* cell lines derived on either side of the date, provided they are otherwise eligible?[28] If, on the contrary, federal funding of embryo research is unacceptable, then it should simply not be allowed at all, regardless of when cell lines were first derived.[29] "It is difficult to see," writes one critic, "what ethical reasoning would commend a policy that takes as its central distinction the time chosen for political convenience to deliver a presidential address."[30]

In response, some supporters of the policy contend that while the cut-off date (August 9, 2001) certainly has no inherent moral significance, it acquires moral meaning by the simple fact that it was the operative date of a newly announced policy that turned on a crucial distinction between the "up-until-now" and the "from-now-on," between past (irrevocable) deeds and future (preventable) ones. That date of announcement was the line between past embryo destruction, which could no longer be undone, and future embryo destruction, which could still be influenced by the federal government's funding rules. If the policy sought to avoid encouraging or offering incentives for any future destruction of human embryos, they argue, then drawing a hard line between past and future would be indispensable. That line could only reasonably be drawn at the moment of the decision's announcement[*] (or before it), since drawing it at some point in the future would create a powerful incentive to quicken the pace of work until the cut-off date arrived. The date, they say, is therefore not a morally arbitrary marker in the context of the policy, but is rather a crucial and unavoidable element of the policy's logic.[31]

---

[*] Only "the now," created by an act of speaking, defines the difference between past and future (tenses and deeds).

B. Unsustainable

A further, and more common, critique of the current policy suggests that its approach cannot be expected to hold over time and that it will fairly quickly prove unsustainable.

One form of this argument suggests that the policy has created a situation in which scientists may make some progress using existing stem cell lines but would then be prohibited from capitalizing on what they have learned and making further progress using stem cell lines not now eligible for federal funds. As the editors of the *Washington Post* put it: "Mr. Bush's compromise policy will be a reasonable one only as long as the existing lines are capable of supporting the research scientists need to perform."[32] Indeed, some have argued that by explicitly encouraging and speaking of the potential medical value of human embryonic stem cell research, the President himself created the circumstances that will make the constraints of the policy unsustainable. "Bush's decision was at its core an endorsement of the promise of human embryonic stem cells and their importance to the fledgling field of regenerative medicine," wrote one critic, and if that is the core message of the decision, then the resulting policy seems insufficient.[33]

Others, arguing on more practical grounds, claim that, regardless of the reasonableness of the principle behind the policy, its effects will prove unbearable for American scientists, who will then force a reconsideration.[34] The limits placed on funding in the current policy, several observers have predicted, will seriously hamper and hold back embryonic stem cell research work in the United States, perhaps causing prominent scientists to leave the country in search of greater support abroad.[35] If that were to occur, it is argued, the policy would prove genuinely damaging to American science, and therefore to the national interest, and would need to be changed.[36]

In response to this specific point, some defenders of the policy have noted that for the time being there has not been news of any notable migration of prominent stem cell researchers to foreign countries, that federal funding *is* now available with no ceiling on its amount, and that American researchers can continue to work with private funds.[37]

But even apart from worries about a "brain-drain" in the field, some have argued that both the lack of funding, and particularly the complexity of the set of conditions under which research using such funding may be conducted, is preventing progress in research and discouraging even private funding in the field,[38] and that the lines made available simply will not be enough,[39] or indeed that they have already proved insufficient.[40] Some argue that the very low number of applications submitted to the NIH for postdoctoral and training projects using the approved stem cell lines provides striking evidence for the chilling effect of the current "in limbo" situation. Some have also suggested that safety issues connected with the way the eligible lines have been derived and developed may make them less suitable for use in human trials and treatments, thus making other cell lines necessary (although presentations before this Council by NIH Director Elias Zerhouni and FDA Commissioner Mark McClellan have contradicted some of these claims).[41] Others, meanwhile, worry that the eligible lines do not offer sufficient genetic diversity to be adequate for research needs or for eventual therapies.[42] As scientists make their case that important work is being hampered, it is argued, the policy will prove unsustainable politically and practically.[43]

A similar argument has also been made in nearly the opposite terms. That is, some have said that if or when ongoing embryonic stem cell research produces a spectacular breakthrough in understanding or treating disease, the pressure to alter the policy would prove unstoppable.[44] This way, whether the future brings announcements of great progress, or whether it brings no news of advances, the result will be pressure for a policy that funds research more broadly.

Those defenders of the policy who have addressed these claims of unsustainability have pointed out that the present policy does not limit the *amount* of federal funds available to the kind of research that may be performed with the approved cell lines. They also point out that much of the pioneering work in mouse stem cell research was done using very few (approximately five) cell lines[45] (though of course the challenge of working with human cell lines is more complex and daunting). But they have also generally pointed to the policy's stated grounding in principle—a principle that would not change in light of scientific advances or delays. In August of 2001, for instance, Health and Human Services Secretary Tommy Thompson told reporters that "neither unexpected scientific breakthroughs nor unanticipated research problems would cause Bush to reconsider" the approach drawn by the policy, because it is based on "a high moral line that this president is not going to cross."[46] As another observer has put it: "The general question is, well, will these cell lines be enough . . . and a non-complicity argument [like the one implied by the policy] will only work if the answer to that question is well, I guess they'll have to be enough."[47] Or, put differently, if the policy is founded primarily in a determination to prevent government funds from encouraging the destruction of human embryos, and only secondarily in a judgment about the value of embryonic stem cell research, then advances in research alone (or the absence of advances alone) would not be sufficient to overturn it. If it is sound before such advances, some argue, it would still be valid after[48]—though again, of course, whether it is right to begin with is itself a point of great contention.

C. Inconsistent

In responding to critiques like those just discussed, defenders of the administration's policy generally point to the principles that define the approach of the policy, as partially laid out in the previous chapter. But some criticism of the policy, directing itself precisely to the claim of consistent adherence to principle, has charged that the policy is morally contradictory, or at least inconsistent, in its own terms.

One common form of the charge of inconsistency concerns the distinction the policy tacitly draws between public and private funding. The current policy addresses itself only to federal funding of embryo research and is silent on the conduct of such research in the private sector. But the source of funding, this line of criticism suggests, could have no bearing on the question of the moral status of human embryos or the propriety of using them in research. If federal funding for research that destroys human embryos is so troublesome, then why should such research be allowed to proceed when privately funded? Acting to restrict one but not the other may be prudent, but it seems inconsistent.[49] Indeed, by implicitly excluding such research from federal guidelines, it may actually encourage more reckless practices and may simply transfer the problem of complicity to the private sphere, where it is even more difficult to monitor and moderate the uses to which human embryos are put. Though these critics understand that the President cannot simply ban embryo research by himself, they argue that he could attempt to convince the Congress to do so, and he could have made such an attempt as an element of his funding policy decision. But he did not—thus bringing into question (for these critics) not only his commitment to the principle articulated by the policy, but also his own view of the moral standing of human embryos, which the policy itself does not make simply apparent.[50] Moreover, critics argue, some prominent defenders of the policy, by making the fact of ongoing private research an element of their defense, might be said to contribute to these doubts about its grounding in consistent principle.[51]

There may be some political or structural reasons for drawing a distinction in federal policy between what is funded and what is permitted. Questions of federalism and other legal realities no doubt enter the picture, and indeed those who oppose the destruction of human embryos are, in many cases, actively seeking prohibitions on all embryo research in individual states.* But, critics point out, they generally say little on the larger question of permissibility at the federal level. By making funding the center of concern, these critics argue, the policy puts into question the importance of preventing embryo destruction more generally, casting some uncertainty over the relation of that policy to one or another position regarding the moral standing of early human embryos.

A related criticism contends that the distinction drawn between research practices in which human embryos are

---

* Lori Andrews detailed both these ongoing efforts and existing legislation in the states in her presentation at the Council's July 2003 meeting and in the accompanying paper. (See Appendix E.)

destroyed and research practices that use the products of previous embryo destruction is itself inconsistent—a distinction without a difference, drawn for political cover.[52] "Pretending that the scientists who do stem cell research are in no way complicit in the destruction of embryos is just wrong, a smoke and mirrors game," writes one critic. "It would be much better to take the issue on directly by making the argument that destroying embryos in this way is morally justified—is, in effect, a just sacrifice to make."[53] Similar objections have also been raised by critics on the other side of the embryo question, who believe that embryo destruction is morally unjustified and that the present policy does not sufficiently distance the federal government from such destruction. "The federal government, for the first time in history, will support research that relies on the destruction of some defenseless human beings for the possible benefit to others," one critic contended in the immediate wake of the August 9, 2001 announcement. "However such a decision is hedged about with qualifications, it allows our nation's research enterprise to cultivate a disrespect for human life."[54]

A further critique of the policy on grounds of inconsistency focuses more particularly on specific elements of the implicit claim to non-complicity, discussed in the previous chapter. If, as many of its advocates argue, the policy takes embryo destruction to be essentially a morally unjustifiable act, and if its provisions aim to make use of the irreversible consequences of that act without in any way encouraging or abetting the act itself, then, critics contend, it is curious that the policy would insist on requiring (in addition) that the eligible stem cell lines must have been obtained from embryos originally intended for reproduction and used with donor consent and without financial inducements. If embryo destruction is in principle a wrong, and if the policy's provisions seek only to keep the federal government from complicity in that wrong, then why should it matter how precisely the wrong was originally committed? The presence of these conditions, it is argued, suggests that the policy is not in fact based in a consistent and principled adherence to the proposition that human embryos should not be destroyed. Indeed, some have argued that this character of the policy suggests that its authors, including the President, may not hold the view that human embryos deserve the same protections as human children or adults. These critics suggest not so much that the current policy is necessarily inconsistent, but that it may only be consistent with a view of human embryos as possessed of intermediate or indeterminate moral standing, rather than a view that holds that human embryos ought simply not to be destroyed for the benefit of others.[55]

In response to these last arguments, some have suggested that the qualifying conditions included in the policy reflect a secondary commitment to long-standing principles of human-subject and embryo research and a recognition of pre-existing standards in such research, rather than a contradiction of the fundamental commitment to avoiding any support for the destruction of nascent human life. The government, they argue, has multiple aims in this area, and these need not undercut each other. In addition to discouraging the creation and use of embryos for purposes other than producing children, one commentator argues, the government also seeks to support the requirement for informed consent to all procedures involving human subjects and to discourage commercial trafficking in human material.[56] Another observer has suggested that the additional conditions are an expression of the fact that some standards for stem cell derivation did exist before August 9, 2001, including those reflected in the Clinton administration's funding guidelines.[57] Even in rejecting the legitimacy of the act of embryo destruction and seeking to discourage it in the future, it is still reasonable to recognize the value of those earlier standards. The policy, it is argued, while establishing a new standard, still takes account of previous standards.[58]

In response to the more general complaints of inconsistency, defenders of the policy, within and outside the administration, have described the present policy as both principled and prudent. The policy, as articulated by these defenders, aims (at least minimally) to uphold and advance the principled conviction that the federal government should not offer support or incentives for the destruction of nascent human life for research. At the same time, they say, it seeks, as much as reasonably possible within the bounds of the principle, to benefit from the results of embryo destruction that has already occurred and can no longer be undone.[59] As

President Bush put it in announcing the policy, "This allows us to explore the promise and potential of stem cell research without crossing a fundamental moral line, by providing taxpayer funding that would sanction or encourage further destruction of human embryos that have at least the potential for life."[60] The policy, argue some of its advocates, aims to put into practice the moral principle of not destroying nascent human life, even if it does not do so on every possible front, or to the greatest possible extent.

Regarding the alleged inconsistency of withholding federal funding but not calling for a ban on all private embryo research, some have pointed out that the Dickey Amendment, under which the President was acting, is itself a "don't fund, don't ban" law. Moreover, they argue that neither statesmanship nor prudence requires that the President do battle with what is settled practice (the free use of embryos in private-sector research) or push zealously against everything to which he is morally opposed, especially in a pluralistic society that is deeply divided on the moral issue. An analogous case may be found in the administration's approach to abortion, a practice that the President says he opposes deeply on moral grounds: the administration supports the legislative ban on federal funding, but has not called for a constitutional amendment that would ban abortion. As President Bush told the pro-life "March for Life" participants in January 2002: "Abortion is an issue that deeply divides our country. And we need to treat those with whom we disagree with respect and civility. We must overcome bitterness and rancor where we find it and seek common ground where we can. But we will continue to speak out on behalf of the most vulnerable members of our society."[61] In a similar way, its defenders contend that the current administration policy on stem cell research funding keeps the public ethical conversation open, may be acceptable to some individuals who hold that human embryos possess an intermediate or unknowable moral standing, and, at the same time, also advances the cause of those who contend that human embryos should be protected from destruction altogether.[62]

The present funding policy has also been defended from the charge of inconsistency on rather different grounds, which, for their advocates, carry different consequences for future federal funding. This defense contends that, if the early embryo is morally equivalent to a child, then the proper moral response would be to ban all future embryo destruction and all stem cell research (public and private) on embryos destroyed after August 9, 2001, and not merely deny it federal funding. Confronted with a practice that involved killing children so that their organs could be used to save the lives of others, advocates of this view contend, no one would simply deny it federal funding while allowing the gruesome practice to continue. But, they point out, there is no reason to attribute to the current funding policy the assumption that the early embryo is morally equivalent to a child. In fact, they contend, the President has never said that early embryos are inviolable, or are persons, or morally equivalent to children. While President Bush explained his funding policy by arguing that "it is unethical to end life in medical research,"[63] he has not sought to prohibit privately-funded embryonic stem cell research. This leads some to conclude that implicit in the President's policy is either the view that the embryo has an intermediate moral status (worthy of serious moral consideration as a developing form of human life) or uncertainty about its moral status. Those who interpret the President's policy in this way point out that, in his address to the nation on stem cell research, he spoke in terms of a human embryo's "potential for life":

> Research on embryonic stem cells raises profound ethical questions, because extracting the stem cell destroys the embryo, and thus destroys its potential for life. Like a snowflake, each of these embryos is unique, with the unique genetic potential of an individual human being.[64]

This, they contend, does not constitute an argument for treating human embryos as possessed of fully human moral standing.[65]

If the present policy is seen to reflect either the intermediate or uncertain view of the moral standing of human embryos, these advocates argue, it is not inconsistent to withhold federal funding, at least for now, when the medical benefits of stem cell research are still speculative, while permitting privately funded embryo research to proceed. They argue that restricting or limiting federal funding is a reasonable way of registering either doubt about the moral status of the embryo, or the moral unease felt by someone who takes nascent human life seriously and does not want it used wantonly, and who therefore wants scientists to prove the promise of this research before permitting them to go further with the support of federal funding. Indeed, they argue, this interpretation is consistent with the President's words in his August 9th address:

> [W]hile we're all hopeful about the potential of this research, no one can be certain that the science will live up to the hope it has generated. . . . Embryonic stem cell research offers both great promise and great peril. So I have decided we must proceed with great care."[66]

Viewing the administration's policy as based on an intermediate or uncertain view of the moral standing of human embryos also makes plausible, in the view of these observers, the fact that the President has neither called for a ban on privately funded embryo research, nor called upon scientists to desist from research on stem cell lines created after August 9, 2001. They contend that it also makes sense of the requirements, discussed above, that stem cell lines created before that date must, in order to qualify for federal funding, be derived from excess embryos created for reproduction, with donor consent, and without financial inducements.

Those who interpret the policy as reflecting an intermediate or uncertain view of the moral status of human embryos argue further that, if embryonic stem cell research vindicates its promise, there would be no categorical reason to prevent a reconsideration of federal funding in the light of medical advances. A further implication is that people may reasonably draw different conclusions about whether this principle justifies federal funding of stem cell lines derived after August 9, 2001 provided they are derived from embryos left over from IVF procedures, donated with consent, and without financial inducement.[67] Although the President has declared that he regards destructive embryo research as unethical without additional qualifications,[68] and although HHS Secretary Tommy Thompson has said that scientific advances would not cause the President to reconsider his policy,[69] those who interpret the President's policy as reflecting an intermediate or uncertain view of the embryo argue that the moral logic of this principle admits the possibility that significant medical breakthroughs could justify a reconsideration.

IV. THE MORAL STANDING OF HUMAN EMBRYOS

Many elements (though, as we shall make clear, not all elements) of the ongoing debate about federal funding of human embryonic stem cell research seem, as we have reviewed them, to turn on a basic disagreement about the nature, character, and moral standing of human embryos. Public debate over the moral standing and appropriate treatment of human embryos has been quite contentious and divisive in recent years. In part, this had to do with its almost inevitable entanglement with the abortion debate, itself a deep and thorny controversy in America. In part, too, this has been connected to the fact that the question of the moral standing of human embryos touches many other fundamental moral and existential questions involving human origins, human dignity, the moral significance of our biology, and its relation to numerous traditional and widely shared moral teachings.[70] Differences of opinion on the moral standing of human embryos often suggest differences on these larger questions of overall worldview.[71] Nonetheless, the question of the moral standing of human embryos itself has been taken by nearly all commentators to be amenable to human reason and argument, and a lively debate has raged despite (or perhaps precisely because of) these widely diverging starting assumptions.

In the public arena, the question of the moral standing of human embryos has often been summed up in the question, "When does (a) human life begin?" This question suggests something of the quandary, although the academic and intellectual debate generally takes a somewhat more nuanced question as its starting point. That question has as its unstated premise the fact that under normal circumstances we regard all born human beings (from newborns through adults) as possessing equal moral worth and meriting equal legal protection. It then reflects upon the ways in which human embryos are similar and different from live-born human individuals, the moral significance of those similarities and differences, and therefore whether embryos should or should not be afforded protections.[72]

The first and most common recourse in seeking an answer to such questions has been human biology, and particularly human embryology. Nearly all participants in the dispute make some reference to biological findings, whether to claim that they teach us about an embryo's essential continuity with and similarity to human beings at other stages of life, or to argue that they reveal profound and morally meaningful discontinuities between embryos and live-born persons.

While we examine these differing contentions, it is crucial to remember—as several commentators in recent years have noted—that the biological findings, however relevant, are not themselves necessarily decisive morally.[73] They may serve better to challenge moral positions founded on erroneous assumptions than to ground some positive moral affirmation or conclusion. For example, a recognition of biological continuity might in some measure undermine the argument that embryo destruction is permissible when certain biological markers or states of development are absent. But it would not by itself show indisputably that embryos are to be treated as simply inviolable.[74] Meanwhile, recognizing the biological significance of some particular point, marker, characteristic, or capacity would not, in itself, imply some decisive moral significance. A description of early embryonic development is necessary though not sufficient to an understanding of the nature and worth of an early embryo.[75] It is not sufficient because any purely biological description requires some interpretation of its anthropological and moral significance before it can function as a guide to action.[76]

With these provisions in mind, we offer the following brief review of developments in the debate over the moral standing of human embryos in the past several years.[*]

---

[*] Readers may find it helpful to consult Notes on Early Human Development (Appendix A).

A. Continuity and Discontinuity

Many participants in the debate take the question of the biological continuity or discontinuity between nascent and later human life to be crucially significant. Some argue that the fundamental organismal continuity from the moment of fertilization until natural death means that no lines can be drawn between embryos and adults. Others argue, on the contrary, that some particular point of discontinuity (or the sum of several such points) marks a morally significant distinction between stages, which difference should guide our treatment of human embryos.

1. The Case for Continuity.

Many of those who seek to defend human embryos base their case on some form of the argument for biological continuity and sameness through time. For example, they argue that a human embryo is an organic whole, a living member of the human species in the earliest stage of natural development, and that, given the appropriate environment, it will, by self-directed integral organic functioning, develop progressively to the next more mature stage and become first a human fetus and then a human infant. Every adult human being around us, they argue, is the same individual who, at an earlier stage of life, was a human embryo. We all were then, as we still are now, distinct and complete human organisms, not mere parts of other organisms.[77]

This view holds that only the very beginning of a new (embryonic) life can serve as a reasonable boundary line in according moral worth to a human organism, because it is the moment marked out by nature for the first visible appearance in the world of a new individual. Before fertilization, no new individual exists. After it, sperm and egg cells are gone—subsumed and transformed into a new, third entity capable of its own internally self-directed development. By itself, no sperm or egg has the potential to become an adult, but zygotes by their very nature do.[78]

Many authors therefore regard the activation of the oocyte (by the penetration of the sperm)[79] or the completion of syngamy (the combining of paternally- and maternally-contributed haploid pro-nuclei to result in a unique diploid nucleus of a developing zygote) as a meaningful marker of the beginning of a new human life worthy of protection.[80] After this point, there is a new genome, in a new individual organism, and there is a zygote (single-celled embryo) already beginning its first cleavage and embarking on its continuous developmental path toward birth.

All further stages and events in embryological development, they argue, are discrete labels applied to an organism that is persistently itself even as it continuously changes in its dimensions, scope, degree of differentiation, and so on. We can learn names for the various stages as if they were static and discrete, but the living and developing embryo is continuously dynamic.[81] More to the point, in the view of these commentators no discrete point in time or development would seem to give any justification for assuming that the embryo in question was one thing at one point and then suddenly became something different (turning, for example, from non-human to human or from non-person to person). None of the biological events (or "points" in processes), they contend, is sufficient to tell us what we are morally permitted or obligated to do to human embryos, in the absence of one or another *additional* premises that shape one's view of these biological events. And if one's guiding premise is that all human persons possess equal moral standing—regardless of their particular powers, size, or appearance—then there are no grounds for denying the earliest human embryo full moral standing as a person. [82]

Some critics of this position argue that it makes too much of mere genetic identity and (uncertain) potential or that it does not make enough of present condition and the significance of development itself.[83] There is more to being human, some observers argue, than possessing a human genome or spontaneous cell division, and it matters that the early human embryo is but a ball of cells, without sentience or sensation and without human form.[84] It matters, too, they argue, that an ex vivo human embryo does not have the potential to develop independently, without further technical intervention.[85] Indeed, some argue that a human embryo in its earliest stages is essentially no different from any human tissue culture in the laboratory,[86] or that, because the ex vivo embryo cannot develop if left to itself, it cannot be thought of as truly continuous with more developed human organisms. It may be, in the description of one observer, not much different from a pile of building materials stored in a warehouse.[87]

Nonetheless, advocates of the argument from continuity suggest that it is dangerous to begin to assign moral worth on the basis of the presence or absence of particular capacities and features, and that instead we must recognize each member of our species from his or her earliest days as a human being deserving of dignified treatment. They contend that a human embryo already has the biological potential needed to enable the exercise, at a later stage of development, of certain functions. Sentience and sensation come in later in the process of development, but their seeds are there right from the beginning. And the fact that an embryo cannot develop outside the body is not an argument for leaving it outside the body. There is, they argue, no clear place to draw a line after the earliest formation of the organism, and so there can be no stark division between the moral standing of nascent human life and that of more mature individuals.[88]

2. The Case for Meaningful Discontinuity and Developing Moral Status.

Many other observers, however, argue that some biologically and morally significant discontinuities do in fact present themselves in the course of early human development. These arguments generally do not simply hinge on biological descriptions—which are, in the absence of analysis, largely devoid of obvious moral significance—but instead begin from some implicit or explicit claim regarding the importance of a particular feature, capacity, form, or function (or the progressive accumulation of these) in defining a developing organism as meaningfully a member of the human race.* Not simply grounded in biology, they appeal also to a moral or even metaphysical claim about the meaning of humanity.[89] They

---

* Also, several traditional religious views of the embryo (among some Jews and some Muslims, for instance) have attributed humanity to the embryo only after a particular point in its development, for example the 40th day.

suggest that the developing human organism might become (at once or progressively) deserving of protection as it becomes able to feel pain, or to exhibit neural activity, or rudimentary features of consciousness, or some elements of the human form, or the capacity to function independently—or as it progressively exhibits more and more of these or other criteria. Until that time, many argue, a developing human deserves some respect because of what it might become, but not protection on par (or nearly so) with that afforded to fully human subjects.[90] They suggest that genetic identity and organismic continuity are not sufficient; what matters is present form and function, more than mere potential.[91]

Several particular putative discontinuities (and combinations of them) have been proposed as candidates for the division between early stages, when a human embryo may be disaggregated for research, and later stages, when it deserves some greater level of protection.

(a) Primitive streak: The most popular candidate for a meaningful point of discontinuity is the appearance of the primitive streak, the earliest visible "structure" that defines the region of the embryo along which the vertebral column will form. The primitive streak generally appears around the 14th day after first cell division. It is taken to indicate the anterior-posterior axis of an embryo (in vertebrates), although recent studies suggest that polarity may be established much earlier, and in fact may be indicated by the point at which the sperm enters the egg.[92]

A principal reason for the importance placed on the primitive streak has to do with the biology of twinning. Prior to the appearance of the primitive streak, an embryo may (rarely, and for unknown reasons) divide completely to form identical twins. Some conclude that individuality must not yet be established, since the embryo might yet become two embryos.[93] Since individuality is essential to human personhood and capacity for moral status, since individuality presumes a definitive single individual, and since the singularity of the embryo is not irrevocably settled prior to the appearance of the primitive streak, they argue that the entity prior to the primitive streak stage lacks definitive individuality and hence also moral status.[94] Critics of this line of reasoning point to the low statistical probability of monozygotic twinning: one in 240 live births (though rates clearly vary among populations in different regions, and precise figures probably cannot be known because some zygotes that split likely later fuse into one, and some singleton births may conceal a twin who died early in gestation). Critics also point to evidence that twinning results, not from an intrinsic drive within the embryo, but from a disruption of the fragile cell dynamics of embryogenesis. Evidence for this, they suggest, may be seen in the increased incidence of monozygotic twinning (up to tenfold in blastocyst transfer) associated with IVF. This suggests, in their view, that twinning is neither a proof of the absence of an integrated individual organism with a drive in the direction of development nor a demonstration ex post facto of the absence of moral worth of the embryo before twinning.[95]

Nonetheless, for this reason, and for others (discussed below) having to do with the formation of the nervous system, the primitive streak has often been taken to be a highly significant marker of embryological development, and many commentators suggest it as a reasonable candidate for a meaningful point of discontinuity. For this reason, many supporters of embryo research regularly propose the 14th day of development as a logical stopping point for permissible embryo research.[96]

(b) Nervous system: A second argument for discontinuity focuses on the developing nervous system. Many observers regard the nervous system as an especially important marker of humanity, both because the human brain is critical for all "higher" human activities, and because the nervous system is the seat of sensation and, especially relevant to this case, the sensation of pain. Proponents of this view hold that before an embryo has developed the capacity for feeling pain (or, in some forms of the argument, before sentience), we cross no crucial moral boundary in subjecting it to destructive research.[97] For some, this is taken to mean that the primitive streak, as the first marker of a future nervous system, is a crucial feature of developing life. For others, only later points of neural development (where pain might plausibly be experienced) are held to be decisive.[98] Critics meanwhile contend that neural development as well as development of other systems (such as the cardio-vascular system) are the natural outcome of the genetic program in action, and should be explained by reference to the previous stage and as leading to the subsequent stage, rather than marking a significant discontinuity.[99] They maintain that the human being is, from the start, an inseparable psycho-physical unity, rather than a pure rationality or consciousness that exists with no meaningful ties to our bodies. From a scientific perspective, such critics hold, there is no meaningful moment when one can definitively designate the biological origins of a human characteristic such as consciousness, because our mind works in and through our body, and the roots of consciousness lie deep in our development. The earliest stages of human development serve as the indispensable and enduring foundations for the powers of freedom and self-awareness that reach their fullest expression in the adult form.[100] Some of those who believe that neural development is crucial, however, argue that the fact of non-sentience and of an inability to experience pain possess great moral significance, quite apart from the question of probable potential.

(c) Human form: Some observers also argue that certain rudimentary features of the human form must be apparent before we can consider a human embryo deserving of protection. In this view, the human form truly signals the presence of a human life in the making and calls upon our moral sentiments to treat the being in question as "one of us."[101] Different versions of this argument appeal to different particular elements (or combinations of elements) of the human form as decisive, but all suggest that a "ball of cells" is not recognizably human and therefore ought not to be treated as *simply* one of us.[102] Some critics of this view argue that humans have different external forms or shapes throughout their lives, and that an organism, particularly in its early stages, should not be judged by its external shape, but rather by its biological constitution, and especially its genetic identity.[103] But adherents of the argument that human form matters contend that genetic identity cannot simply be decisive of moral worth.

These various particular cases for discontinuity (and these are not the only ones that have been propounded over the years) are not mutually exclusive. Indeed, many of them point to more than one particular element of early human development as finally decisive of moral standing. But they share the belief that moral status accrues only at some later stage of the developing human organism. Their claim, in the broadest terms, is that in its earliest stages a human embryo is not yet simply a human being or a human person, and that it need not be treated as though it were.[104] Human development, they contend, is an essential element in any understanding of human life, and an organism at the earliest stages of that process is not to be treated the same as one much farther along.[105] There are developmental differences, and these differences matter, in ways to be determined by human choice and understanding, as well as by a grasp of the biological facts.[106]

Critics of this view contend that while it is certainly true that human beings at different stages of development are not to be treated the same (as children are not given the responsibilities of adults), the crucial treatment here at issue is destructive treatment. No human being, at any stage, they argue, should simply be destroyed for research, and the "use" of an embryo for research, no matter how valuable one deems the research to be, could not amount to treatment of that embryo as "deserving some degree of respect." The degree of respect granted in destroying the embryo would be zero, they contend.[107]

Nonetheless, the case for developing moral status, as articulated by a great number of participants in the policy debates of the past several years, often results in an expression of what has sometimes been termed the "special respect" approach to human embryos: an embryo in its earliest stages is not accorded the full moral standing of a human person, but it is nonetheless regarded as deserving some degree of respect and is treated as more than a mere object or collection of somatic cells in tissue culture. In practice, adherents of this view tend to accept the use of early human embryos in medically valuable research under some circumstances, but they seek to apply some scrutiny to the reasons for which embryos will be used, the circumstances under which those embryos are obtained, and other relevant factors. Several bodies advising the federal government on human embryo research over the years have expressed this view, to varying degrees. The Ethics Advisory Board to the Department of Health, Education and Welfare concluded in 1979 that the early human embryo deserves "profound respect" as a form of developing human life (though not necessarily "the full legal and moral rights attributed to persons").[108] The NIH Human Embryo Research Panel agreed in 1994 that "the preimplantation human embryo warrants serious moral consideration as a developing form of human life," though the panel argued that this did not mean that research should be prohibited.[109] In 1999, the National Bioethics Advisory Commission (NBAC) cited broad agreement in society that "human embryos deserve respect as a form of human life,"[110] but, like its predecessors, did not recognize the embryo as having the full rights of a human person. The special respect position has been held by its advocates to be consistent with a range of possible particular policies on embryo research, including fairly restrictive ones, and indeed could be held consistent even with an outright restriction on the destruction of human embryos.*[111] To consider the embryo "inviolable" (and therefore not a mere utility to be instrumentally used), it is not necessary to presuppose its moral equality with a child, an adult or even a later stage gestating fetus. There may be increasing moral obligations (and natural sentiments) associated with a deepening relationality that extend moral duty without in any way implicitly eroding an imperative of protection at earlier stages of developing life. Most of those who have articulated the special respect position in the public debate have, however, tended to then argue that some research

---

* This is partly due to the fact that "*special* respect," and "*intermediate* moral status," are rather vague terms, and embrace a very wide range of degrees of "specialness" and "intermediacy."

on embryos should be permitted within certain boundaries, even if they have not always agreed on the permissible extent or the appropriate boundaries.

B. Special Cases and Exceptions

Some arguments in favor of permitting and funding embryo research, grounded in the "special respect" approach, do not in fact explicitly (or exclusively) rely on arguments about discontinuity or a biologically grounded view of developing moral status. Instead, or in addition, they rely upon questions of embryo viability and potential, and they are aimed at exploring unique circumstances to address and perhaps resolve questions of the moral standing of *certain particular* human embryos.

1. IVF "Spares."

Much of the debate surrounding embryonic stem cell research has focused on the use of cryogenically preserved IVF embryos, left over from assisted reproduction procedures and stored, perhaps indefinitely, in the freezers of IVF clinics. One recent study suggests there are hundreds of thousands of such embryos in the United States alone, though only a very small percentage of them (less than 3 percent, approximately 3,000 or more) has ever been donated for research.[112] Although all were produced with reproductive intent and were stored for further reproductive efforts should the first attempt fail, most of these frozen embryos may never be claimed by the original egg and sperm donors for use in assisted reproduction procedures. They are almost certain to remain frozen and, eventually, to die without developing further. Although there have been some efforts to build interest in adopting such embryos, such adoption, some commentators argue, is very unlikely to become a large-scale phenomenon or to affect the fate of most of these stored embryos.[113] Under the present funding policy, if these frozen embryos were donated for research and embryonic stem cells were derived from them, research on the resulting cells would not be eligible for federal funding. Many observers argue that it should be: Since *these* particular embryos are almost certainly destined to die without ever developing, it would be appropriate to at least redeem some possible good from their existence and unavoidable demise.[114]

Some people have pushed the point further. Since the vast majority of the (huge number of) cryopreserved embryos will almost certainly not be adopted, and since many may not be viable even if they were to be transferred to a woman's uterus, a few observers have argued that, practically speaking, the frozen embryos are virtually all already lost.[115] To be sure, they are not already dead, but they are in a so-called "terminal situation" from which no rescue is practically possible. In view of this situation, one commentator proposes extending to these embryos the principle that sometimes, he argues, permits the killing of innocents. That is, killing may be morally permissible in cases where the person will soon die for other unavoidable reasons and where there is another person who can somehow be rescued through or as a result of such a normally impermissible act of killing (thus, as he puts it, "nothing more is lost"). He admits that the case of cryopreserved embryos stretches the application of the "nothing is lost" principle beyond its previous uses, because the embryos in question are alive and at risk of death *only* because of human choices and designs specifically directed toward them. The principle is also stretched because the lives that might someday be saved through today's embryo deaths are quite remote. The potential lives saved are those of unspecified future persons with diseases that might be treated by therapies that as yet do not exist and may or may not exist in the future. However, against the weight of all these *ifs*, which some find formidable, there is the present fact that (like the embryos used to create stem cell lines derived before August 9th of 2001) the cryopreserved embryos are already here, with little or no prospect of rescue—they are, in this observer's description, already lost.[116]

Presumably, if destruction of "spare embryos" for human embryonic stem cell research were generally agreed to be permissible through this "nothing is lost" principle, it could be federally funded, subject to such routine secondary considerations as the need for free and informed consent by donors.

Yet this argument, not surprisingly, has met with opposition. Some critics have claimed that it employs circular moral reasoning. The embryos, they argue, are in a "terminal situation" because of human choice and design; thus to then decide that, since they are going to die anyway, they may as well be put to good use is to ignore the moral implications of the original decision to create and freeze them. Critics argue, moreover, that when thinking about our responsibilities to those who are soon to die, we would normally say that it makes a considerable moral difference whether we simply accept their dying or whether we positively embrace it as our aim.[117] Yet some proponents of using IVF spares argue that the present situation is best understood as a forced choice between two regrettable alternatives for the final disposition of stored embryos (whether donated for research or abandoned). One choice may then be justified as the lesser evil. Even if one deems the original decisions leading to the creation and storage of these embryos questionable, the embryos exist, and the earlier decisions cannot be undone.[118]

Some have also worried about the possibility of a "slippery slope," by which the uses of "spare" IVF embryos under this justification might open the door to their wider use in experiments in natural embryogenesis, toxicological studies or chimerizations, or perhaps their development in an artificial (or natural) endometrium,[119] (though the reasonableness of "slippery slope" arguments is often disputed).[120] Other critics point out that the "nothing is lost" principle is not permitted to govern decisions regarding lethal experiments on the terminally ill, on death-row inmates, or even on fetuses slated for abortion.[121]

A further issue involves the question of whether accepting the "nothing is lost" principle for already existing embryos condones in principle the use of future excess embryos, or whether the principle actually requires efforts to prevent the creation or storage of "excess" embryos in the future.[122] Further, this application of the principle relates only to embryos originally created for the purpose of reproduction but not transferred to initiate a pregnancy. Should their use in research be accepted, however, it is not clear that it would be possible to differentiate between embryos created originally for reproduction and extra embryos created with an eye to research uses, since the process of producing them would be identical (though consideration of the extra risks involved for the woman egg donor could act as a counter against any large-scale embryo-creation-for-research).[123]

Other observers, however, begin from the presence of cryogenically preserved embryos but extend further the argument justifying their use in research. They argue that there is good reason to use embryos for research, not only because a situation some judge tragic already exists but because donating embryos is an act of beneficence and using them is a social obligation incurred by their gift.[124] This approach would have us start by recognizing that, in the current situation, there is a set of embryos (in IVF clinics) none of which will ever enter a uterus, or even a (still hypothetical) artificial uterus.[125] These embryos, in one commentator's term, are "unenabled" and have no potential for full development. Since there is no possible way for such embryos to develop, there is no "possible person" to whom any "unenabled" embryo corresponds; therefore using them in research involves no loss of possible life.[126]

Critics of this approach argue that in effect it allows the moral standing of any given embryo to be decided simply by those responsible for it. Thus, whether a given embryo has moral standing depends only on whether it has a practical prospect of developing, as evidenced by whether it will be transferred to a uterus. But whether it enters a uterus depends on human choices, so the moral standing of a given embryo depends on human choices. If we choose to withhold essential enabling conditions for an embryo's development, these critics argue, then that embryo will lack not moral standing but merely the opportunity to develop. Its intrinsic nature, including its *own* potential to develop (given needed conditions), has not changed.[127]

Nonetheless, in the view of a number of commentators, the fact that so many frozen embryos already exist (even if only a small percentage of them are donated for research) changes the balance of duties and respect owed to any single ex vivo embryo. Several observers have argued that the presence of these frozen embryos, with little or no chance of attaining birth, creates special circumstances in which the use of embryos in research (whether destroying them in the course of research or allowing them to die and subsequently using them)[128] is independent of any final judgment regarding the moral standing or worth of the ordinary human embryo, as such.[129]

2. Natural Embryo Loss.

Some authors tie their moral arguments regarding the use of embryos in research to the fact of the high rate of embryo loss under natural conditions of sexual intercourse and unassisted reproduction. They argue that directed destruction of ex vivo embryos for the purpose of research would not result in greater embryonic loss than occurs in this natural process and would result in far greater benefits for humanity.[130] The rate of natural embryo loss after conception in unassisted human reproduction (taking in losses both before and after implantation), though not accurately known, is thought to be high, some suggest as high as 80 percent.[131] Moreover, the fact of natural loss is now fairly well known, so that persons who engage in unprotected intercourse, whether seeking to reproduce or not, are knowingly bringing about the conception of many embryos that will die. We generally do not regard this embryo loss as unacceptably tragic, and we do not engage in great efforts to avert it or to find ways to diminish it (beyond research to prevent miscarriage, for instance). For this reason, these commentators argue, using artificially created embryos for purposes of research would not destroy a greater portion of those embryos than ordinarily die in natural unassisted reproduction.[132]

Moreover, they suggest, the high rate of natural embryo loss should call into question the views of those who believe that early-stage human embryos merit equal treatment with human children and adults. If so many die in the natural course of things, why do we not treat natural procreation as a great fountain of tragedy and carnage? The natural rate of embryo loss, and most people's failure to mourn its consequences, they suggest, should teach us something about the limited significance of human embryos in the earliest stages. One observer adds that the same logic should diminish our concerns about creating extra human embryos for research, as long as sufficient embryos are created for implantation to keep the chances of survival for any given embryo as good as the chances in natural reproduction.[133] Another argues further that human embryonic stem cell research might actually raise the probability of longer survival for all humans, including embryonic ones, and is therefore a case of permissible taking of life, even on the assumption that the embryos are persons.[134]

Opponents of this view, however, have argued that natural deaths of embryos and the deaths caused by intentional efforts to destroy ex vivo embryos for research are not morally equivalent. There are many things that happen naturally that we are not therefore justified in doing deliberately. Indeed, the rate of natural loss of live-born human beings is 100 percent, but that does not justify their killing. And, they argue, even if one were permitted to analogize the deaths of frozen embryos in vitro to the embryonic wastage in vivo, in neither case were the embryos lost destined or created for anything other than their procreative end. In contrast, they argue, using embryos for research bears no relation to their natural direction or trajectory. Critics also argue that the character of our reaction to the natural embryonic death does not justify our practice of destructive embryo research. For they believe that a creature's moral worth is not dependent on the emotional reaction of others to its death.[135] The absence of moral sentiment does not imply an absence of moral obligation (nor a right of adverse intervention in a naturally developing human life). Moreover, critics contend, it is not clear how many of these "natural losses" were in fact failures of the fertilization process, so that there was never a unified, integrated organism in the first place, and thus never the loss of a human embryo. It is also unclear, they argue, how much of the supposedly "natural" loss rate is actually due to contingent and changeable factors, such as environmental pollution, pesticides, or endometrial problems, and so is not simply an unavoidable fact of embryonic existence.[136]

3. Prediction of Non-Viability of Embryos.

Some people, hoping to get around the moral dilemma of destroying even "unenabled" embryos, seek to identify a subset of embryos that might reasonably be regarded in advance as *non-viable*. One proposal has involved the possibility that cloned human embryos, created by somatic cell nuclear transfer, may prove to be non-viable or unable to develop beyond a certain point (biological evidence that this may be the case is presented by Rudolph Jaenisch in Appendix N) or may even, by their nature and origins, simply not constitute the equivalent of human embryos. If this turns out to be true, it is further argued, it might be possible to use them without arousing some of the ethical dilemmas that accompany the use of otherwise potentially viable human embryos.[137] Others point to a possible sub-group of those embryos currently frozen in storage as potentially non-viable. Although those embryos are not yet dead and, if thawed, would still exhibit some cellular function, some would be unlikely to survive even were transfer attempted. Since IVF procedures usually produce more embryos than can be transferred at one time, goes the argument, the clinicians choose for transfer those among the available embryos that look "the best" and seem most likely to survive and develop—so that those that are frozen are those deemed less likely to develop. Moreover, by applying similar judgment to the unimplanted embryos, we might identify those that would be least likely to survive even under the most favorable circumstances. These embryos might then reasonably be regarded as non-viable and therefore available for research since their use will not disrupt a potential life.[138]

There has not been much direct reaction to this view in the ongoing ethical debates. Yet some observers have noted, in other contexts, that the techniques used to identify which IVF embryos are more or less likely to develop successfully— estimates based usually on visual assessment of the embryos—have never been proven effective or even tested to ascertain their validity.[139] Moreover, some argue, the true viability of cloned human embryos or of cryogenically stored embryos could not be determined in advance without attempting to implant them.[140]

4. Creation of Non-Viable Embryo-Like Artifacts.

Others, seeking to bypass altogether the issue of viability, propose the possibility of creating a biological entity that cannot rightly be called a living organism, yet that has the generic organic powers necessary to produce embryonic stem cells. They suggest that somatic cell nuclear transfer, or a similar technique, could be used to create an entity that lacks, by design, the qualities and capabilities essential to be designated a human life in process. By intentional alteration of the somatic cell nuclear components or the cytoplasm of the oocyte into which they are transferred, researchers may be able to construct an "artifact" that is biologically (and morally) more akin to cells in tissue culture, but could still provide a source of functional human embryonic stem cells.[141] Proponents of this innovation aim to shift the ethical debate from the question of whether or when a human embryo should be considered a human being to the question of which organized structures and potentials constitute the minimal criteria for considering an entity *to be* a human embryo.[142]

Absent all the essential elements (including a full complement of chromosomes, proper genetic sequence and chromatin configuration, and cytoplasmic structures and transcription factors), advocates of this proposal argue, there can be no integrated whole, no organism, and hence no human embryo. By technically constructing biological entities lacking these essential elements yet bearing the partial organic potential often found in failures of fertilization, they suggest it may be possible to procure embryonic stem cells without producing an organismal or embryonic entity that can meaningfully be designated a being with moral standing.[143]

Proponents argue that there is a natural biological precedent for such an entity lacking the qualities and characteristics of an organism yet capable of generating cells with the character of embryonic stem cells. Teratomas are germ cell tumors that generate all three germ layers as well as more advanced cells, tissues, and partial limb and organ primordia. Yet these chaotic, disorganized, and nonfunctional masses lack entirely the structure and dynamic character of an organism. Likewise, failures of fertilization due to abnormal complements or chromatin configurations (imprinting) of the chromosomes may still proceed along partial trajectories of organic growth without being meaningfully designated as organismal entities.

These natural examples of "partial generative potential" (described by some as 'pseudo-embryos'), together with other observations of early embryonic process, have led to a diverse array of suggestions for ways that embryonic stem cells may be produced without raising the moral issues involved in the creation and destruction of human embryos. These suggestions include the use of aneuploidies, polyploidies, viable cells from embryos in arrested development, parthenotes, and chimeras of human nuclear material and animal oocytes. Each presents its own particular technical challenges and raises unique and unfamiliar moral considerations. The scientific prospects for such projects remain largely unexplored in humans, though some animal work has shown promise, and proponents argue that they are within the reach of current technology.[144]

The crucial principle in all such efforts, proponents argue, is the preemptive nature of the intervention: undertaken at a stage before the transition to organismal status. They contend that just as we have learned that neither genes, nor cells, nor even whole organs define the locus of human moral standing, in this era of developmental biology we will come to recognize that cells and tissues with "partial generative potential" may be used for medical benefit without a violation of human life or dignity. Moreover, they argue, the moral distinctions essential to discern and define the categories of organism, embryo, and human being will be critical to progress in scientific research involving embryonic stem cells, chimeras, and laboratory studies of fertilization and early embryogenesis. These advances in developmental biology, they contend, will depend on clarifying these categories and defining the moral boundaries in a way that at once defends human dignity while clearing the path for scientific progress.[145]

This proposal has drawn criticism on several fronts. First, critics suggest, it would require significant research to ensure that the procedure reliably produced the desired sort of "non-embryonic" entity yet also still yielded normal human embryonic stem cells, and such research might itself be morally problematic. Second, this proposal raises a series of further scientific and ethical questions, including those regarding the minimal degree of "partial generative potential" for an entity to be considered an organism, and for an entity to be considered a human embryo. They point further to the risk of creating entities that are so ambiguous as to leave their moral standing in serious doubt, at least for those people who believe that the early stages of human life have at least some moral worth. Finally, proposals to use human oocytes raise moral concern regarding the source of supply, in this case as in the larger arena of in vitro fertilization and experimentation.[146]

Although this approach has never been tested in humans, animal experiments suggest it has potential, and it has begun to play a part in the debates over the moral standing of human embryos and the permissibility of embryo research more generally.

V. SOCIETAL SIGNIFICANCE AND PUBLIC RESPONSIBILITY

While the bulk of the public debate surrounding embryonic stem cell research has been directed to the question of the moral standing of human embryos, some commentators have raised a number of other crucial and serious concerns. They have argued that the debate suffers from focusing too narrowly on questions of the standing of human embryos, when other issues—including the duties and responsibilities of those who engage in embryonic stem cell research, the implications of such uses of nascent life for our society (rather than just for the embryos themselves), the significance of the public debate, and a series of other issues—also bear heavily on the subject, and may illuminate it in ways at least as significant.[147]

Some authors, including some who do *not* believe that human embryos should simply be treated as inviolable persons, have argued that the instrumentalization of human embryos—the seeds of the next generation—might tend to coarsen the sensibilities of our society toward future generations, and toward human life in general, quite apart from the effects on the embryos themselves.[148] Others also argue that by setting down the path laid out by human embryonic stem cell research, we open the way for other, and more troubling, techniques and developments. Since human suffering and disease will never come to an end, they suggest, the resort to extreme and potentially exploitative methods is unlikely to find a logical stopping point. Today, they argue, scientists want to use only the earliest embryos; but what will happen when it turns out that later-stage embryos are even more valuable in developing treatments for disease?[149]

Critics of this view have generally argued that it fails to offer a sufficient ground for impeding the promise and potential of medical treatments that might result from embryonic stem cell work. These critics see medical research as a central moral duty, and they argue that a society that prevents such research undermines not only medical progress, but the moral progress of the community as a whole. While instrumentalization and moral coarsening are real worries, these critics argue, long-term fears of a "slippery slope" do not justify renouncing the potential of today's research. Future difficulties, they say, can be faced if and when they arrive in earnest.[150]

Other observers have raised concerns related to the ethical and policy debate itself, rather than to the specifics of one technique or another, or of one funding policy or another. Some, for instance, have argued that what is needed in the human embryonic stem cell research debate is not only an exchange of views about the substantive issues (though that is surely crucial) but also some sense of the appropriate democratic process for deliberation and for establishing appropriate public policy on such a profoundly contentious matter. The embryonic stem cell debate, it is argued, offers a valuable opportunity to think through the ways in which the American polity debates the most contentious moral issues.[151] Some even suggest that such matters should be removed from the political process altogether, and left in the hands of a regulatory body specifically charged with monitoring and decision-making authority.[152]

Other observers worry that the promise of embryonic stem cell research to bring swift or immediate cures to those who are now ailing has been oversold. They point out that, two decades ago, similar claims of rapid cures were made on behalf of fetal tissue transplantation research, but have not as yet been realized (though, of course, the danger of unfulfilled hope always looms over medical research). Such talk, some observers have argued, tends to raise false hopes, and thus does a genuine disservice to the sick and disabled.[153] Worse, some regard it as cruel and immoral to exploit the hopes and fears of the desperately ill and their families, especially when— as several scientists in the field have remarked—it is not very likely that stem cell based remedies will be available for most people now suffering from the potentially targeted diseases. This moral concern is tied directly to the longstanding bioethical principle of truth-telling, which obliges physicians to inform their patients honestly about their condition and prognosis and encourages researchers to be truthful in their assessments of the potential for new treatments and cures, whether or not what they have to say is what patients and their loved ones want to hear. "It is misleading to suggest that stem cells will bring cures," one writer observes, "particularly cures for patients now coping with the serious diseases the research targets."[154]

A related concern, raised by some of the same commentators, involves what has been termed "the disproportionate emphasis on stem cell research in contemporary health politics."[155] The prominence of this debate in American politics, they argue, may tend to distract us from the fact that many Americans, and even more people elsewhere, lack very basic healthcare and have no access to those medical tools and techniques that already exist and that raise no profound controversy.[156] The concern for justice, and for the proper setting of priorities, they argue, requires us to see this line of research in its proper perspective. Because federal resources for research are limited, decisions must be made about which areas should receive high (and low) priority, and decisions must be made about how much to devote to research and how much to devote to programs that provide proven health care to patients. These commentators suggest that this view does not mean that stem cell research should not be funded, but only that we must keep in mind that funds are also needed for many other approaches to fighting suffering and disease.[157] Critics of this view, however, respond that an excessive preoccupation with *existing* health care needs can jeopardize new medical research and medical progress, and that today's "basic medicine" was once experimental research.[158]

Meanwhile, others have argued that the debate has been too narrow in another way. Any federal policy, they suggest, must take note not only of the potential promise of embryonic stem cell science, but also possible alternatives to such research, alternatives less morally troubling to many

Americans. Many opponents of public funding for human embryonic stem cell research, for instance, have argued that more attention should be paid, and more resources devoted, to *adult* stem cell research, which raises few of the moral difficulties present in the embryonic stem cell debates[159] (though, for some, it still does raise a number of ethical difficulties).[160] Such work, they contend, might even make embryonic stem cell research (or, at the very least, publicly funded embryonic stem cell research)[161] unnecessary, if it proves sufficiently useful.*

In response, however, others point out that adult stem cell research already receives about ten times the amount of federal funding apportioned to human embryonic stem cell research. Critics also argue that embryonic stem cells may possess unique advantages over adult stem cells, just as (in some circumstances) the opposite may be the case, and that therefore both avenues, as well as research using stem cells derived from fetal tissue, should be pursued simultaneously.[162] And they contend that opponents of embryonic stem cell research have oversold the promise of adult stem cells so that the public might come to see embryonic stem cell research as unnecessary.[163] In the next chapter, we examine some of the scientific facts regarding adult and embryonic stem cells, but in the context of the ethical and political debates, the distinctions between them have been quite important and prominent.

In these ways, the controversial possibility of scientific alternatives, as well as concerns about the health of American culture and democracy, the honesty of a political debate that touches on the hopes and fears of many who are suffering, and the bigger picture of health-care politics all impinge upon the question of federal funding of human embryonic stem cell research.

---

* In 1999, the National Bioethics Advisory Commission stated that, "In our judgment, the derivation of stem cells from embryos remaining following infertility treatments is justifiable only if no less morally problematic alternatives are available for advancing the research," though the commission did *not* conclude that adult stem cells were sufficiently shown to offer such an alternative. (National Bioethics Advisory Commission, *Ethical Issues in Human Stem Cell Research*, Washington, D.C.: Government Printing Office, 1999, p. 53.)

VI. CONCLUSION

Participants in the public debate surrounding human embryonic stem cell research and the administration's funding policy have addressed themselves to many complicated and difficult ethical matters: the character of the moral question at issue; the nature, object, ethical assumptions, and premises of the policy itself; the vexing question of what is owed to the early human embryo; and a number of other related concerns. As we have seen, strong and powerfully argued views have been presented on various sides of each of these questions. For now, neither side to the debate seems close to fully persuading the other of the truth it thinks it sees. But the rich and growing ethical debates do suggest the possibility of progress toward greater understanding of the issues, and toward more informed public decisionmaking, as all parties to the deliberation appreciate better just what is at stake, not only for them or their opponents, but indeed for all of us.

ENDNOTES

[1] "Press Conference by President Bush and Italian Prime Minister Berlusconi," as made available by the White House Press Office, July 23, 2001.

[2] See, for instance, Weissman, I. L., "Stem Cells—Scientific, Medical and Political Issues," *New England Journal of Medicine* 346(20): 1576-1579 (2002).

[3] See, for instance, Green, R., "Determining Moral Status," *American Journal of Bioethics* 2(1): 28-29 (2002).

[4] See, for instance, Orr, R., "The Moral Status of the Embryonal Stem Cell: Inherent or Imputed?" *American Journal of Bioethics* 2(1): 57-59 (2002).

[5] See, for instance, Doerflinger, R., "Ditching Religion and Reality," *American Journal of Bioethics* 2(1): 31 (2002).

[6] See, for instance, Latham, S., "Ethics and Politics," *American Journal of Bioethics* 2(1): 46 (2002).

[7] See, for instance, Marquis, D., "Stem cell research: The Failure of Bioethics," *Free Inquiry* 22(1): Winter 2002.

[8] See, for instance, the presentation of Richard Doerflinger before the Council on June 12, 2003. A transcript of that presentation is available on the Council website at www.bioethics.gov.

[9] See, for instance, FitzGerald, K., "Questions Concerning the Current Stem Cell Debate," *American Journal of Bioethics* 2(1): 50-51 (2002).

[10] See, for instance, Haseltine, W., "Regenerative Medicine: A Future Healing Art," *Brookings Review*, Winter 2003.

[11] See, for instance, Stolberg, S., "Scientists Urge Bigger Supply of Stem Cells," *The New York Times*, September 11, 2001, p. A1.

[12] See, for instance, Perry, D., "Patients' Voices: The Powerful Sound in the Stem Cell Debate," *Science* 287: 5457, 1423; and the presentation of Irving Weissman before the Council on February 13, 2002, available on the Council website at www.bioethics.gov.

[13] This point was taken up by the Council in its report *Human Cloning and Human Dignity: An Ethical Inquiry* (July 2002), Chapter 6, pp. 167-168.

[14] One useful description of this approach to the issue is Zoloth, L., "Jordan's Banks, A View from the First Years of Human Embryonic Stem Cell Research," *American Journal of Bioethics* 2(1): 7 (2002).

[15] Callahan, D., *What Price Better Health: The Hazards of the Research Imperative.* Los Angeles, CA.: University of California Press, 2003; also see Daniel Callahan's presentation before the Council on July 24, 2003, available on the Council's website at www.bioethics.gov.

[16] FitzGerald, K., *op. cit.*, pp. 50-51.

[17] For examples, see Weiss, R., "Legal Barriers to Human Cloning May Not Hold Up," *The Washington Post*, May 23, 2001.

[18] See, for instance, Elias, P., "States, Feds May Clash over Cloning Rules," *Genomics and Genetics Weekly*, April 12, 2002, p. 24.

[19] This, for instance, was the line of reasoning of the American Bar Association when, in August of 2002, it issued a resolution supporting limited therapeutic research using cloned human embryos. The report of the ABA's Section of Individual Rights and Responsibilities (available from the ABA) lays out the case.

[20] See, for instance, Russo, E., "Policy Questions and Some Answers," *The Scientist*, April 15, 2003.

[21] R. Alta Charo, of the University of Wisconsin Law School, quoted in Coyle, M., "The Clone Zone," *The National Law Journal*, May 15, 2002.

[22] Alan Meisel of the University of Pittsburgh School of Law, quoted in Coyle, M., "The Clone Zone," *The National Law Journal*, May 15, 2002.

[23] Berkowitz, P., "The Meaning of Federal Funding," a paper commissioned by the President's Council on Bioethics and included as Appendix F of this report.

[24] See: *Maher v. Roe* 432 U.S. 464 (1977); *Harris v. McRae* 448 U.S. 297 (1980); *Rust v. Sullivan* 500 U.S. 173 (1991).

[25] Berkowitz, P., "The Meaning of Federal Funding," a paper commissioned by the President's Council on Bioethics, and included as Appendix F of this report. See also Khushf, G., and Best, R., "Stem Cells and the Man on the Moon: Should We Go There from Here?" *American Journal of Bioethics* 2(1): 38 (2002).

[26] FitzGerald, K., *op. cit.*, pp. 50-51.

[27] Doerflinger, R., *op. cit.*, p. 31.

[28] Childress, J., "Federal Policy Toward Embryonic Stem cell research," *American Journal of Bioethics* 2(1): 34 (2002).

[29] Kahn, J., "Missing the mark on stem cells," *Bioethics Examiner* 5:3, Fall 2001, p. 1.

[30] Murray, T., "Hard Cell," *The American Prospect,* September 24, 2001.

[31] See, for instance, Lefkowitz, J., "The Facts on Stem Cells," *The Washington Post*, October 30, 2003, p. A23; and Bush, G.W., "Stem Cell Science and the Preservation of Life," *The New York Times*, August 12, 2001, p. D13.

[32] "How many lines," *The Washington Post*, August 31, 2001, p. A22.

[33] Daley, G., "Cloning and Stem Cells—Handicapping the Political and Scientific Debates," *New England Journal of Medicine* 349(3): 211-212 (2003).

[34] Kennedy, D., "Stem Cells: Still Here, Still Waiting," *Science* 300: 865 (2003).

[35] See, for instance, Brickley, P., "Scientists Seek Passports to Freer Environments," *The Scientist* 15:36 (2001).

[36] See, for instance, Weissman, I., "Stem Cells—Scientific, Medical and Political Issues," *New England Journal of Medicine* 346(20): 1578-1579 (2002).

[37] See, for instance, Weise, E., "USA's Stem Cell Scientists Fear a Research Brain-Drain," *USA Today*, May 12, 2003, p. 6D.

[38] Clark, J., "Squandering Our Technological Future," *The New York Times*, August 30, 2001, p. A19. See also the presentation of Thomas Okarma, President and CEO of Geron Corporation, before the Council on September 4, 2003. The full transcript of Okarma's presentation may be found on the Council's website at www.bioethics.gov.

[39] Robertson, J., "Crossing the Ethical Chasm: Embryo Status and Moral Complicity," *American Journal of Bioethics* 2(1): 33 (2002).

[40] See, for instance, Pizzo, P., "Remove Obstacles to Stem cell research," *San Jose Mercury-News*, June 19, 2003; and Holden, C. and Vogel, G., "'Show Us the Cells' U.S. Researchers Say," *Science* 297: 923 (2002).

[41] The most notable discussion of safety concerns has been Dawson, L., et al., "Safety Issues in Cell-Based Intervention Trials" *Fertility and Sterility* 80: 5, 1077-1085, 2003. We must note, however, that the assessment of the subject presented in that article relies upon the assumption that "all of [the approved] lines were derived using mouse feeder layers." (p. 1078). In his presentation before the Council on September 4, 2003, NIH Director Elias Zerhouni flatly contradicted this assertion, telling the Council that "there are at least those, which is about 16 lines, I believe, that have not been exposed to either mouse or human cell—human feeder cell lines." (The transcript of Zerhouni's presentation is available on the Council's website at www.bioethics.gov.) Those *eligible* unexposed lines, however, are not presently among the lines *available* to researchers, and it cannot be known in advance if or when they might become available or whether they will prove useable. In addition, the FDA has stated that even in the case of those lines that were developed with mouse feeders, "FDA does not intend that the agency's regulation of xenotransplantation will preclude the use of these hES cell lines." (Food and Drug Administration, "Guidance for Industry - Source Animal, Product, Preclinical, and Clinical Issues Concerning Use of Xenotransplantation Products in Humans," April 3, 2003.) FDA Commissioner Mark McClellan also made that clear in his presentation before the Council, available at the Council's website.

[42] Faden, et al., "Public Stem Cell Banks: Considerations of Justice in Stem Cell Research and Therapy" *Hastings Center Report* 33:6 (2003).

[43] See, for instance, Capron, A., "Stem Cell Politics: The New Shape to the Road Ahead," *American Journal of Bioethics* 2(1): 35-36 (2002).

[44] Robertson, J., *op. cit.*, pp. 33-34.

[45] See, for instance, the testimony of HHS Secretary Tommy Thompson before the Senate Health, Education, Labor and Pensions Committee, September 5, 2001; and Lefkowitz, J., "The Facts on Stem Cells," *The Washington Post*, October 30, 2003, p. A23.

[46] Brownstein, R., "Bush Won't Budge on Stem Cell Position, Health Secretary Says," *Los Angeles Times*, August 13, 2001, p. A9.

[47] This comment was made by Council Member Gilbert Meilaender in Council discussion on September 4, 2003. A transcript of that discussion is available on the Council's website at www.bioethics.gov.

[48] See, for instance, FitzGerald, K., *op. cit.*, pp. 50-51.

[49] Khushf, G., and Best, R., *op. cit.*, p. 38.

[50] See, for instance, the discussion of the Council in its September 4, 2003 meeting, particularly the argument elucidated by Council Member Michael Sandel. A transcript of that session is available on the Council website at www.bioethics.gov.

[51] See, for instance, Stolberg, S., *op. cit.*

[52] See, for instance, Robertson, J., "Ethics and Policy in Embryonic Stem cell research," *Kennedy Institute of Ethics Journal* 9(2): 109-136 (1999); Spike, J. "Bush and Stem cell research: An Ethically Confused Policy," *American Journal of Bioethics* 2(1): 46 (2003); and Capron, A., *op. cit.*, p. 36.

[53] The statement was made by bioethicist Glen McGee, in reference to the Clinton-era proposed NIH rules on funding, quoted in Spanogle, J., "Transforming Life," *The Baylor Line*, Winter 2000, p. 30.

[54] The remark was made by Bishop Joseph A. Fiorenza, President of the U.S. Conference of Catholic Bishops, in a news release ("Catholic Bishops Criticize Bush Policy on Embryo Research," made available by the United States Conference of Catholic Bishops, August 9, 2001).

[55] This argument arose in the course of Council discussion on September 4, 2003 (see particularly the comments of Council Members Michael Sandel and Charles Krauthammer in that discussion). A transcript of that session is available on the Council's website at www.bioethics.gov.

[56] *Ibid.*, see particularly the comments of Council Members Mary Ann Glendon and Gilbert Meilaender in that discussion.

[57] For these guidelines, see Appendix D.

[58] *Ibid.*, see particularly the comments of Council Member William Hurlbut in that discussion.

[59] Some reasons to regard public funding of activities in a light different from mere permission are proposed in Khushf, G., and Best, R., *op. cit.*, p. 38.

[60] "Remarks by the President on Stem Cell Research," as made available by the White House Press Office, August 9, 2001.

[61] "President's Phone Call to March for Life Participants," as made available by the White House Press Office, January 22, 2002.

[62] See Council discussion on September 4, 2003, especially comments by Council Members Charles Krauthammer, Francis Fukuyama, Gilbert Meilaender, and Leon

Kass. A transcript of that session is available on the Council's website at www.bioethics.gov.

[63] Bush, G.W., "Stem Cell Science and the Preservation of Life," *The New York Times*, August 12, 2001, p. D13.

[64] "Remarks by the President on Stem Cell Research," as made available by the White House Press Office, August 9, 2001.

[65] See, for instance, Council discussion on September 4, 2003 (particularly the comments of Council member Michael Sandel and invited guest Peter Berkowitz in that discussion). A transcript of that session is available on the Council's website at www.bioethics.gov.

[66] "Remarks by the President on Stem Cell Research," as made available by the White House Press Office, August 9, 2001.

[67] See, for instance, Council discussion on September 4, 2003 (particularly the comments of Council Members Charles Krauthammer and James Wilson in that discussion). A transcript of that session is available on the Council's website at www.bioethics.gov.

[68] For instance, in discussing research on cloned human embryos, President Bush said, "Research cloning would contradict the most fundamental principle of medical ethics, that no human life should be exploited or extinguished for the benefit of another. Yet a law permitting research cloning, while forbidding the birth of a cloned child, would require the destruction of nascent human life." ("Remarks by the President on Human Cloning Legislation," as made available by the White House Press Office, April 10, 2002); and in explaining his stem cell funding policy, the President wrote: "While it is unethical to end life in medical research, it is ethical to benefit from research where life and death decisions have already been made." (Bush, G.W., "Stem Cell Science and the Preservation of Life," *The New York Times*, August 12, 2001, p. D13.)

[69] As noted previously, in August of 2001 Health and Human Services Secretary Tommy Thompson told reporters that "neither unexpected scientific breakthroughs nor unanticipated research problems would cause Bush to reconsider." (See, Brownstein, R., "Bush Won't Budge on Stem Cell Position, Health Secretary Says," *Los Angeles Times*, August 13, 2001, p. A9).

[70] One useful account of these issues is Cohen, E., "Of Embryos and Empire," *The New Atlantis* 2: 3-16 (2003).

[71] See, for instance, London, A., "Embryos, Stem Cells, and the 'Strategic' Element of Public Moral Reasoning," *American Journal of Bioethics* 2(1): 56 (2002).

[72] For examples of this way of proceeding—both in arguments supporting embryo research, and arguments opposing it—see, among numerous other sources, the Council's July 2002 report, *Human Cloning and Human Dignity: An Ethical Inquiry*, Chapter 6.

[73] See, for instance, McCartney, J., "Embryonic Stem cell research and Respect for Human Life: Philosophical and Legal Reflections," *Albany Law Review* 65: 597-624 (2002).

[74] See, for instance, Orr, R., *op. cit.*, pp. 57-58.

[75] See, for instance, Steinberg, D., "Can Moral Worthiness Be Seen Using a Microscope?" *American Journal of Bioethics* 2(1): 49 (2002).

[76] See, for instance, Green, R., *op. cit.*, p. 20.

[77] See, for instance, Doerflinger, R., *op. cit.*, pp. 31-33; Orr, R., *op. cit.*, pp. 57-58; and the personal statements of Council Members William Hurlbut and Robert George and Alfonso Gómez-Lobo, in the Council's July 2002 report *Human Cloning and Human Dignity: An Ethical Inquiry*, among others.

[78] See, for instance, Orr, R., *op. cit.*, pp. 57-58.

[79] See, for instance, the personal statement of Council Member William Hurlbut, in the Council's July 2002 report *Human Cloning and Human Dignity: An Ethical Inquiry*.

[80] See, for instance, Sullivan, D., "The Conception View of Personhood: A Review," *Ethics and Medicine* 19(1): 11-33 (Spring 2003).

[81] A form of this argument was presented by some Members of the Council in the Council's July, 2002 report *Human Cloning and Human Dignity: An Ethical Inquiry*, chapter 6, and in several of the personal statements appended to that report.

[82] See, for instance, Steinberg, D., *op. cit.*, p. 49.

[83] See, for instance, the transcript of Council discussion on October 17, 2003, especially the comments of Council Member Elizabeth Blackburn. The full transcript is available on the Council's website at www.bioethics.gov. Also see Green, R., *op. cit.*, pp. 20-30.

[84] See, for instance, Lanza, R., et al., "The Ethical Validity of Using Nuclear Transfer in Human Transplantation," *Journal of the American Medical Association*, 284(24): 3175-3179 (2000). Also, see the views expressed by some Members of this Council in our July 2002 report *Human Cloning and Human Dignity: An Ethical Inquiry*, chapter 6, and in the personal statements of Council Members Elizabeth Blackburn, Michael Gazzaniga, and Janet Rowley.

[85] See, for instance, the transcript of Council discussion on April 25, 2002, especially the comments of Council Member Daniel Foster. The full transcript is available on the Council's website at www.bioethics.gov.

[86] See, for instance, the position articulated in "Position Number Two" of the "Moral Case for Cloning for Biomedical Research" presented by some Members of the Council in the Council's July, 2002 report *Human Cloning and Human Dignity: An Ethical Inquiry*, Chapter 6; and the personal statement of Council Member Michael Gazzaniga, appended to that report.

[87] See, for instance, the comments of Council Member Michael Gazzaniga (and response from other Members) during Council discussion on February 14, 2002. The full transcript is available on the Council's website at www.bioethics.gov.

[88] See, for instance, Orr, R., *op. cit.*, pp. 57-59; Callahan, S., "Zygotes and Blastocysts: Human enough to protect," *Commonweal*, June 14, 2002; the position articulated by several Members of this Council in our July 2002 report *Human Cloning and Human Dignity: An Ethical Inquiry*, Chapter 6; and the personal statements of Council

Members Robert George and Alfonso Gómez-Lobo, and William Hurlbut in that same report.

[89] See, for instance, Steinberg, D., *op. cit.*, pp. 49-50; and Green, R., *op. cit.*, p. 20.

[90] See, for instance, Meyer, M. J., and Nelson, L. J., "Respecting What We Destroy: Reflections on Human Embryo Research," *Hastings Center Report* 31(1): 16-23 (2001).

[91] See, for instance, Strong, C., "The Moral Status of Preembryos, Embryos, Fetuses, and Infants," *The Journal of Medicine and Philosophy* 22(5): 457-478 (1997).

[92] Pearson, H., "Your Destiny, from Day One," *Nature* 418: 14-15 (2002).

[93] See, for instance, McCartney, J., *op. cit.*, pp. 601-602; Brogaard, B., "The moral status of the human embryo: the twinning argument," *Free Inquiry* Winter 2002; Meyer, M. J., *op. cit.*, p. 18. The question of twinning—pro and con—was also taken up by the Council in its July 2002 report, *Human Cloning and Human Dignity: An Ethical Inquiry,* Chapter 6.

[94] See, for instance, Green , R., *op. cit.*, pp. 21-22; Lanza, R., *op. cit.*, p. 3177.

[95] See, for instance, Ashley, B., and Moraczewski, A., "Cloning, Aquinas, and the Embryonic Person," *National Catholic Bioethics Quarterly* 1(2): 189-201 (2001); Orr, R., *op. cit.*, p. 58; Marquis, D., *op. cit.*; and Lori Andrews quoted in Green, R., *op. cit.*, p. 22. The question of twinning—pro and con—was also taken up by the Council in its July 2002 report, *Human Cloning and Human Dignity: An Ethical Inquiry,* Chapter 6, and in the personal statement of Council Member William Hurlbut, attached to that report.

[96] See, for instance, National Bioethics Advisory Commission, *Ethical Issues in Human Stem cell research*, Washington, D.C.: Government Printing Office, 1999, p. 7; and the position articulated by several Members of this Council in our July 2002 report *Human Cloning and Human Dignity: An Ethical Inquiry*, Chapter 6, among many others.

[97] See, for instance, Lanza, R., *op. cit.*, p. 3177.

[98] See, for instance, Strong, C., *op. cit.*, p. 467.

[99] See, for instance, the views expressed by some Members of this Council in our July 2002 report *Human Cloning and Human Dignity: An Ethical Inquiry*, Chapter 6, and in the personal statement of Council Members Robert George and Alfonso Gómez-Lobo.

[100] See, for instance, the personal statement of Council Member William Hurlbut, in the Council's July 2002 report *Human Cloning and Human Dignity: An Ethical Inquiry.*

[101] See, for instance, Wilson, J., "On Abortion," *Commentary* January 1994; and the position articulated by several Members of this Council in our July 2002 report *Human Cloning and Human Dignity: An Ethical Inquiry*, Chapter 6.

[102] See, for instance, Green, R., *op. cit.*, pp. 20-30; and the position articulated by several Members of this Council in our July 2002 report *Human Cloning and Human Dignity: An Ethical Inquiry*, Chapter 6.

[103] See, for instance, the views expressed by some Members of this Council in our July 2002 report *Human Cloning and Human Dignity: An Ethical Inquiry*, Chapter 6, and in the personal statement of Council Members Robert George and Alfonso Gómez-Lobo.

[104] See, for instance, National Institutes of Health, *Report of the Human Embryo Research Panel*, Bethesda, MD.: NIH, 1994; and the position articulated by several Members of this Council in our July 2002 report *Human Cloning and Human Dignity: An Ethical Inquiry*, Chapter 6.

[105] See, for instance, McGee, G., "The Idolatry of Absolutizing in the Stem Cell Debate," *American Journal of Bioethics* 2(1): 53-54 (2002).

[106] See, for instance, Green, R., *op. cit.*, p. 20.

[107] See, for instance, the views expressed by some Members of this Council in our July 2002 report *Human Cloning and Human Dignity: An Ethical Inquiry*, Chapter 6, and in the personal statements of Council Members Gilbert Meilaender, William Hurlbut, Robert George and Alfonso Gómez-Lobo.

[108] "Report of the Ethics Advisory Board," 44 *Fed. Reg.* 35,033-35,058 (June 18, 1979) at 35,056.

[109] NIH, *op. cit.*, *Report of the Human Embryo Research Panel*, p. 2.

[110] National Bioethics Advisory Commission (NBAC), *Ethical Issues in Human Stem Cell Research* Bethesda, MD.: Vol. I, p. ii; cf. p. 2 (September 1999).

[111] See, for instance, Lebacqz, K., "On the Elusive Nature of Respect," in Holland, S., Lebacqz K., and Zoloth, L., eds., *The Human Embryonic Stem Cell Debate*, Cambridge, MA.: MIT Press, 2001, pp. 149-162; and the personal statement of Council Member Rebecca Dresser, appended to the Council's July 2002 report *Human Cloning and Human Dignity: An Ethical Inquiry*.

[112] Hoffman, D.I., et al., "Cryopreserved Embryos in the United States and Their Availability for Research," *Fertility and Sterility* 79: 1063-1069 (2003).

[113] See, for instance, Spike, J., *op. cit.*, p. 45.

[114] See, for instance, the testimony of Michael West before the Labor, HHS, and Education Subcommittee of the Senate Appropriations Committee, December 4, 2001, among others.

[115] Outka, G., "The Ethics of Stem cell research," a paper presented to the Council for its April 2002 meeting, and available on the Council website at www.bioethics.gov. A slightly revised version appeared in the *Kennedy Institute of Ethics Journal*, 12:2, pp. 175-213.

[116] *Ibid.*

[117] See, for instance, "The Stem Cell Sell," *Commonweal*, August 17, 2001; Meilaender, G., "Spare Embryos," *The Weekly Standard*, August 26, 2002; and Council discussion of Gene Outka's paper, in its April 2002 meeting, a transcript of which is available on the Council website at www.bioethics.gov.

[118] See, for instance, McCartney, J, *op. cit.*, p. 615.

[119] Council Member William Hurlbut, in Council communication.

[120] See, for instance, the comments of Council Member James Wilson during Council discussion on April 25, 2002. The full transcript is available on the Council's website at www.bioethics.gov.

[121] A federal law, 42 U.S.C. § 289, is directed specifically against any such lethal experimentation on these populations.

[122] This concern is raised in Outka, G., "The Ethics of Stem Cell Research," a paper presented to the Council for its April 2002 meeting, and available on the Council website at www.bioethics.gov. Also see the Council discussion of that paper, available on the Council website; and see Meilaender, G., *op. cit.*

[123] See, for instance, the transcript of Council discussion on April 25, 2002, available on the Council's website at www.bioethics.gov.

[124] Guenin, L. M., "Morals and Primordials," *Science* 292: 1659-1660 (2001).

[125] Spike, J., *op. cit.*, p. 45.

[126] Guenin, L., *op. cit.*

[127] See, for instance, Orr, R., *op. cit.*, p. 58.

[128] Mahowald, M. and Mahowald A., "Embryonic Stem Cell Retrieval and a Possible Ethical Bypass," *American Journal of Bioethics* 2(1): 42-43 (2002).

[129] See, for instance, Zoloth, L., *op. cit.*, pp. 4-5.

[130] See, for instance, Harris, J., "The ethical use of human embryonic stem cells in research and therapy," in Burley, J. and Harris, J., eds., *A Companion to Genetics: Philosophy and the Genetic Revolution*, Oxford: Basil Blackwell, 2001; and Savulescu, J., "The Embryonic Stem Cell Lottery and the Cannibalization of Human Beings," *Bioethics* 16(6): 508-529 (2002).

[131] This subject was discussed in a Council session on January 16, 2003. When asked, in the course of a presentation before the Council, about the rate of natural embryo loss, Dr. John Opitz (Professor of Pediatrics, Human Genetics, and Obstetrics/ Gynecology, School of Medicine, University of Utah) responded: "Estimates range all the way from 60 percent to 80 percent of the very earliest stages, cleavage stages, for example, that are lost." The transcript of this session is available on the Council website at www.bioethics.gov.

[132] Spike, J., *op. cit.*, p. 45.

[133] Harris, J., "Stem Cells, Sex & Procreation," a paper presented to the American Philosophical Society, March 29, 2003.

[134] Savulescu, J., *op. cit.*, pp. 508-529.

[135] See, for instance, Orr, R., *op. cit.*, pp. 57-59.

[136] See, for instance, Council discussion in its January 2003 meeting, a transcript of which is available on the Council website at www.bioethics.gov.

[137] See, for instance, the comments of Rudolph Jaenisch before the Council in its July 2003 meeting, and Dr. Jaenisch's accompanying paper (included in Appendix N of this report). See also the personal statement of Council Member Paul McHugh, appended to the Council's July 2002 report *Human Cloning and Human Dignity: An Ethical Inquiry*.

[138] See, for instance, Zoloth, L., *op. cit.*, pp. 4-5.

[139] See, for instance, Mandavilli, A., "Fertility's New Frontier Takes Shape in the Test Tube," *Nature Medicine* 9: 1095 (2003).

[140] See, for instance, Council discussion with Rudolph Jaenisch at the Council's July 2003 meeting, and discussion taken up in the Council's July 2002 report *Human Cloning and Human Dignity: An Ethical Inquiry,* Chapter 6.

[141] See, for instance, the personal statement of Council Member William Hurlbut, in the Council's July 2002 report *Human Cloning and Human Dignity: An Ethical Inquiry;* the transcripts of Council discussions on June 20, 2002 and July 24, 2003, available on the Council's website at www.bioethics.gov; and Weiss, R., "Can Scientists Bypass Stem Cells' Moral Minefield?" *The Washington Post*, December 14, 1998, p. A3.

[142] See, for instance, the transcript of Council discussion on July 24, 2003, available on the Council's website at www.bioethics.gov.

[143] See, for instance, the transcript of Council discussion on June 20, 2002, available on the Council's website at www.bioethics.gov.

[144] See remarks by Professor Rudolph Jaenisch before the Council's July 24, 2003 meeting. A transcript of that session is available on the Council's website at www.bioethics.gov.

[145] See, for instance, the personal statement of Council Member William Hurlbut, in the Council's July 2002 report *Human Cloning and Human Dignity: An Ethical Inquiry.*

[146] See, for instance, a number of arguments raised in Weiss, R., "Can Scientists Bypass Stem Cells' Moral Minefield?" *The Washington Post*, December 14, 1998, p. A3; and the transcript of Council discussion on June 20, 2002, available on the Council's website at www.bioethics.gov.

[147] Lauritzen, P., "Report on the Ethics of Stem cell research," a commissioned paper prepared at the request of the Council and included in Appendix G of this report. See also the Council's July 2002 report *Human Cloning and Human Dignity: An Ethical Inquiry,* Chapter 6.

[148] See, for instance, Kass, L., "The Meaning of Life - In the Laboratory," *The Public Interest*, Winter 2002. Also see the Council's July 2002 report *Human Cloning and Human Dignity: An Ethical Inquiry,* Chapter 6.

[149] This case is articulated by several Members of the Council in the Council's July 2002 report *Human Cloning and Human Dignity: An Ethical Inquiry,* Chapter 6. See also, "Biotechnology: A House Divided," *The Public Interest*, Winter 2003.

[150] This case is articulated, for instance, by several Members of the Council in the Council's July 2002 report *Human Cloning and Human Dignity: An Ethical Inquiry,* Chapter 6.

[151] Strong, C., "Those Divisive Stem Cells: Dealing with Our Most Contentious Issues," *American Journal of Bioethics* 2(1): 39-40 (2002).

[152] See, for instance, the Council's July 2002 report *Human Cloning and Human Dignity: An Ethical Inquiry,* Chapters 6 and 8, available on the Council's website at www.bioethics.gov.

[153] See, for instance, Dresser, R., "Embryonic Stem Cells: Expanding the Analysis," *American Journal of Bioethics* 2(1): 40-41 (2002); and FitzGerald, K., *op. cit.,* pp. 50-51.

[154] Dresser, R., *op. cit.,* p. 41.

[155] *Ibid.*

[156] See also FitzGerald, K., *op. cit.,* pp. 50-51.

[157] Dresser, R., *op. cit.,* p. 41.

[158] See, for example, the discussion and presentations at the Council's July 2003 meeting, available on the Council's website at www.bioethics.gov.

[159] See, for instance, Orr, R., *op. cit.,* pp. 57-58; "A review of the National Institute of Health's Guidelines for Research Using Human Pluripotent Stem Cells," *Issues in Law and Medicine* 3(17): 293; and the testimony of Representative David Weldon before the Technology and Space Subcommittee of the Senate Commerce, Science and Transportation Committee, January 29, 2003.

[160] Lauritzen, P., "Report on the Ethics of Stem cell research," a commissioned paper prepared at the request of the Council and included in Appendix G of this report.

[161] See, for instance, Oldham, R., "Stem Cells: Private Sector Can Do It Better," *The Wall Street Journal,* August 28, 2001, p. A14.

[162] See, for example, the discussion and presentations at the Council's April 2002 meeting, available on the Council's website at www.bioethics.gov; and Verfaillie, C., "Multipotent Adult Progenitor Cells: An Update," a commissioned paper prepared at the request of the Council and included in Appendix J of this report.

[163] See, for example, the discussion and presentations at the Council's July 2003 meeting, available on the Council's website at www.bioethics.gov.

4

Recent Developments in Stem Cell Research and Therapy

Research using human and animal <u>stem cells</u>[*] is an extremely active area of current biomedical inquiry. It is contributing new knowledge about the pathways of normal and abnormal cell differentiation and organismal development. It is opening vistas of new cell transplantation therapies for human diseases. Although the availability of a variety of human stem cells is relatively recent—the isolation of human <u>embryonic stem cells</u> was first reported only in 1998—much is happening in both publicly funded and privately funded research centers around the world. It is difficult for anyone to stay abreast of all the results now rapidly accumulating.

To help us fulfill our mandate to "monitor stem cell research," the President's Council on Bioethics asked several experts to survey the recent published scientific literature and to contribute articles on various areas of stem cell research to this report (see articles by Drs. Gearhart,[1] Ludwig and Thomson,[2] Verfaillie,[3] Prentice,[4] Itescu,[5,6] and Jaenisch[7] in the Appendices). These reviews and the present chapter emphasize peer-reviewed, published work with human stem cells through July 2003. Interested readers should also consult the wide variety of other review articles that have appeared.[8]

This chapter should be read in conjunction with the commissioned review articles cited above. It draws on their find-

---

[*] In this chapter, technical terms that are defined in the Glossary are underlined when they are used for the first time.

ings, as well as on the Council's own monitoring activities, but it makes no attempt to summarize all the complexity of stem cell research or the vast array of results. Rather we offer here some general observations and specific examples that might help non-scientist readers understand the overall state of present human stem cell research, its therapeutic promise, and some of the problems that need to be solved if the research is to yield sound knowledge and clinical benefit. To that end, we highlight the importance of well-characterized, stable preparations of stem cells for obtaining reproducible experimental results, and we identify several problems that must be solved before these requirements can be fully met. This chapter then describes, by way of illustration and example, some of the better-characterized adult and embryonic stem cells. It also indicates some of the specific investigations that are being conducted with their aid. Finally, it considers how human stem cells are being used to explore their potential for treating disease, using experiments in animal models of Type-1 diabetes as an example, and it points out some of the difficulties that must be overcome before stem cell-based remedies may be available to treat human diseases.

We confine our attention here to newly identified types of human stem cells and their potential use in research and future medical treatment. Accordingly, we do not consider those stem cell types that are already well established in medical practice and research. Specifically, we will not examine those preparations of bone marrow cells that have been clinically used for some years to treat various forms of anemia and cancer.[9] Neither will we deal with hematopoietic (blood-forming) stem cells that have been isolated and purified from bone marrow and are now being intensively studied.[10] Although these developments lie beyond the scope of this report, the demonstrated usefulness of these cells for research and therapy encourages many researchers to expect similar benefits from the newer stem cells that we shall consider here.

I. STEM CELLS AND THEIR DERIVATIVES

The adult human body, and all its differentiated cells, tissues, and organs, arise from a small group of cells contained within the early underline{embryo} at the underline{blastocyst stage} of its development. During in vivo embryonic development, these cells, constituting the underline{inner cell mass (ICM)}, will divide and differentiate in concert with each other and with the whole of which they are a part, eventually producing the specialized and integrated tissues and organs of the body. But when embryos are grown [using in vitro fertilization (IVF)] in a laboratory setting, these ICM cells may be removed and isolated, and under appropriate conditions some will proliferate in vitro and become embryonic stem cell lines.

These embryonic stem cells are capable of becoming many different types of differentiated cells if stimulated to do so in vitro [see endnote 2 for references]. However, it is not yet clear that the cells that survive the in vitro selection process to become embryonic stem cells have *all* of the same biological properties and potentials as the ICM cells of the blastocyst.[7] In particular, it is not known for certain that human embryonic stem cells in vitro can give rise to *all* the different cell types of the adult body.[*]

As noted in the Introduction to this report, stem cells are a diverse class of cells, which can now be isolated from a variety of embryonic, fetal, and adult tissues. Stem cells share two characteristic properties: (1) unlimited or prolonged *self-renewal* (that is, the capacity to maintain a pool of stem cells like themselves), and (2) *potency for differentiation*, the potential to produce *more differentiated* cell types—usually

---

[*] It is also not known whether stem cells, either human or animal, when cultured in vitro apart from the embryonic whole from which they were originally derived, will function in all respects like cells do when they act as parts of a developing organic whole.

more than one and, in some cases, many.* When stem cells head down the pathway toward differentiation, they usually proceed by first giving rise to a more specialized kind of stem cell (sometimes called "precursor cells" or "progenitor cells"), which can in turn either proliferate through self-renewal or produce fully specialized or differentiated cells (see Figure 1).

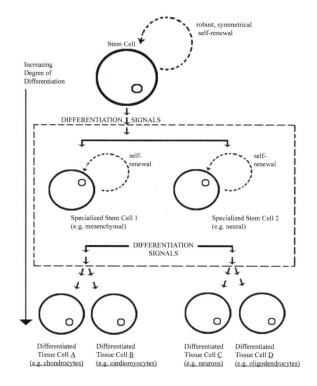

Figure 1. Schematic Diagram of Some Stages in Cell Differentiation

At the top of the figure is an undifferentiated stem cell; in the central box are more "specialized" stem cells (or "precursor cells" or "progenitor cells"); at the bottom are various differentiated cells that are derived from the specialized stem

---

* Some stem cells, however, give rise to only one type of specialized cell. For example, one type of stem cell found in the epidermis (skin) apparently gives rise only to keratinoctyes (cells that produce the protein keratin, found in hair and nails).

cells. Dashed arrows indicate symmetrical (in the sense that both the daughter cells are stem cells) cell divisions that produce more stem cells (self-renewal). Solid arrows indicate asymmetric cell divisions that produce more differentiated daughter cells. (There may also be self-renewal with asymmetric division—not shown here—in which one daughter cell initiates a differentiation pathway while the other remains a stem cell.) Differentiation signals can be supplied by both soluble proteins and by specific, cell-surface binding sites. Some of the specialized stem cells inside the dashed box, for example, mesenchymal stem cells, can be isolated from tissues after birth and correspond to *adult* stem cells. Scientists are currently investigating whether, at least in some cases, the process can be reversed, that is, whether specialized cells may, on appropriate signals, *dedifferentiate* to become precursor or even fully undifferentiated stem cells.

The terminology used to describe different stem cell types can be confusing. As used in this chapter, *stem cells* are self-renewing, cultured cells, grown and preserved in vitro, that are capable—upon exposure to appropriate signals—of differentiating themselves into (usually more than one) specialized cell types. Stem cells may be classified either according to their *origins* or according to their *developmental potential*.

Stem cells may be obtained from various sources: from embryos, from fetal tissues, from umbilical cord blood, and from tissues of adults (or children). Thus, depending on their *origin*, stem cell preparations may be called adult stem cells,* embryonic stem cells, embryonic germ cells, or fetal stem cells. *Adult stem cells* [see (4)] are cells derived from various tissues or organs in humans or animals that have the two characteristic properties of stem cells (self-renewal and potency for differentiation). *Embryonic stem cells* (ESCs) [see (2)] are derived from cells isolated from the inner cell mass of early embryos. *Embryonic germ cells* (EGCs) [see (1)] are stem cells

---

* As already noted in Chapter 1, "adult stem cells" is something of a misnomer. The cells are not themselves "adult." As non-embryonic stem cells, they are, however, partially differentiated and many of them are multipotent. (See discussion in the text that follows shortly.)

derived from the primordial germ cells of a fetus. *Fetal stem cells* (not further discussed in this chapter, but included for the sake of completeness) are derived from the developing tissues and organs of fetuses; because they come (unlike EGCs) from already differentiated tissues, they are (like adult stem cells) "non-embryonic," and may be expected to behave as such.

Depending on their *developmental potential*, cells may be called pluripotent, multipotent, or unipotent. Cells that can produce *all* the cell types of the developing body, such as the ICM cells of the blastocyst, are said to be *pluripotent*. The somewhat more specialized stem cells, of the sort found in the developed organs or tissues of the body, are said to be *multipotent* if they produce more than one differentiated tissue cell type, and *unipotent* if they produce only one differentiated tissue cell type.

We introduce in this chapter an additional term: *stem cell preparation*. A *stem cell preparation* is a population of stem cells, prepared, grown, and preserved under certain conditions. Because different laboratories (or even the same one) can have different preparations of the same type of stem cell, it is important to recognize the potential differences between particular preparations of embryonic stem cells.[*] It will sometimes be important to call attention to this fact, by speaking of a "*preparation* of ES cells" (or a preparation of adult stem cells) rather than of "ES cells," pure and simple. We will use the term "stem cell preparations" when we are speaking of a diverse group of stem cell cultures, when we are speaking of stem cell cultures that contain an admixture of other types of cells, or when the developmental homogeneity of the stem cells in the population has not been defined.

Adult and embryonic stem cell populations have also been called "stem cell lines." In the past, the term "cell line" denoted a cell population (usually of cancer cells containing abnormal chromosome numbers or structure, or both) that could

---

[*] Embryonic stem cell cultures prepared from different embryos of a single inbred mouse strain are more likely to have closely similar biological properties than will ESC cultures from genetically different individual human beings.

grow "indefinitely" in vitro. Embryonic and some adult stem cell preparations are capable of prolonged growth beyond 50 population doublings in vitro while retaining their characteristic stem cell properties and initially with no change in the chromosome numbers and structure. It is not yet known whether any preparation of human ES cells (generally believed to be much longer-lived than adult stem cells) will continue to grow "indefinitely," without undergoing genetic changes.

Under the influence of various cell-differentiation signals, embryonic stem cells differentiate into numerous distinct types of more specialized cells. Some of these are specialized stem cells that can also self-renew, while retaining their ability also to differentiate into multiple cell types. Recent research has led to the isolation of an increasing number of adult (non-embryonic) stem cells (dashed box area of Figure 1) from such tissues as bone marrow (for example, hematopoietic and mesenchymal stem cells), brain (for example, neural stem cells) and other tissues [see (4)]. Although these stem cell preparations differ from one another in their future fates, they tend to be grouped together (especially in the public policy debates) under the name "adult stem cells," even though they may have been obtained from children or even from umbilical cord blood obtained at the time of childbirth.

Subsequent exposure to additional differentiation signals can cause these specialized stem cells to differentiate further, so that they finally give rise to the variety of differentiated cells that make up the adult body (labeled A-D in Figure 1). At each stage of the differentiation process, specific sets of genes are expressed (or "turned on") and other sets are repressed (or "turned off"), to produce the specific proteins that give each cell its distinctive properties. At each stage along the way, proteins called transcription factors play key roles in determining which sets of genes are expressed and repressed, and therefore what sort of a cell the newly differentiated cell will become.

II. REPRODUCIBLE RESULTS USING STEM CELL PREPARATIONS AND THEIR DERIVATIVES

A major goal of scientific research is the acquisition of reliable knowledge based on experiments that yield reproducible results. Reproducible results are possible only if the materials used in experiments remain constant and stable. To obtain reproducible results in experiments using stem cells, it is essential to produce, preserve, characterize, and continually re-characterize preparations of stem cells in ways that increase the likelihood that the cells used to repeat experiments will remain unchanged—a technically challenging task. The tendency of stem cells in vitro to differentiate spontaneously into more specialized cells makes the task of obtaining homogeneous and stable stem cell preparations especially challenging, and much basic research is needed to learn how to control the fate of these cells. Failure to control the cells may yield experimental results that are difficult or impossible to reproduce. The following more specific observations make clear the dimensions of this difficulty.

A. Initial Stem Cell Preparations Can Contain Multiple Cell Types

Isolation of *adult* stem cells from source tissues such as bone marrow, brain, or muscle initially yields a heterogeneous cell preparation. The initial preparation contains the several cell types found in the source tissue, and it may also include red blood cells, white blood cells, and (possibly) circulating stem cells, owing to the presence of blood flowing through the tissue in question. Initial mixtures of cells may then be treated in various ways to remove unwanted contaminating cells, thereby increasing the proportion of *stem cells* in the preparation. But seldom, if ever, does one produce an adult stem cell preparation that is 100 percent stem cells, unless the adult stem cell preparation has been "single-cell cloned" in vitro (see below).

The way in which human *embryonic* stem cells have been produced from ICM cells also raises a question about the "species homogeneity" of the initial cell preparations. In the past, human *embryonic* stem cells were isolated and maintained by in vitro growth on top of irradiated (so that they no longer divide) "feeder layers" of mouse cells. It is thought that the feeder cells secrete factor(s) that enable the stem cells to divide while maintaining a relatively undifferentiated state. Although the mouse cells have been treated to prevent their cell division, should any of them happen to survive, human embryonic stem cells prepared in this way *may* contain some viable mouse cells.[*] More recently, several groups have shown that it is possible to grow ESCs on feeder layers of human cells, including fibroblasts obtained from skin biopsies, or without any feeder cell layer at all.[11] One way to be certain that human embryonic stem cell preparations do not contain any mouse feeder cells is through "single cell cloning" (see below).

B. Genetically Homogenous Stem Cells through Single Cell Cloning

Some preparations of stem cells growing in vitro have been "single cell cloned," that is, grown as a population derived from a *single* stem cell. By placing a cylinder over a single cell located with a microscope, scientists are able to isolate within the cylinder all the progeny produced by subsequent cell divisions beginning from this single cell. The result is a stem cell preparation in which all the cells are descended from the original single cell. The cells within the cylinder are then harvested and grown to greater numbers in vitro, and the resulting stem cell preparation is said to be "single cell cloned." The stem cells within a "single cell cloned" population are, at least to begin with, *genetically* homogeneous because they are all derived from the same original cell. Some of the ESC preparations produced prior to August 9, 2001 have been "single cell cloned."[12]

---

[*] The issue of possible mouse virus contamination is dealt with in Section F, below.

C. Expansion in Vitro, Preservation, and Storage

Reproducible results require that preparations of stem cells, even if genetically homogenous when first isolated, remain stable over time and during preservation. This, too, is not a simple matter with stem cells, despite the fact that the self-renewal characteristic of human embryonic and adult stem cells enables them—unlike differentiated cells from many human tissues—to be grown in large numbers in vitro while maintaining their essential stem cell characteristics. After such expansion, many, presumably identical, vials of the cells can be frozen and preserved at very low temperatures. Frozen stem cell preparations can later be thawed and grown again in vitro to produce larger numbers of cells.

As with all dividing cells, stem cells are subject to a very small but definite chance of mutation during DNA replication; *thus, prolonged growth in vitro could introduce genetic heterogeneity into an originally homogeneous population.* During this process of repeated expansion and preservation, subtle changes in the growth conditions or other variables may give rise to "selective pressures" that can increase the heterogeneity in a stem cell preparation by favoring the multiplication of advantaged cell variants in the population. It is not known at present how many of the 78 human ESC preparations, designated as eligible for federal funding under the current policy, have developed genetic variants that may make them unsuitable for further research.

Whether several cycles of freezing and thawing change the phenotypic characteristics of stem cell preparations needs detailed study. However, the practical advantages of preserving stem cell preparations by freezing are too large to ignore. Such preservation makes it possible to repeat an experiment many times with a very similar stem cell preparation. It would also make it possible, should stem cell based therapies be developed in the future, to treat multiple patients with a common, well-characterized cell preparation derived from a single initial stem cell sample.

D. Chromosome Changes

In addition to the possible loss of homogeneity in stem cell preparations owing to variability in growth conditions or to freezing and thawing, there is the possibility of variation being introduced during the processes of growth and cell division. Normal human stem cells (like all human somatic cells) have 46 chromosomes. During the copying of chromosomal DNA and the separation of daughter chromosomes at cell division, rare mistakes occur that lead to the formation of abnormal chromosomes or maldistribution of normal ones. Cells with abnormal chromosomes or chromosome numbers can progress to malignancy, so retention of the normal human chromosome number and structure is an essential characteristic of useful human stem cell preparations. The most studied preparations of human stem cells generally have normal human chromosome numbers and structure.[3,13,*] Nevertheless, vigilance is needed, for even a small number of chromosomally abnormal cells could end up causing cancer in future clinical trials of stem cell based therapies.

E. Developmental Heterogeneity of Stem Cell Preparations

The in vitro growth conditions and the presence of specific chemicals or proteins, or both, in the culture medium can influence the differentiation pathway taken by stem cells as they start to differentiate. Thus, even initially homogeneous, "single cell cloned" stem cell preparations may become *developmentally* heterogeneous over time, with respect to the

---

* As of November 2003, reports were available about the chromosome patterns of only 21 out of the 78 ESC preparations designated as eligible for federal funding; 11 of the 12 preparations currently available as of that time had their chromosome patterns characterized, and they appear normal. However, a recent publication, presenting results from two different laboratories, reports abnormalities in chromosome number and structure in some samples of three different human ESC preparations. Two of these ESC preparations are among the preparations currently available for federal funding. [Draper, J.S., et al., "Recurrent gain of chromosomes 17q and 12 in cultured human embryonic stem cells," *Nature Biotechnology* December 7, 2003, advance online publication.]

percentage of cells in the preparation that are in one or another differentiated state. For example, a stem cell preparation after growth in vitro under specific conditions might contain 75 percent fully differentiated (insulin-producing) cells and 25 percent partially differentiated cells. The biological properties of the fully differentiated cells and the partially differentiated cells are likely to be different. If such a cell preparation is used in research, or transplanted into an animal model of human disease and a biological effect is observed, one must do additional experiments to determine whether the effect was due to the fully differentiated cells or to the partially differentiated cells (or perhaps to both acting together) in the now mixed preparation.

F. Microbial Contamination

Stem cell preparations originally isolated from humans and expanded in vitro may also be variably contaminated with human viruses, bacteria, fungi, and mycoplasma. ESC preparations isolated using mouse feeder cell layers might also be contaminated with mouse viruses. Specific tests need to be performed on the source tissue and periodically on the resulting stem cell preparations to rule out the presence of these contaminants. Some of these contaminants can also multiply when stem cells are grown in vitro, and their presence can influence the results obtained when stem cell preparations are used in subsequent experiments. The presence of such contaminants can also potentially affect the reproducibility of the results of experiments in which stem cell preparations are studied in vivo in experimental animals.

In summary, there are numerous challenges to obtaining and preserving the uniform and stable preparations of stem cells necessary for reliable research and, eventually, for safe and effective possible therapies. Researchers must address multiple factors in order to maximize the probability of obtaining reproducible results with human stem cell preparations. Human stem cell preparations that are

- "single cell cloned," with a normal chromosome structure and number, *and*

- stored as multiple samples that are preserved at very low temperature, *and*

- compared in experiments where cells from the same lot of frozen material are used, *and*

- well-characterized as to the absence of cellular, viral, bacterial, fungal, and mycoplasma contaminants, *and*

- tested to determine the proportion of stem cells and various differentiated cells in the cell preparation used in the experiments,

are most likely to yield experimental results that will be reproducible. Preparations with these properties will be the most useful both in basic research and in investigations of possible clinical applications.

III. MAJOR EXAMPLES OF HUMAN STEM CELLS

In this section we discuss major examples of human stem cells that meet many of the criteria listed above. Among human *adult* stem cells, we focus on mesenchymal stem cells (MSCs),[4] multipotent adult progenitor cells (MAPCs),[3] and neural stem cells, and among human *embryonic* stem cells, on ESC[2] and EGC[1] cells. For information on the wide variety of other human stem cell preparations isolated from adult tissues, see reference (4) (Appendix K).

Further research on some of these other adult stem cell preparations may demonstrate that they can also be "single cell cloned," expanded considerably by growth in vitro with retention of normal chromosome structure and number, and preserved by freezing and storage at low temperatures. At that point, it would be very important to compare the properties of these other adult stem cells, and the more differentiated cells that can be derived from them, with the already characterized human embryonic and adult stem cell preparations.

A. Human Adult Stem Cells

1. Human Mesenchymal Stem Cells.

Bone marrow contains at least two major kinds of stem cells: hematopoietic stem cells,[10] which give rise to the red cells and white cells of the blood, and mesenchymal stem cells,* which can be reproducibly isolated and expanded in vitro and that can differentiate in vitro into cells with properties of cartilage, bone, adipose (fat), and muscle cells.[14]

The characteristics (morphology, expressed proteins, and biological properties) of these cells have been somewhat difficult to specify, because they appear to vary depending upon the in vitro culture conditions and the specific cell preparation.[15] However, there is a recent report indicating that MSCs, if isolated using three somewhat different methods, give rise to stem cell preparations whose properties are very similar to one another.[16] Using dual antibody staining and fluorescence-activated cell sorting, Gronthos and colleagues[17] isolated human MSCs in almost pure form and expanded them substantially in vitro. Thus, human MSC preparations isolated in different laboratories by different methods may have similar but not identical properties.

A molecular analysis of genes expressed in a single-cell-derived colony of MSCs provided evidence for the activity of genes also turned on in bone, cartilage, adipose, muscle, hematopoiesis-supporting stromal, endothelial, and neuronal cells.[15] These results are surprising in that MSCs derived from a single cell appear to be expressing genes associated with multiple major cell lineages. It is possible that different cells within the colony had already entered into distinct differentiation pathways, resulting in a *developmentally*

---

* The terms "stromal stem cells," "mesenchymal stem cells," and "mesenchymal progenitor cells" have all been used by different authors to describe these cells.

heterogeneous population composed of several different cell types.

Mesenchymal stem cells are important for research and therapy for several reasons. First, because they can be differentiated in vitro into multiple cell types, they make possible detailed research on the molecular events underlying differentiation into bone,[18] cartilage, and fat cell lineages. Second, they have recently been shown to support the in vitro growth of human *embryonic* stem cells.[19] Thus, they could replace the mouse feeder cells used previously, obviating the need to satisfy FDA requirements for xenotransplantation, should the ESCs or their derivatives ever be used in human clinical research or transplantation therapy. Third, clinical studies are already underway in which MSCs are co-transplanted with autologous hematopoietic stem cells into cancer patients to replace their blood cell-forming system, destroyed by radiation or high dose chemotherapy.[20] It is believed that the MSCs will support the repopulation of the bone marrow by the injected hematopoietic stem cells.

In addition, injecting allogeneic MSCs (MSCs from a genetically different human donor) may also prove valuable in modulating the immune system to make it more accepting of foreign tissue grafts [see Itescu review, reference (5)]. Finally, MSCs have the potential for cell-replacement therapies in injuries involving bone, tendon, or cartilage and possibly other diseases. They are, in fact, already being tested as experimental therapies for osteogenesis imperfecta,[21] metachromatic leukodystrophy, and Hurler syndrome.[22] These last two studies are of great interest, since *allogeneic* MSCs were used and no serious adverse immune reactions were noted.

2. Multipotent Adult Progenitor Cells (MAPCs).

Verfaillie and coworkers recently described the isolation of MAPCs from rat, mouse, and human bone marrow [see (3) and references cited therein]. Like MSCs, MAPCs can also be differentiated in vitro into cells with the properties of cartilage, bone, adipose, and muscle cells. In addition, there is evidence for the in vitro differentiation of human MAPCs into functional, hepatocyte-like cells,[23] a potential that has not so far been shown for MSCs. There is increasing interest in MAPCs, both as potential precursors of multiple differentiated tissues and, ultimately, for possible autologous transplantation therapy.

The relationship between human MSCs and the human MAPCs described by Verfaillie and coworkers [see (3)] needs to be clarified by further research. Both kinds of cells are isolated from bone marrow aspirates as cells that adhere to plastic. Each can be differentiated in vitro into cells with cartilage, bone, and fat cell properties. They express several of the same cell antigens, but are reported to differ in a few others.[3] MAPCs have to be maintained at specific, low cell densities when grown in vitro, otherwise they tend to differentiate into MSCs.[3] It remains important that the isolation and properties of MAPCs be reproduced in additional laboratories.

3. Human Neural Stem Cells.

The nervous system is made up of three major types of cells, neurons or nerve cells proper, and two kinds of supporting or glial cells (oligodendrocyte, astrocyte). Stem cells capable of differentiating into one or more of these neural cell lineages can be isolated from brain tissue (particularly the olfactory bulb and lining of the ventricles)[24,25] and grown in vitro. In the presence of purified growth-factor proteins, the population of cells can be expanded by growth in vitro as round clumps of cells called neurospheres. However, many neurospheres grown in culture are developmentally heterogeneous in that they contain more than one neural cell type, and the number of self-renewing cells is frequently low (less than five percent).[26]

Although neural stem cells are still insufficiently understood, they are already proving valuable in basic research on neural development. The ability to grow reproducible neural stem cells in vitro has facilitated identification of important neural stem cell growth factors and their cellular receptors. For example, human neural stem cells from the developing human brain cortex, expanded in culture in the presence of leukemia inhibitory factor (LIF), allowed growth of a self-renewing neural stem cell preparation for up to 110 population doublings. Withdrawal of LIF led to decreased expression of about 200 genes,[27] which were specifically identified through use of "gene chips" manufactured by Affymetrix. These genes are presumably involved in promoting or preserving the stem cell's capacity for self-renewal in the undifferentiated state. The number and specificity of the molecular changes characterized in these experiments powerfully illustrate the usefulness of neural and other stem cell preparations in basic biomedical research.

Human neural stem cells are also being injected into animals to test their effects on animal models of human neurological disease. To track the fate of the introduced human cells, they must first be modified or "marked" in ways that permit their specific detection.* Marked human neural stem cells are easily tracked after they are injected into experimental animals, making it possible to determine whether they survive and migrate following injection. Studies of this type have provided evidence that human neural cells can migrate extensively in the brain after injection.[28] In addition, such cells can be injected into animal models of human diseases such as intracerebral hemorrhage and Parkinson Disease (PD) to study their effect on the progression of the disease.[29] Although human neural stem cells may not yet be as well characterized as MSCs or ESCs, they are being actively studied with the hope that they can be used in future treatments for devastating neurological diseases such as Alzheimer Disease and PD.

4. Adult Stem Cells from Other Sources.

Prentice [see (4)] has summarized a large amount of recent information on preparations of stem cells isolated from amniotic fluid, peripheral blood, umbilical cord blood, umbilical cord, brain tissue, muscle, liver, pancreas, cornea, salivary gland, skin, tendon, heart, cartilage, thymus, dental pulp, and adipose tissue. Studies of many of the stem cell preparations from these sources are just getting started, and further work is needed to determine their biological properties and their relatedness to other stem cell types. In some cases, the long-

---

* Stem cell preparations are frequently transduced in vitro with foreign genes that, when expressed, produce readily visualized proteins, such as Green Fluorescent Protein (GFP).

term expandability in vitro of these stem cells has not been demonstrated. Yet, the demonstration that they can be isolated from such tissue compartments in animals should spur the search for similar human stem cell types.

As Prentice also reports,[4] many attempts have already been made using various preparations of adult stem cells to influence or alter the course of diseases in animal models. Despite the fact that the stem cell preparations used are not well characterized, and reproducible results have yet to be obtained, preliminary findings are sometimes encouraging. It is of course not yet clear whether the injected cells are functioning as stem cells, fusing with existing host cells, or stimulating the influx of the host's own stem cells into the target tissue.* But, if reproduced, these preliminary findings may point the way to future therapies, even in the absence of precise knowledge of the mechanism(s) of cellular action.

B. Human Embryonic Stem Cells

1. Human Embryonic Stem Cells (ESCs).

Human embryonic stem cells have been isolated from the inner cell masses of blastocyst-stage human embryos in multiple laboratories around the world.† There is great interest in understanding the properties of these cells because they hold out the promise of being able to be differentiated into a large number of different cell types for possible cell therapies, as contrasted with the more limited number of cell types available by differentiation of specific adult stem cell preparations. As of July 2003, 12 ESC preparations (up from 2 such preparations a year earlier) out of a total of 78 "eligible" preparations of human ESCs were available for shipment to recipients of U.S. federal research grants.‡ The review by

---

* In a recent review article on adult stem cell plasticity, Raff [see (8)] discusses the phenomenon of spontaneous cell fusion masquerading as cell plasticity.

† According to published reports, laboratories in Australia, Britain, China, India, Iran, Israel, Japan, Korea, Singapore, Sweden, and the United States have isolated ESC preparations.

‡ For current information on available and eligible ESC preparations see http://stemcells.nih.gov/registry/index.asp.

Ludwig and Thomson[2] lists more than 40 peer-reviewed human ESC primary research papers that have been published since the initial publication in 1998.

Although isolated from different blastocyst-stage human embryos in laboratories in different parts of the world, ESCs have a number of properties in common. These include the presence of common cell surface antigens (recognized by binding of specific antibodies), expression of the enzymes alkaline phosphatase and telomerase, and production of a common gene-regulating transcription factor known as Oct-4. At least 12 different preparations of ESCs have been expanded by growth in vitro, frozen and stored at low temperature, and at least partially characterized.[13] Some of these ESC preparations have been "single-cell cloned."

Human ESCs have been differentiated in vitro into neural (neurons, astrocytes, and oligodendrocytes), cardiac (synchronously contracting cardiomyocytes), endothelial (blood vessels), hematopoietic (multiple blood cell lineages), hepatocyte (liver cell), and trophoblast (placenta) lineages.[2] In the case of neural and cardiac lineages, similar results have been obtained in different laboratories using different preparations of ESCs, thus fulfilling the "reproducible results" criterion described above. For other lineages, the results described have not yet been reproduced in another laboratory.

2. Embryonic Germ Cells.

Human embryonic germ cells are isolated from the primordial germ tissues of aborted fetuses. Gearhart[1] has summarized the results of recent research with human and mouse EG cells. One study focused on regulation of imprinted genes in EG cells: it showed "that general dysregulation of imprinted genes will not be a barrier to their (EG cell) use in transplantation studies."[30*] In addition, Kerr and coworkers[31] showed that cells derived from human EG cells, when introduced into the cerebrospinal fluid of rats, became

---

* Previous work had shown that variation in imprinted gene expression was observed in cloned mice, and that it might be partly responsible for their subtle genetic defects. So it was reassuring that the pattern of imprinted gene expression appeared to be normal in EG cells.

extensively distributed over the length of the spinal cord and expressed markers of various nerve cell types. Rats paralyzed by virus-induced nerve-cell loss recovered partial motor function after transplantation with the human cells. The authors suggested that this could be due to the secretion of transforming growth factor-α and brain-derived growth factor by the transplanted cells and subsequent enhancement of rat neuron survival and function.

Until recently, work with human EG cells came primarily from one laboratory. Recently the isolation and properties of human EG cells have been independently confirmed.[32] Because human EG cells share many (but not all) properties with ESCs, these cells offer another important avenue of inquiry.

3. Embryonic Stem Cells from Cloned Embryos (Cloned ESCs).

Although it has yet to be accomplished in practice, somatic cell nuclear transfer (SCNT) could create cloned human embryos from which embryonic stem cells could be isolated that would be genetically virtually identical to the person who donated the nucleus for SCNT: hence *cloned* ESCs [see (7)]. In theory, using such cloned embryonic stem cells from individual patients might provide a way around possible immune rejection (see below), though in practice this could require individual cloned embryos for each prospective patient—a daunting task. And clinical uses might require a separate FDA approval for every single cloned stem cell line or its derivatives.

The ability to produce cloned mouse stem cells and genetically modify them in vitro has made possible an experiment demonstrating the potential of cloned *human* embryonic stem cells in the possible future treatment of *human* genetic diseases. Rideout et al.[33] used a mutant mouse strain that was deficient in immune system function. They produced a cloned mouse *embryonic* stem cell line carrying the mutation, and then specifically repaired that gene mutation in vitro. The repaired cloned stem cell preparation was then differentiated in vitro into bone marrow precursor cells. When these precursor cells were injected back into the genetically mutant mice, they produced partial restoration of immune system function.

Production of cloned *human* embryonic stem cell preparations remains technically very difficult and ethically controversial. Recently however, Chen and coworkers[34] have reported that fusion of human fibroblasts with enucleated rabbit oocytes in vitro leads to the development of embryo-like structures from which cell preparations with properties similar to human embryonic stem cells can be isolated. This work needs to be confirmed by repetition in other laboratories.

In addition, further work is needed to decisively settle the question of whether rabbit (or human egg donor) mitochondrial DNA and rabbit (or human egg donor) mitochondrial proteins persist in the embryonic stem cell preparations. Persistence of these foreign mitochondrial proteins in these human ESC-like preparations could possibly increase the probability of immune rejection of the cloned cells, thus limiting their clinical application, although the immune reaction might not be as severe as that to foreign proteins produced under the direction of chromosomal genes. The presence of foreign or aberrant mitochondria also carries the risk of transmitting mitochondrial disease (caused by defects in mitochondrial DNA) that could be detrimental to the cells and to the recipient into whom they might eventually be transplanted.

IV. BASIC RESEARCH USING HUMAN STEM CELLS

Human stem cells are proving useful in basic research in several ways. They are useful in unraveling the complex molecular pathways governing human differentiation. For example, because ESCs can be stimulated in vitro to produce more differentiated cells, this transition can be studied in greater detail and under better-controlled conditions than it can be in vivo. In the best circumstances, these differentiated cells can be grown as largely homogeneous cell populations, and their gene expression profiles can be compared in detail.

Also, stem cell preparations can be used to produce populations of specialized cells that are not easily obtained in other ways. In one case, for example, this approach has provided large quantities of human trophoblast-like cells that have not been previously available.[35] In addition, cultures of differentiated cells derived from stem cells could be used to test new drugs and chemical compounds for toxicity and mutagenicity.[36] As experience with these differentiated derivatives of human ESCs grows, it may become possible to reduce or eliminate the use of live animals in such testing protocols.

In the near future, the differentiated state of various human cell types will be characterized not just by a few biological markers, but by the pattern and levels of expression of hundreds or thousands of genes. Integration of this knowledge with the catalog of all human genes produced during the Human Genome Project will gradually give us knowledge of which genes are key regulators of human development and which genes are central to maintaining the stem cell state.[37] Increased understanding of the molecular pathways of human cell differentiation should eventually lead to the ability to direct in vitro differentiation along pathways that yield cells useful in medical treatment. In addition, when the normal range of gene expression patterns is known, researchers can then determine which genes are expressed abnormally in various diseases, thus increasing our understanding of and ability to treat these diseases.

A group of stem cell researchers has recently outlined a set of important research questions that, once answered, will greatly enhance our understanding of human embryonic stem cells and their potential fates and possible uses.[38] They include the following:

- What is the most effective way to isolate and grow ESCs?
- How is the self-renewal of ESCs regulated?
- Are all ESC lines the same?
- How can ESCs be genetically altered?
- What controls the processes of ESC differentiation?
- What new tools are needed to measure ESC differentiation in vitro and in vivo?

V. HUMAN STEM CELLS AND THE
TREATMENT OF DISEASE

A major goal of stem cell research is to provide healthy differentiated cells that, once transplanted, could repair or replace a patient's diseased or destroyed tissues. In pursuit of this goal, one likely approach would start by isolating stem cells that could be expanded substantially in vitro. A large number of the cultivated stem cells could then be stored in the frozen state, extensively tested for safety and efficacy as outlined above, and used as reproducible starting material from which to prepare differentiated cell preparations that will express the needed beneficial properties when they are transplanted into patients with specific diseases or deficiencies.

To make more concrete both the potential of this approach and the obstacles it faces, we will summarize, as a case study example, some current information on the properties of cells derived from human stem cell populations that have been used in an animal model of Type-1 diabetes. But before doing so, we discuss an obstacle to any successful program of stem cell-based transplantation therapy: the problem of immune rejection of the transplanted cells.

A. Will Stem Cell-Based Therapies Be Limited by Immune Rejection?

Much of the impetus for human stem cell research comes from the hope that stem cells (or, more likely, differentiated cells derived from them) will one day prove useful in cell transplantation therapies for a variety of human diseases. Such cell transplantation would augment the current practice of whole organ transplantation. To the extent that the healing process works with in vitro derived cells, the need for organ donors and long waiting lists for organ donation might be reduced or even eliminated.

Will the recipient (patient) accept or reject the transplanted human cells? In principle, the problem might seem avoidable altogether: adult stem cells could be obtained from each individual patient needing treatment. They could then be grown or modified to produce the desired (underline autologous and hence rejection-proof) transplantable cells. But the logistical difficulties in processing separate and unique materials for each patient suggest that this approach may not be practical. The cost and time required to produce sufficient numbers of well-characterized cells suitable for therapy suggest that it will be cells derived from one or another unique stem cell line that will be used to treat many (genetically different) individual patients (_allogeneic_ cell transplantation).

When allogeneic organ or tissue transplantation is currently done using, for example, bone marrow, kidney, or heart, powerful immunosuppressive drugs—carrying undesirable side effects—must be used to prevent immunological rejection of the transplanted tissue.[5] Without such immunosuppression, the patient's T-lymphocytes and natural killer (NK) cells recognize surface molecules on the transplanted cells as "foreign" and attack and destroy the cells. Also, in whole organ transplantation, _donor_ T-lymphocytes and NK cells, entering the recipient with the transplanted organ, can also destroy the tissues of the transplant _recipient_ (called "graft versus host" disease).

Are the differentiated derivatives of human stem cells as likely to incite immune rejection, when transplanted, as are solid organs? Do their surfaces carry those protein antigens that will be recognized as "foreign"? Experiments have been done to examine human ESC and MSC preparations growing in vitro for the expression of surface molecules known to play important roles in the immune rejection process. Drukker and coworkers[39] showed that _embryonic_ stem cells in vitro express very low levels of the immunologically crucial major histocompatibility complex class I (MHC-I) proteins on their cell surface. The presence of MHC-I proteins increased moderately when the ESCs became differentiated, whether in vitro or in vivo. A more pronounced increase in MHC-I antigen expression was observed when the ESCs were exposed to gamma-interferon, a protein produced in the body during immune reactions. Thus, under some circumstances, human

ESC-derived cells can express cell surface molecules that could lead to immune rejection upon allogeneic transplantation.

Similarly, Majumdar and colleagues showed that human *mesenchymal* stem cells in vitro express multiple proteins on their cell surfaces that would enable them to bind to, and interact with, T-lymphocytes. They also observed that gamma-interferon increased expression of both human leukocyte antigen (HLA) class I and class II molecules on the surface of these MSCs.[40] These results indicate that it will probably not be possible to predict, solely on the basis of in vitro experiments, the likelihood that transplanted allogeneic MSCs would trigger immune rejection processes in vivo.

Many further studies in this area are badly needed. At this time there is insufficient information to determine which, if any, of the approaches to get around the rejection problem will eventually prove successful.

B. Case Study: Stem Cells in the Future Treatment of Type-1 Diabetes?

1. The Disease and Its Causes.

The human body converts the sugar glucose into cell energy for heart and brain functioning, and indeed, for all bodily and mental activities. Glucose is derived from dietary carbohydrates, is stored as glycogen in the liver, and is released again when needed into the bloodstream. A protein hormone called insulin, produced by the beta cells in the islets of the pancreas, facilitates the entrance of glucose from the bloodstream into the cells, where it is then metabolized. Insulin is critical for regulating the body's use of glucose and the glucose concentration in the circulating blood.

The body's failure to produce sufficient amounts of insulin results in diabetes, an extremely common metabolic disease affecting over 10 million Americans, often with widespread and devastating consequences. In some five to ten percent of cases, known as Type-1 diabetes (or "juvenile diabetes"), the disease is caused by "autoimmunity," a process in which the body's immune system attacks "self."[*] T-lymphocytes attack the patient's own insulin-producing beta cells in the pancreas. Eventually, this results in destruction of ninety percent or so of the beta cells, resulting in the diabetic state.

With a deficiency or absence of insulin, the blood glucose becomes elevated and may lead to diabetic coma, a fatal condition if untreated. Chronic diabetes, both Type-1 and the much more common Type-2 diabetes (which is not autoimmune, but largely genetic), causes late complications in the retina, kidneys, nerves, and blood vessels. It is the leading cause of blindness, kidney failure, and amputations in the U.S. and a major cause of strokes and heart attacks.

Type-1 diabetes is a devastating, lifelong condition that currently affects an estimated 550,000-1,100,000 Americans,[41] including many children. It imposes a significant burden on the U.S. healthcare system and the economy as a whole, over and above the disabilities and impairments borne by individual sufferers. Recent estimates suggest that treatment of all forms of diabetes costs Americans a total of $132 billion per year.[42] At 5-10 percent of all diabetes cases, the costs of Type-1 diabetes can be estimated as $6.5-$13 billion per year.

2. Current Therapy Choices and Outcomes.

The current treatment of Type-1 diabetes consists of insulin injections, given several times a day in response to repeatedly measured blood glucose levels. Although this treatment is life-prolonging, the procedures are painful and burdensome, and in many cases they do not adequately control blood glucose concentrations. Whole pancreas transplants can essentially cure Type-1 diabetes, but fewer than 2,000 donor pancreases become available for transplantation in the U.S. each year, and they are primarily used to treat patients who also need a kidney transplant. Like all recipients of donated organs, pancreas transplant recipients must continuously take powerful drugs

---

[*] Normally the immune system protects against infectious and toxic agents and surveys for cancer cells with the intent of destroying them but does not attack one's own tissues. There are many other autoimmune diseases, such as some forms of thyroiditis and lupus erythematosis.

to suppress the immunological rejection of the transplanted pancreas.

In addition to treatment with whole pancreas transplantation, small numbers of Type-1 diabetes patients have been treated by transplantation of donor pancreatic islets into the liver of the patient coupled with a less intensive immunosuppressive treatment (the Edmonton protocol).[43] Expanded clinical trials of this procedure are currently underway. Scientists are also evaluating methods of slowing the original autoimmune destruction of pancreatic beta cells that produces the disease in the first place.

Whole pancreas and islet cell transplants ameliorate Type-1 diabetes, but there is nowhere near enough of these materials to treat all in need. To overcome this shortage, people hope that human stem cells can be induced—at will and in bulk—to differentiate in vitro into functional pancreatic beta cells, available for transplantation. Of course, it would still be crucial to prevent immunological destruction of the newly transplanted stem cell-derived beta cells.

3. Stem Cell Therapy for Type-1 Diabetes?

Initial experiments in mice suggested that insulin-producing cells could be obtained from mouse embryonic stem cells following in vitro differentiation.[44] Can this approach be extended to human stem cells? A number of attempts have been made, with promising initial findings, yet they are not easily evaluated, partly because the criteria for characterizing the cells are not standardized. In a recent paper, Lechner and Habener provided a list of six criteria to define the characteristics of pancreas-derived "beta-like" cells that could be potentially useful in treatment of Type-1 diabetes.[45]

We have used those criteria to facilitate assessment of the current state of progress toward development of functional "beta-like" cells that might eventually be tested in Type-1 diabetes patients. Table 1 summarizes and compares the properties of human cell preparations recently produced in research seeking this objective by Abraham et al.,[46] Zulewski et al.,[47] Assady et al.,[48] Zhao et al.,[49] and Zalzman et al.,[50] and tested in mouse models of human diabetes.

Table 1: Comparison of Insulin-Producing Cells Derived from Human Stem Cells

References	Cell Source: Clonally Isolated / Marked?	Beta-cell markers	Ultrastructural Examination to Ensure Endogenous Insulin Production	Glucose-responsive Insulin Secretion?	In Vivo Studies	Tumorigenicity?
Abraham et al, 2002 (46); Zulewski et al, 2001 (47)	Clonally isolated adult stem cells (derived from adult pancreatic islets)	PDX-1 (+) CK-19 (+)	Insulin mRNA(+); Insulin protein (+); No ultra-structural examination	Not assessed	None	Not assessed
Assady et al, 2001(48)	Clonally isolated embryonic stem cells	PDX-1 (-); GK (+); GLUT-2 (+)	Insulin mRNA (+) Insulin protein (+); No ultrastructural examination; possible insulin uptake from serum	No	None	Not assessed
Zhao et al, 2002 (49)	Uncloned cadaver islets (cultured in vitro)	CK-19 (+)	Preproinsulin mRNA (+); Insulin protein (+);electron microscopyinsul-in secretory granuoles (+)	Yes	High blood glucose concentrations reversed in STZ/SCID mice	Not assessed
Zalzman et al, 2003 (50)	Cloned fetal liver cells: immortalized with human telomerase and transduced with rat PDX-1	Human and rat PDX-1 (+); GK (-); GLUT-2 (-)	Insulin mRNA (+); Insulin protein (+); No ultra- structural examination	Yes	High blood glucose concentrations reversed in STZ/NOD-SCID mice; high blood glucose returned upon graft removal	No tumors at 3 months after transplantation

Beta-cell-specific markers: PDX-1: (a.k.a IPF-1), a regulatory gene important for beta-cell function; Glucokinase (GK), an enzyme that detects high levels of glucose and modulates insulin release; GLUT-2, a protein associated with glucose-responsive insulin secretion. CK-19 is a marker for pancreatic duct cells. Insulin production criteria: synthesis of messenger RNA for insulin or preproinsulin; tests for the presence of insulin protein; and ultrastructural studies (electron microscopy) to determine the presence of typical insulin secretory granules. In addition, the glucose-responsiveness of insulin production and release, an essential characteristic of normal beta-cell function, was assessed in a number of the studies described above. Both mouse models of Type-1 diabetes used mice that had a condition known as Severe Combined Immunodeficiency (SCID) and were treated with streptozotocin (STZ), a drug that induces selective destruction of the insulin-producing cells. The mice in the Zalzman study were also born with a form of mouse diabetes, and are called Non-Obese Diabetic (NOD) mice.

As the results described in Table 1 indicate, cells derived from some human stem cells transplanted into specific strains of mice mimicking major aspects of Type-1 human diabetes[51] were able to reverse high blood glucose concentrations. Although these results are encouraging, the transplant rejection question remains unanswered because the likely immune rejection of the transplanted human cells was prevented in these experiments by using special strains of immunodeficient mice that lack the capacity to recognize and attack foreign cells.

No tumors were observed in the transplanted mice, but the experiments were terminated after about three months, an insufficient time for much tumor development to occur. Because many Type-1 diabetes patients are children and because a largely effective therapy (insulin injection) is currently available, the introduction of islet cell transplant therapy will need a high degree of certainty that the introduced cells or their derivatives will not become malignant over the course of the patient's life. Stringent tests of the cancer-causing potential of candidate cell preparations will be required, including multi-year studies in animals that live longer than mice or rats. Long-term follow-up of children and adult patients who had received bone marrow transplants many years ago has revealed an increased risk of severe neurologic complications[52] and a variety of types of cancer.[53]

C. Therapeutic Applications of Mesenchymal Stem Cells (MSCs)

Before stem cell based therapies are used to treat human diseases, they will have to gain approval through the Food and Drug Administration (FDA) regulatory process. The first step in this process is filing an Investigational New Drug (IND) application. As of July 2003, four IND applications have been filed for clinical applications of mesenchymal stem cells. The disease indications include: (1) providing MSC support for peripheral blood stem cell transplantation in cancer treatment, (2) providing MSC support for cord blood transplantation in cancer treatment, (3) using MSCs to stimulate regeneration of cardiac tissue after acute myocardial infarction (heart attack), and (4) using MSCs to stimulate regeneration of cardiac tissue in cases of congestive heart failure. The first two applications are currently in Phase II of the regulatory process, with pivotal Phase III trials scheduled to begin in 2004.[54]

D. Evaluating the Different Types of Stem Cells

A major unresolved issue at present involves the therapeutic potential of human adult stem cells compared with embryonic stem cells. The answer may well be different for different diseases and for patients of different ages. For example, in treating an elderly patient with Parkinson's Disease, the use of adult stem cells may be appropriate even if these cells may have a more limited number of cell divisions remaining. On the other hand, treating a child with Type I Diabetes, one may want to use embryonic stem cells because of their potentially greater longevity, or other factors. The only valid way to resolve these questions is by instituting rigorous therapeutic trials which test the efficacy of the different types of stem cells in treating a variety of different diseases to determine their comparative efficacy. Clearly, such trials would be a long-term endeavor, since it would take years to obtain answers to these very critical questions.

VI. PRIVATE SECTOR ACTIVITY

In the United States, much of the basic research on animal stem cells and human *adult* stem cells has been publicly funded. Yet before 2001, research in the U.S., using human ESCs could only be done in the private sector (the locus also of much research on animal and human adult stem cells). The current state of knowledge about human ESCs (and also about human MSCs) reflects pioneering and on-going stem cell research funded by the private sector in the U.S.[54,55] For example, the work that led to the 1998 reports of the first isolation of both ESCs and EGCs, was funded by Geron Corporation. Embryonic and adult stem cell research is today vigorously pursued by many companies and supported by several private philanthropic foundations,[56] and the results of some of this research have been published in peer-reviewed journals.[57]

Private sector organizations have pursued and been awarded patents on the stem cells themselves and methods for producing and using them to treat disease. As noted above, at least one company (Osiris Therapeutics) has protocols under review at the FDA for clinical trials with MSCs. It seems likely that private sector companies will continue to play large roles in the future development of stem cell based therapies.

VII. PRELIMINARY CONCLUSIONS

While it might be argued that it is too soon to attempt to draw *any* conclusions about the state of a field that is changing as rapidly as stem cell research, we draw the following preliminary conclusions regarding the current state of the field.

Human stem cells can be reproducibly isolated from a variety of embryonic, fetal, and adult tissue sources. Some human stem cell preparations (for example, human ESCs, EGCs, MSCs, and MAPCs) can be reproducibly expanded to substantially larger cell numbers in vitro, the cells can be stored frozen and recovered, and they can be characterized and compared by a variety of techniques. These cells are receiving a large share of the attention regarding possible future (non-hematopoietic) stem cell transplantation therapies.

Preparations of ESCs, EGCs, MSCs, and MAPCs can be induced to differentiate in vitro into a variety of cells with properties similar to those found in differentiated tissues.

Research using these human stem cell preparations holds promise for: (a) increased understanding of the basic molecular process underlying cell differentiation, (b) increased understanding of the early stages of genetic diseases (and possibly cancer), and (c) future cell transplantation therapies for human diseases.

The case study of developing stem cell-based therapies for Type-1 diabetes illustrates that, although insulin-producing cells have been derived from human stem cell preparations, we could still have a long way to go before stem cell-based therapies can be developed and made available for this disease. This appears to be true irrespective of whether one starts from human embryonic stem cells or from human adult stem cells. The transplant rejection problem remains a major obstacle, but only one among many.

Human mesenchymal stem cells are currently being evaluated in pre-clinical studies and clinical trials for several specific human diseases.

Much basic and applied research remains to be done if human stem cells are to achieve their promise in regenerative medicine.[58] This research is expensive and technically challenging, and requires scientists willing to take a long perspective in order to discover, through painstaking research, which combinations of techniques could turn out to be successful. Strong financial support, public and private, will be indispensable to achieving success.

ENDNOTES

[1] Gearhart, J., "Human Embyronic Germ Cells: June 2001-July 2003. The Published Record," Paper prepared for the President's Council on Bioethics, July 2003. [Appendix H]

[2] Ludwig, T. E. and Thomson, J. A., "Current Progress in Human Embryonic Stem Cell Research," Paper prepared for the President's Council on Bioethics, July 2003. [Appendix I]

[3] Verfaillie, C., "Multipotent Adult Progenitor Cells: An Update," Paper prepared for the President's Council on Bioethics, July 2003. [Appendix J]

[4] Prentice, D., "Adult Stem Cells," Paper prepared for the President's Council on Bioethics, July 2003. [Appendix K]

[5] Itescu, S., "Stem Cells and Tissue Regeneration: Lessons from Recipients of Solid Organ Transplantation," Paper prepared for the President's Council on Bioethics, June 2003. [Appendix L]

[6] Itescu, S., "Potential Use of Cellular Therapy For Patients With Heart Disease," Paper prepared for the President's Council on Bioethics, August 2003. [Appendix M]

[7] Jaenisch, R., "The Biology of Nuclear Cloning and the Potential of Embryonic Stem Cells for Transplantation Therapy," Paper prepared for the President's Council on Bioethics, July 2003. [Appendix N]

[8] See, among others, Bianco, P., et al., "Bone marrow stromal cells: nature, biology and potential applications," *Stem Cells* 19: 180-192 (2001); Martinez-Serrano, A., et al., "Human neural stem and progenitor cells: in vitro and in vivo properties, and potential for gene therapy and cell replacement in the CNS," *Current Gene Therapy* 1: 279-299 (2001); Nir, S., et al., "Human embryonic stem cells for cardiovascular repair," *Cardiovascular Research* 58: 313-323 (2003); Raff, M., "Adult stem cell plasticity: fact or artifact?" *Annual Review of Cell and Developmental Biology* 19: 1-22 (2003).

[9] Storb, R., "Allogeneic hematopoietic stem cell transplantation – Yesterday, today and tomorrow," *Experimental Hematology* 31: 1-10 (2003).

[10] Kondo, M., et al., "Biology of Hematopoietic Stem Cells and Progenitors: Implications for Clinical Application," *Annual Review of Immunology* 21: 759-806 (2003) and references cited therein.

[11] Xu, C., et al., "Feeder-free growth of undifferentiated human embryonic stem cells," *Nature Biotechnology* 19: 971-974 (2001); Richards, M., et al., "Human feeders support prolonged undifferentiated growth of human inner cell masses and embryonic stem cells," *Nature Biotechnology* 20: 933-936 (2002); Amit, M., et al., "Human Feeder Layers for Human Embryonic Stem Cells," *Biology of Reproduction* 68: 2150-2156 (2003); Richards, M., et al., "Comparative Evaluation of Various Human Feeders for Prolonged Undifferentiated Growth of Human Embryonic Stem Cells," *Stem Cells* 21: 546-556 (2003).

[12] Amit, M., et al., "Clonally derived Human Embryonic Stem Cell Lines Maintain Pluripotency and Proliferative Potential for Prolonged Periods of Culture," *Developmental Biology* 227: 271-278 (2000); Amit, M. and Itskovitz-Eldor, J., "Derivation and spontaneous differentiation of human embryonic stem cells," *Journal of Anatomy* 200: 225-232 (2002).

[13] Carpenter, M. K., et al., "Characterization and Differentiation of Human Embryonic Stem Cells," *Cloning and Stem Cells* 5: 79-88 (2003).

[14] Pittenger, M. F. et al., "Multilineage potential of adult human mesenchymal stem cells," *Science* 284: 143-147 (1999); Pittenger, M., et al., "Adult mesenchymal stem cells: Potential for muscle and tendon regeneration and use in gene therapy," *Journal of Musculoskeletal and Neuronal Interactions* 2: 309-320 (2002).

[15] Tremain, N., et al., "MicroSAGE Analysis of 2,353 Expressed Genes in a Single-Cell Derived Colony of Undifferentiated Human Mesenchymal Stem Cells Reveals mRNAs of Multiple Cell Lineages," *Stem Cells* 19: 408-418 (2001).

[16] Lodie, T. A., et al., "Systematic analysis of reportedly distinct populations of multipotent bone marrow-derived stem cells reveals a lack of distinction," *Tissue Engineering* 8: 739-751 (2002).

[17] Gronthos, S., et al., "Molecular and cellular characterization of highly purified stromal stem cells derived from bone marrow," *Journal of Cell Science* 116: 1827-1835 (2003).

[18] Qi, H., et al., "Identification of genes responsible for osteoblast differentiation from human mesodermal progenitor cells," *Proceedings of the National Academy of Sciences of the United States of America* 100: 3305-3310 (2003).

[19] Cheng, L., et al., "Human adult marrow cells support prolonged expansion of human embryonic stem cells in culture," *Stem Cells* 21: 131-142 (2003).

[20] Koc, O. N., et al., "Rapid hematopoietic recovery after coinfusion of autologous-blood stem cells and culture-expanded marrow mesenchymal cells in advanced breast cancer patients receiving high-dose chemotherapy," *Journal of Clinical Oncology* 18: 307-316 (2000).

[21] Horwitz, E. M., et al., "Isolated allogeneic bone marrow-derived mesenchymal cells engraft and stimulate growth in children with osteogenesis imperfecta: Implications for cell therapy of bone," *Proceedings of the National Academy of Sciences of the United States of America* 99: 8932-8937 (2002).

[22] Koc, O. N., et al., "Allogeneic mesenchymal stem cell infusion for treatment of metachromatic leukodystrophy (MLD) and Hurler syndrome (MPS-IH)," *Bone Marrow Transplantation* 30: 215-222 (2002).

[23] Schwartz, R. E., et al., "Multipotent adult progenitor cells from bone marrow differentiate into functional hepatocyte-like cells," *Journal of Clinical Investigation* 109: 1291-1302 (2002).

[24] Pagano, S. F., et al., "Isolation and characterization of neural stem cells from the adult human olfactory bulb," *Stem Cells* 18: 295-300 (2000).

[25] Liu, Z., and Martin, L. J., "Olfactory bulb core is a rich source of neural progenitor and stem cells in adult rodent and human," *Journal of Comparative Neurology* 459: 368-391 (2003).

[26] Pevny, L., and Rao, M. S., "The stem-cell menagerie," *Trends in Neurosciences* 26: 351-359 (2003).

[27] Wright, L. S., et al., "Gene expression in human neural stem cells: effects of leukemia inhibitory factor," *Journal of Neurochemistry* 86: 179-195 (2003).

[28] See, for example, Englund, U., et al., "Transplantation of human neural progenitor cells into the neonatal rat brain: extensive migration and differentiation with long-distance axonal projections," *Experimental Neurology* 173: 1-21 (2002); Chu, K., et al., "Human neural stem cells can migrate, differentiate, and integrate after intravenous transplantation in adult rats with transient forebrain ischemia," *Neuroscience Letters* 343: 129-133 (2003).

[29] See, for example, Jeong, S., et al., "Human neural cell transplantation promotes functional recovery in rats with experimental intracerebral hemorrhage," *Stroke* 34: 2258-2263 (2003); Liker, M., et al., "Human neural stem cell transplantation in the MPTP-lesioned mouse," *Brain Research* 971: 168-177 (2003).

[30] Onyango, P., et al., "Monoallelic expression and methylation of imprinted genes in human and mouse embryonic germ cell lineages," *Proceedings of the National Academy of Sciences of the United States of America* 99: 10599-10604 (2002).

[31] Kerr, D. A., et al., "Human Embryonic Germ Cell Derivatives Facilitate Motor Recovery of Rats with Diffuse Motor Neuron Injury," *The Journal of Neuroscience* 23: 5131-5140 (2003).

[32] Turnpenny, L., et al., "Derivation of Human Embryonic Germ Cells: An Alternative Source of Pluripotent Stem Cells," *Stem Cells* 21: 598-609 (2003).

[33] Rideout, W., et al., "Correction of a genetic defect by nuclear transplantation and combined cell and gene therapy," *Cell* 109: 17-27 (2002); Tsai, R. Y. L., et al., "Plasticity, niches and the use of stem cells," *Developmental Cell* 2: 707-712 (2002); For political and legislative aspects of the debate relative to these articles, see Daly, G., "Cloning and Stem Cells—Handicapping the Political and Scientific Debates," *New England Journal of Medicine* 349: 211-212 (2003).

[34] Chen, Y., et al., "Embryonic stem cells generated by nuclear transfer of human somatic nuclei into rabbit oocytes," *Cell Research* 13: 251-264 (2003).

[35] Xu, R. H., et al., "BMP4 initiates human embryonic cell differentiation to trophoblast," *Nature Biotechnology* 20: 1261-1264 (2002).

[36] Rohwedel, J., et al., "Embryonic stem cells as an in vitro model for mutagenicity, cytotoxicity, and embryotoxicity studies: present state and future prospects," *Toxicology In Vitro* 15: 741-753 (2001).

[37] Sato, N., et al., "Molecular signature of human embryonic stem cells and its comparison with the mouse," *Developmental Biology* 260: 404-413 (2003); Ramalho-Santos, M., et al., "'Stemness': Transcriptional Profiling of Embryonic and Adult Stem Cells," *Science* 298: 597-600 (2002); Ivanova, N. B., et al., "A Stem Cell Molecular Signature," *Science* 298: 601-604 (2002).

[38] Brivanlou, A. H., et al., "Stem cells. Setting standards for human embryonic stem cells," *Science* 300: 913-916 (2003).

[39] Drukker, M., et al., "Characterization of the expression of MHC proteins in human embryonic stem cells," *Proceedings of the National Academy of Sciences of the United States of America* 99: 9864-9869 (2002).

[40] Majumdar, M. K., et al., "Characterization and functionality of cell surface molecules on human mesenchymal stem cells," *Journal of Biomedical Science* 10: 228-241 (2003).

[41] American Diabetes Association, "Facts and Figures," http://diabetes.org/main/info/facts/facts.jsp (accessed June 23, 2003).

[42] Hogan, P., et al., "Economic Costs of Diabetes in the US in 2002," *Diabetes Care* 26: 917-932 (2003).

[43] Ryan, E. A., et al., "Clinical outcomes and insulin secretion after islet transplantation with the Edmonton protocol," *Diabetes* 50: 710-719 (2001).

[44] Soria, B., et al., "Insulin-secreting cells derived from embryonic stem cells normalize glycemia in streptozotocin-induced diabetic mice," *Diabetes* 49: 157-162 (2000); Lumelsky, N., et al., "Differentiation of embryonic stem cells to insulin-secreting structures similar to pancreatic islets," *Science* 292: 1389-1394 (2001); Hori, Y., et al., "Growth inhibitors promote differentiation of insulin-producing tissue from embryonic stem cells," *Proceedings of the National Academy of Sciences of the United States of America* 99: 16105-16110 (2002).

[45] Lechner, A. and Habener, J. F., "Stem/progenitor cells derived from adult tissues: potential for the treatment of diabetes mellitus," *American Journal of Physiology - Endocrinology and Metabolism* 284: E259-266 (2003). The criteria that these authors outlined were as follows:

- The stem or progenitor cell should be clonally isolated or marked; "enrichment" of a certain cell type alone is not sufficient.
- In vitro differentiation to a fully functional beta cell should be unequivocally established. Insulin expression per se does not make a particular cell a beta cell. The expression of other markers of beta cells (e.g. Pdx1/Ipf1, GLUT2, and glucokinase) or other endocrine islet cells should be demonstrated.
- Ultrastructural studies should confirm the formation of mature endocrine cells by identification of characteristic insulin secretory granules.
- The in vitro function of endocrine cells, differentiated from stem cells, should be reminiscent of the natural counterparts. For beta cells, this would imply a significant glucose-responsive insulin secretion, adequate responses to incretin hormones and secretagogues, and the expected electrophysiological properties.
- In vivo studies in diabetic animals should demonstrate a reproducible and durable effect of the stem/progenitor-derived tissue on the attenuation of the diabetic phenotype. It should also be demonstrated that removal of the stem cell-derived graft after a certain period of time leads to reappearance of the diabetes.
- For future clinical use, the tumorigenicity of stem/progenitor tissue should be determined.
- Additionally, immune responses toward the transplanted cells should be examined.

[46] Abraham, E. J., et al., "Insulinotropic hormone glucagons-like peptide-1 differentiation of human pancreatic islet-derived progenitor cells into insulin-producing cells," *Endocrinology* 143: 3152-3161 (2002).

[47] Zulewski, H., et al., "Multipotential Nestin-Positive Stem Cells Isolated From Adult Pancreatic Islets Differentiate Ex Vivo Into Pancreatic Endocrine, Exocrine and Hepatic Phenotypes," *Diabetes* 50: 521-533 (2001).

[48] Assady, S., et al., "Insulin production by human embryonic stem cells," *Diabetes* 50: 1691-1697 (2001).

[49] Zhao, M., et al., "Amelioration of streptozotocin-induced diabetes in mice using human islet cells derived from long-term culture in vitro," *Transplantation* 73: 1454-1460 (2002).

[50] Zalzman, M., et al., "Reversal of hyperglycemia in mice using human expandable insulin-producing cells differentiated from fetal liver cells," *Proceedings of the National Academy of Sciences of the United States of America* 100: 7253-7258 (2003).

[51] For a useful summary of the advantages and limitations of rodent models of diabetes see: Atkinson, M. A. and Leiter, E. H., "The NOD mouse model of type 1 diabetes: As good as it gets?" *Nature Medicine* 5: 601-604 (1999).

[52] Faraci, M., et al., "Severe neurologic complications after hematopoietic stem cell transplantation in children," *Neurology* 59: 1895-1904 (2002).

[53] Baker, K. S., et al., "New Malignancies After Blood or Marrow Stem-Cell Transplantation in Children and Adults: Incidence and Risk Factors," *Journal of Clinical Investigation* 21: 1352-1358 (2003).

[54] Pursley, W. H., Presentation at the September 4, 2003, meeting of the President's Council on Bioethics, Washington, D.C., available at www.bioethics.gov.

[55] Okarma, T., Presentation at the September 4, 2003, meeting of the President's Council on Bioethics, Washington, D.C., available at www.bioethics.gov.

[56] See presentations from the Juvenile Diabetes Research Foundation International and the Michael J. Fox Foundation at the September 4, 2003, meeting of the President's Council on Bioethics, Washington, D.C., available at www.bioethics.gov.

[57] See, for example, Carpenter, M. K., et al., "Characterization and Differentiation of Human Embryonic Stem Cells," *Cloning and Stem Cells* 5: 79-88 (2003), and Pittenger, M. F. et al., "Multilineage potential of adult human mesenchymal stem cells," *Science* 284: 143-147 (1999), and Pittenger, M. F., et al., "Adult mesenchymal stem cells: Potential for muscle and tendon regeneration and use in gene therapy," *Journal of Musculoskeletal and Neuronal Interactions* 2: 309-320 (2002).

[58] Daley, G. Q., et al., "Realistic Prospects for Stem Cell Therapeutics," *Hematology* American Society for Hematology Education Program: 398-418 (2003).

GLOSSARY*

Adipose tissue: A type of connective tissue that stores fat.

Adult stem cell: An undifferentiated cell found in a differentiated tissue that can renew itself and (with certain limitations) differentiate to yield all the specialized cell types of the tissue from which it originated. (NIH)

Allogeneic cell transplantation: Transplantation of cells from one individual to another of the same species.

Amniotic fluid: Fluid that fills the innermost membrane, the amnion, that envelopes the developing embryo or fetus.

Amnion: Innermost of the extra-embryonic membranes enveloping the embryo in utero and containing the amniotic fluid. (SMD)

Aneuploid: Having an abnormal number of chromosomes. (SMD)

Angiogenesis: Development of new blood vessels. (SMD)

Antigen: A substance that, when introduced into the body, stimulates the production of protein molecules called antibodies that can bind specifically to the substance.

Astrocyte: A type of nerve cell that has supportive and metabolic functions rather than signal conduction.

Autologous: In transplantation, referring to a graft in which the donor and recipient areas are in the same individual. (SMD)

---

* Definitions marked "(CR)" are from the Council's report on human cloning (*Human Cloning and Human Dignity: An Ethical Inquiry,* Washington, D.C.: Government Printing Office, 2002). Definitions marked "(NIH)" are from the National Institutes of Health on-line stem cell glossary at http://stemcells.nih.gov (accessed September 5, 2003). Definitions marked "(NRC)" are from the National Research Council report, Stem Cell Research and the Future of Regenerative Medicine (Washington, D.C.: National Research Council, 2001). Definitions marked "(SMD)" are from Stedman's Medical Dictionary.

Autosome: Any chromosome other than a sex chromosome, that is, any chromosome other than an X or a Y. (SMD)

Bacteria: Any of numerous unicellular microorganisms, existing either as free living organisms or as parasites, and having a broad range of biochemical, often pathogenic properties.

Blastocyst: (a) Name used for an organism at the blastocyst stage of development. (CR) **(b)** A preimplantation embryo of about 150 to 200 cells. The blastocyst consists of a sphere made up of an outer layer of cells (the trophectoderm), a fluid-filled cavity (the blastocoel), and a cluster of cells on the interior (the inner cell mass). (NIH)

Blastocyst stage: An early stage in the development of embryos, when (in mammals) the embryo is a spherical body comprising an inner cell mass that will become the fetus surrounded by an outer ring of cells that will become part of the placenta. (CR)

Bone marrow: The soft, fatty, vascular tissue that fills most bone cavities and is the source of red blood cells and many white blood cells.

Cardiomyoctes: Heart muscle cells.

Cartilage: A type of connective tissue that is firm but resilient. It is found in joints and also as supportive structure, for example in the ears.

Cell culture: Growth of cells in vitro on an artificial medium for experimental research. (NIH)

Cerebrospinal fluid: A blood serum-like fluid that bathes parts of the brain and the interior cavity of the spinal cord.

Chromosomes: Structures inside the nucleus of a cell, made up of long pieces of DNA coated with specialized cell proteins, that are duplicated at each mitotic cell division. Chromosomes thus transmit the genes of the organism from one generation to the next. (CR)

Clone: A line of cells that is genetically identical to the originating cell; in this case, a stem cell. (NIH)

Cord blood: Blood in the umbilical cord and placenta.

Cornea: Transparent tissue at the front of the eye.

Cryopreserved embryos: Embryos, generally those produced by in vitro fertilization exceeding the number that can be transferred for uterine implantation, that have been frozen.

Culture medium: The broth that covers cells in a culture dish, which contains nutrients to feed the cells as well as other growth factors that may be added to direct desired changes in the cells. (NIH)

Dental pulp: The soft part inside a tooth, containing blood vessels and nerves.

Diploid: Refers to the full complement of chromosomes in a somatic cell, distinct for each species (forty-six in human beings). (CR)

Diploid human cell: A cell having forty-six chromosomes. (CR)

Ectoderm: Upper, outermost layer of a group of cells derived from the inner cell mass of the blastocyst; it gives rise to skin nerves and brain. (NIH)

Edmonton protocol: A procedure (developed in Canada) for transplanting pancreatic islet cells to the liver of a patient with Type I diabetes.

Embryo: (a) In humans, the developing organism from the time of fertilization until the end of the eighth week of gestation, when it becomes known as a fetus. (NIH) (b) The developing organism from the time of fertilization until significant differentiation has occurred, when the organism becomes known as a fetus. An organism in the early stages of development. (CR)

Embryonic germ cells: Cells found in a specific part of the embryo/ fetus called the gonadal ridge that normally develop into mature gametes. (NIH)

Embryonic stem cells: Primitive (undifferentiated) cells from the embryo that have the potential to become wide variety of specialized cell types. (NIH)

Embryonic stem cell line: Embryonic stem cells, which have been cultured under in vitro conditions that allow proliferation without differentiation for months to years. (NIH)

Endoderm: Lower layer of a group of cells derived from the inner cell mass of the blastocyst; it gives rise to lungs and digestive organs. (NIH)

Endometrium: The mucous membranes lining the uterus.

Endothelial: relating to a flat layer of cells lining the heart, for example, or blood vessels. (SMD)

Epidermal growth factor: A cell messenger protein that has effects including stimulation of epidermal development, in newborn animals it hastens eyelid-opening and tooth-eruption (SMD)

Ex vivo: Outside the body, frequently the equivalent of "in vitro."

Fate (of cell progeny): The normal outcome of differentiation of a cell's progeny.

Feeder layer: Cells used in co-culture to maintain pluripotent stem cells. Cells usually consist of mouse embryonic fibroblasts. (NIH)

Fertilization: The process whereby male and female gametes unite. (NIH)

Fetus: A developing human from usually two months after conception to birth. (NIH)

Fibroblast: A stellate (star-shaped) or spindle-shaped cell with cytoplasmic processes present in connective tissue, capable of forming collagen fibers. (SMD)

Gamete: A reproductive cell (egg or sperm). (CR)

Gamma-interferon: A type of small protein with antiviral activity, made by T lymphocytes.

Gastrulation: The process whereby the cells of the blastocyst are translocated to establish three germ layers. Also sometimes used to mark the end of the blastocyst stage and the beginning of the next stage of embryonic development. (Based on SMD)

Gene: A functional unit of heredity that is a segment of DNA located in a specific site on a chromosome. A gene directs the formation of an enzyme or other protein. (NIH)

Genome: The total gene complement of a set of chromosomes. (SMD)

Germ cells (or primordial germ cells): A gamete, that is, a sperm or egg, OR a primordial cell that can mature into a sperm or egg. (NRC)

Germ layers: The three initial tissue layers arising in the embryo—endoderm, mesoderm, and ectoderm—from which all other somatic tissue-types develop. (NRC)

Gonad: An organ that produces sex cells (testes or ovaries). (SMD)

Gonadal ridges: Embryonic structures arising in humans at about five weeks, eventually developing into gonads (either testes or ovaries).

Green Fluorescent Protein: A protein naturally occurring in some animals including jelly fish that spontaneously fluoresces. It can be used as a noninvasive marker in living cells by attaching it to different proteins and then letting it fluoresce so as to track the cell.

Haploid human cell: A cell such as an egg or sperm that contains only twenty-three chromosomes. (CR)

Hematopoietic stem cell: A stem cell from which all red and white blood cells develop. (NIH)

Hepatocyte: Liver cell.

Histocompatible: The immunological characteristic of cells or tissue that causes them to be tolerated by another cell or tissue; that allows some tissues to be grafted effectively to others. (NRC)

Hurler syndrome: A heritable condition involving deficiency of an enzyme (alpha-L-iduronidase), leading to abnormal accumulations of materials inside cells, then resulting in abnormal development of cartilage and bone and other systems. (SMD)

ICM cells: Cells from the inner cell mass, a population of cells inside the blastula that give rise to the body of the new organism rather than to the chorion or other supporting structures.

Immunodeficient: Unable to develop a normal immune response to, for example, a foreign substance.

Immunosuppressive drugs: Drugs that prevent or interfere with the development of an immunologic response. After a transplant, immunosuppressive drugs are usually necessary in order to prevent the recipient from rejecting the transplant.

Implantation: The attachement of the blastocyst to the uterine lining, and its subsequent embedding there. (Based on SMD)

In vitro fertilization (IVF): The union of an egg and sperm, where the event takes place outside the body and in an artificial environment (the literal meaning of "in vitro" is "in glass"; for example, in a test tube). (CR)

Inner cell mass: The cluster of cells inside the blastocyst. These cells give rise to the embryonic disk of the later embryo and, ultimately, the fetus. (NIH)

Karyotype: The chromosome characteristics (number, shape, etc) of an individual cell or cell line, usually presented as a systematized array in pairs. (SMD)

Leukemia inhibitory factor: A cell messenger protein originally noted for inhibition of mouse M1 myeloid leukemia cells that also has effects including inhibiting differentiation to maintain stem cells.

Lineage: The descendants of a common ancestor.

Long-term self-renewal: The ability of stem cells to renew themselves by dividing into the same non-specialized cell type over long periods (many months to years) depending on the specific type of stem cell. (NIH)

Lymphocyte: A motile cell formed in tissues such as the lymph nodes, that functions in the development of immunity.

Meiosis: A special process of cell division comprising two nuclear divisions in rapid succession that result in four cells (that will become gametes) with the haploid number of chromosomes. (Based on SMD)

Mesenchymal stem cells: Cells from the immature embryonic connective tissue. A number of cell types come from mesenchymal stem cells, including chondrocytes, which produce cartilage. (NIH)

Mesoderm: Middle layer of a group of cells derived from the inner cell mass of the blastocyst; it gives rise to.bone, muscle, and connective tissue. (NIH)

Metachromatic leukodystrophy: A heritable metabolic disorder, usually of infancy, characterized by myelin loss and other abnormalities of the white matter of the nervous system, leading to progressive paralysis and mental retardation or dementia. (SMD)

Mitochondria: Small, energy-producing organelles inside cells.

Mitochondrial DNA: Genetic material inside the mitochondria. Essentially all the mitochondria of an individual come from the cytoplasm of the egg, so all mitochondrial DNA is inherited through the maternal line.

Mitochondrial proteins: Proteins that are part of the mitochondria.

Mitosis: Cell division, resulting in two cells that each have the diploid number of chromosomes and are just like the original cell.

Morphology: Configuration or structure, shape.

Mutagenicity: Tendency to promote mutations, that is, genetic alterations.

Multipotent: As applied to stem cells, the ability to differentiate into at least two, more differentiated descendant cells.

Multipotent adult progenitor cells (MAPC): Cells isolated from bone marrow that can be differentiated into cells with characteristics of cartilage, fat, and bone.

Mycoplasma: A general category of microorganisms that shares some characteristics of bacteria.

Natural killer cell: A cell type of the immune system that destroys tumor cells and cells infected with some types of organisms.

Olfactory bulb: A part of the brain involved in detecting and discriminating among different smells.

Oligodendrocyte: A type of neuroglia, that is, a particular type of cell that is part of the nervous system with supportive and metabolic functions rather than signal conduction, this type forms the myelin sheath around nerve fibers.

Oocytes: Egg cells.

Osteogenesis imperfecta: A large and miscellaneous group of conditions of abnormal fragility and plasticity of bone, with recurring fractures on trivial trauma. (SMD)

Pancreas: An organ of the digestive system that secretes the hormones insulin and glucagon, as well as digestive enyzymes.

Pancreatic beta cells: Cells of the pancreas (located in pancreatic islets, or islets of Langerhans) that produce insulin.

Parkinson disease: A neurological syndrome usually resulting from deficiency of the neurotransmitter dopamine . . . ; characterized by rhythmical muscular tremors . . . (SMD)

Phenotypic characteristics: The genetically and environmentally determined physical characteristics of an organism.

Placenta: The oval or discoid spongy structure in the uterus from which the fetus derives it nourishment and oxygen. (NRC)

Pluripotent: having great developmental plasticity, as a *pluripotent* stem cell. Cells that can produce *all* the cell types of the developing body, such as the ICM cells of the blastocyst, are said to be *pluripotent*.

Polarity: The property of having two opposite poles, sides or ends (for example, humans have left-right polarity, also front-back polarity and head-tailward polarity).

Population doublings: The number of times cells growing in vitro have increased the total number of cells by a factor of 2 compared to the initial number of cells.

Primitive streak: A band of cells appearing in the embryo at the start of the third week of development, that marks the axis along which the spinal chord develops.

Primordial germ cell: A gamete, that is, a sperm or egg, OR a primordial cell that can mature into a sperm or egg. (NRC)

Salivary gland: One of several pairs of glands in the mouth that secrete saliva.

Skin biopsy: Process of removing tissue, in this case skin, from living patients for diagnostic examination, or the tissue specimen obtained by that process. (SMD)

"Single-cell cloned": A procedure pertaining to cells in vitro in which the descendants of a single cell are physically isolated from other cells growing in a dish, and then expanded into a larger population.

Somatic cell nuclear transfer (SCNT): A method of cloning: transfer of the nucleus from a donor somatic cell into an enucleated egg to produce a cloned embryo.

Somite: One of the longitudinal series of segments into which the body of many animals (including vertebrates) is divided. (Merriam-Webster on line)

Stem cells: Stem cells are undifferentiated multipotent precursor cells that are capable both of perpetuating themselves as stem cells and of undergoing differentiation into one or more specialized types of cells. (CR)

Stromal: Relating to the stroma of an organ or other structure, that is, its framework, usually of connective tissue, rather than its specific substance. (SMD)

Syngamy: The coming together of the egg and sperm at fertilization.

Thymus: An organ of the developing immune system, active mainly in childhood.

T-lymphocyte: A cell type of the immune system that matures in the thymus and is responsible for cell-mediated immunity.

Type-1 diabetes: A form of insulin dependent diabetes, usually becoming evident in childhood, resulting from an autoimmune reaction that destroys the pancreatic beta cells, so that the body cannot produce its own insulin. In those cases where the condition is not apparent until adulthood, it is called latent autoimmune diabetes of adulthood (LADA).

Transcription factors: Specialized proteins that bind to specific sites on DNA and turn on or turn off the expression of different sets of genes.

Trophoblast: The extraembryonic tissue responsible for implantation, developing into the placenta, and controlling the exchange of oxygen and metabolites between mother and embryo. (NIH)

Twinning: Development of monozygotic twins, that is, when a very early embryo separates into two pieces, each of which continues development, so that two embryos actually come from one zygote.

Uterine: Pertaining to the uterus.

Virus: A submicroscopic pathogen composed essentially of a core of DNA or RNA enclosed by a protein coat, able to replicate only within a living cell.

Xenotransplantion: A transplant of tissue from an animal of one species to an animal of another species.

Zygote: The diploid cell that results from the fertilization of an egg cell by a sperm cell. (CR)

Appendix A.

Notes on Early Human Development

The term "embryo" refers to an organism in the early stages of its development. In humans, the term is traditionally reserved for the first two months of development. After that point, the term "embryo" is replaced by the term "fetus," which then applies until birth. Some authors further reserve the term "embryo" for the organism only after it has implanted and established its placental connection to the pregnant woman. Similarly many also reserve the term "pregnancy" for the state of the woman only after implantation. At the beginning of the individual's development, the entity is a single cell. After two months, it has limbs, distinct fingers and toes, internal development, and countless cells. So the term "embryo" applies to an individual throughout a vast range of developmental change. This document is a description of early human development, with emphasis on those events or structures that have figured most prominently in recent discussions of research using human embryos or their parts, especially for stem cell research.[1]

Development has fascinated centuries of observers, as they pursued deeper understanding of the stability of species characteristics at least from one generation to the next, as well as the uniqueness of each offspring. Uniqueness is especially marked in sexually reproducing organisms, that is, organisms where the genetic make-up of the offspring comes from a combination of maternal and paternal DNA, because a new genome is formed in each instance of conception. The stability reflects inheritance connecting one generation with the past and future members of its line.

Organisms and the processes of their development have evolved. As a result, the development of any organism has a species-specific pattern, but also shares many of the same developmental processes with other species related from its evolutionary origins. Many of the processes discussed here are common not just to all humans, or to all mammals, but to all vertebrates. In some cases they are shared even with invertebrates as well.

The process whereby a new individual of the species comes into being has been at the center of too many deep inquiries to list here, let alone discuss in the depth they deserve. But even in this short document it is important to note one question that is related to the connection of one generation to the next and previous generations.

That is, how are we to understand the apparent directedness of development, following a complex network of pathways from a single cell to a multi-system, free-living, and even conscious being? This process occurs in a reliable pattern time after time, but also is sufficiently resilient to perturbations that developing entities can recover from significant disturbances. For example, at early stages of development an embryo may divide (or be cut) completely in half, and then each half recovers to form an entire offspring, resulting in identical twins.

Different notions of purposive directedness, functional explanation, and even vital forces have been invoked to explain development. One of the insights, from the relation of development to evolution, is that the development of an individual reflects the fact that it is descended from individuals that reproduced successfully and, like its forebears whose DNA it inherited, its development reflects their past survival with their particular characteristics. This legacy of ancestral success at survival is manifested in the new organism's apparent directedness toward development along lines that enhance its own survival. Even very early embryos follow patterns of differentiation in the progeny of different cells. These patterns, in embryology, are called the *fate* of the progeny of a cell. The fate of the progeny of the newest single cell embryo is maximally broad—if it survives it will give rise to every type of cell of the species. But as the embryo becomes multicellular, its cells specialize and, in the absence of artificial perturbation, their progeny have increasingly specialized fates as well.

The evolved events and processes of development include some that reflect distant relations, such as the yolk sac that is conserved in placental mammals, including human beings. Other events or processes exhibit the evolution of more specific characteristics. In animals such as human beings, the specialized and complex membraneous structures that form the connection between the individual body of the pregnant woman and the developing individual body of the offspring begin to arise in the first week. Human embryos implant in the uterine wall starting at about the sixth day after conception, so of course they must arrive in the uterus with membranes capable of participating in that bond. They do not have a fully formed placenta at such an early stage, nor is the uterine wall unilaterally ready, but rather the contact of embryo and endometrium initiates complementary development finally resulting in the fully developed placenta. One way to look at it is that the early embryo's very structure points to the future, showing its overall developmental fate to be connected to the maternal body. Another perspective is that this process reflects the past survival of many generations. In both senses, no moment of development can be understood in isolation from the context of the organism's reflection of its predecessors in evolution, and its directed differentiation toward its future functioning.

I. GERM CELLS

For the beginning of an embryo, one can look both at the newly fertilized egg, and also further back, to embryos of the previous generation. The beginning of an individual is, of course, the union of egg and sperm, specifically the union of DNA in the nucleus of each, so as to form a new complete genome. But the egg and sperm in turn develop from primordial germ cells that were themselves developed when the parents of the new individual were embryos. This description starts at that point (Figure 1).

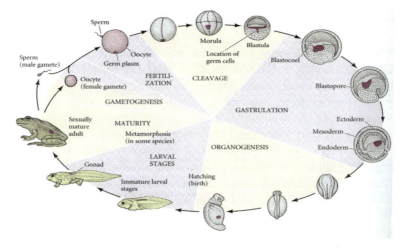

Figure 1: Developmental cycle, here of a frog. Note the continuity of germ plasm. [Figure 2.1, page 26, in Gilbert, S. *Developmental Biology.* 6th Edition. Sunderland, Mass.: Sinauer Associates Inc., 2000. Figure reproduced with permission of Sinauer Associates.]

The primordial germ cells are the cells that will give rise to either ova or sperm. They are large cells with some distinctive characteristics that make it possible to track them in development. Note that in Figure 1, they are highlighted throughout the life cycle of the animal. Primordial germ cells appear in embryonic development prior to the formation of the gonads (ovaries in female, or testes in a male). In humans and other mammals, the primordial germ cells actually develop first in the yolk sac. In either sex, the primordial germ cells migrate in through the developing gut of the embryo and then populate the new gonads of whichever type. In humans, the primordial germ cells first appear by the end of the fourth week of development, and begin their migration to the gonads. The primordial germ cells share certain characteristics with embryonic stem cells, including self-renewal and pluripotency. Primordial germ cells have been recovered from fetuses that were aborted (for reasons unrelated to research) and cell lines have been established from them, the progeny of which showed characteristics of multiple different types of cells.[2]

After the primordial germ cells populate the gonads, some continue to divide by mitosis, producing more like themselves. The primordial germ cells are diploid, meaning that they have all the normal chromosomes of the organism in pairs. In humans, this means that they have 22 pairs of autosomes, and one pair of sex chromosomes, or 46 total. *Mitosis* is the name of the process whereby the cell replicates its DNA and then divides equally to result in two cells, each cell including an entire complement of DNA just like the first cell before the division (in humans, that is the 46 total chromosomes mentioned) (Figure 2).

But if a cell is to become an ovum or sperm ready to combine with a gamete of the complementary type to produce a new organism (at first a zygote) containing the normal number of chromosomes, it must undergo a special type of cell division whereby each gamete acquires only half the diploid number. Each mature ovum or sperm must include only 23 single (not paired) chromosomes. Mature ova or sperm cells are haploid, indicating that their 23 chromosomes in their nuclei are unpaired (and after they combine, then the resulting single cell the zygote is again diploid). The process whereby the diploid primordial germ cells develop into haploid gametes is called *meiosis* (Figure 3). Mitosis is part of the life cycle of any cell, but meiosis or meiotic division occurs only in the development of haploid ova and sperm from diploid primordial germ cells. The process itself appears as though the cell nucleus is undergoing two rounds of mitosis, but omits the step of replicating DNA on the second cycle. In the "first round," the differentiating primordial germ cell replicates its DNA, and then in the "second round" it divides again (without another replication). In the second division, the pairs of chromosomes separate, leaving each of the new cells with just one copy of each of the 22 (in humans) autosomes and just one sex chromosome.

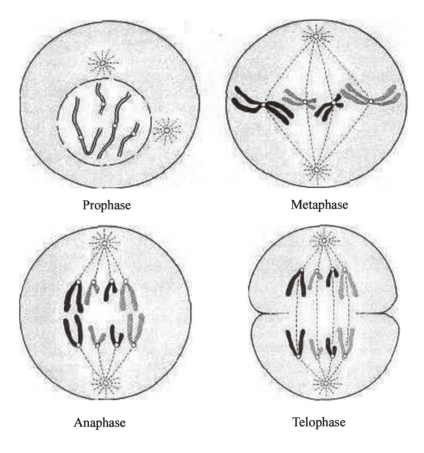

Prophase   Metaphase

Anaphase   Telophase

Figure 2: Schematic summary of the principal stages in mitotic cell division, simplified to show the movement of just two pairs of chromosomes. [Figure 3-4, page 61, in Carlson, B.M. *Patten's Foundations of Embryology*. 4th Edition. New York: McGraw-Hill, 1981. Figure(s) reproduced with permission of the McGraw-Hill Companies.]

This process does not always occur flawlessly. Errors, such as failure of the chromosomes to separate properly, sometimes produce new cells that have the wrong number of chromosomes, a condition called aneuploidy (that is, not the true number). One cell may have an extra copy of one of the chromosomes while the other cell is missing a copy. Such a condition can be detected in the lab by

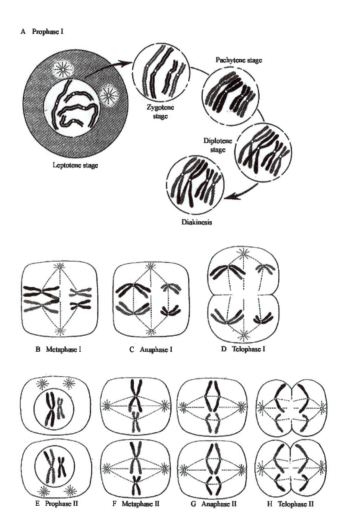

Figure 3: Schematic summary of the major stages of meiosis in a generalized germ cell, simplified to showing movement of two pairs of chromosomes at the start. [Carlson, 4th ed., Figure 3-6, p. 64.]

collecting some cells when they are about to go through mitosis so their chromosomes can be stained and be spread out so their number and appearance can be examined. A normal set of chromosomes produces a characteristic picture (22 recognizable pairs and a pair of sex chromosomes) called the normal karyotype. If a cell is aneuploid, it will produce an abnormal karyotype picture. If the aneuploid cell becomes an egg or sperm and is then involved in a conception, the embryo is also aneuploid. Aneuploidies are not uncommon events in germ cell development, but aneuploid *survival* is uncommon; nearly all aneuploidies are fatal very early in development.

II. FERTILIZATION AND CLEAVAGE

Like the word "embryo," the word "conception" refers to a series of events or processes, not an instantaneous occurrence. Human development begins after the union of egg and sperm cells during a process known as fertilization. Fertilization itself comprises a sequence of events that begins with the contact of a sperm cell with an egg cell and ends with the fusion of their two pronuclei (each containing 23 chromosomes) to form a new diploid cell, called a *zygote*. Fertilization normally occurs in the *ampulla* of the uterine tube 12-24 hours after ovulation (Figure 4).

Before that, however, sperm must travel through the vagina and the cervix, through the uterus, and then up the uterine tube. Smooth muscle contractions in the uterine tubes as well as ciliary activity (waving of hair-like structures) of the tube's lining both are important in the transport of sperm up, and of the ovum into and then down, the uterine tube. Many more sperm, on the order of tens, or even hundreds, of millions, are ejaculated than reach the ovum. Those sperm that do come into the vicinity of the ovum must get through the material covering the ovum (the corona radiata and the zona pellucida) and finally contact and bind to the ovum's membrane, by means of specialized structures in the head of the sperm cell. When a sperm does get into the ovum, then the ovum membrane changes so that other sperm cannot enter. Meanwhile, the sperm cell in the egg is also undergoing changes and its specialized structures fall away. The haploid nuclei of both the sperm and the egg are now called male and female pronuclei. Both swell, as their densely packed DNA loosens up prior to replication, and they also migrate toward the center of the ovum. Then their nuclear membranes disintegrate and the paternally and maternally contributed chromosomes pair up, an event called *syngamy*. In this integration, the diploid

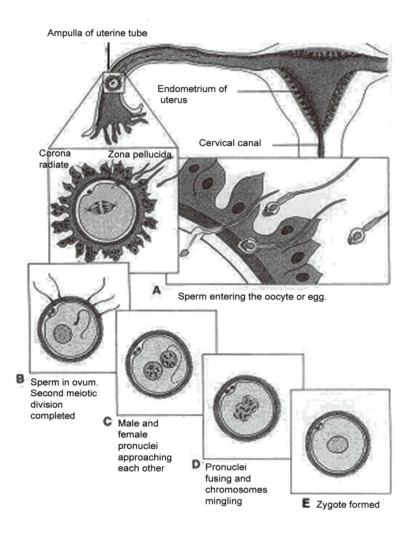

Ampulla of uterine tube

Endometrium of uterus

Cervical canal

Corona radiate

Zona pellucida

A Sperm entering the oocyte or egg.

B Sperm in ovum. Second meiotic division completed

C Male and female pronuclei approaching each other

D Pronuclei fusing and chromosomes mingling

E Zygote formed

Figure 4A-E: Steps in the process of fertilization. The sequence of events begins with contact between a sperm and a secondary oocyte (a mature egg) in the ampulla of the uterine tube, and ends with formation of a zygote. [A-E: Fig. 1-1, page 3, in Moore, *Essentials of Human Embryology*, 1988, with permission from Elsevier.]

Figure 4F: (see following page) Shows fertilization, syngamy [from www.visembryo.com by Mouseworks, Inc.]

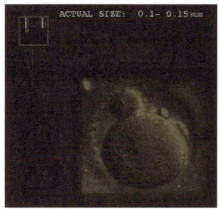

ACTUAL SIZE: 0.1- 0.15mm

Figure 4F: Fertilization, Syngamy

chromosome number is restored, and a new complete genome comes into being. The result of syngamy is an entity with an individual genome. Further, if all goes well, it is an entity that is capable of developing into a fully formed individual of the species. The fertilized egg is now called a zygote. It is at this point already entering the first stage of its first mitotic division, and beginning cleavage (Figure 5).

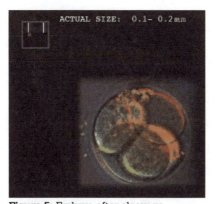

ACTUAL SIZE: 0.1- 0.2mm

Figure 5: Embryo after cleavage.
[www.visembryo.com by Mouseworks, Inc.]

Like other vertebrates, humans have polarity in three dimensions (head-tail, or back-front, and left-right). Establishing polarity is one of the most basic manifestations of emerging specialization. But the egg is roughly spherical, and it is not readily apparent how polarity is established. Although it had been shown long ago that the point of sperm entry determines the plane of first cleavage (and thus subsequent ones) in amphibian eggs, mammals were believed until recently to remain spherically symmetrical until later in development. Recent data on mammalian zygotes, however, suggests that the point of sperm entry may similarly determine the cleavage plane.[3] Even the first two cells resulting from the first cleavage may have different propensities, which persist through the next divisions as the progeny of one cell tend to become the body of the offspring and progeny of the other cell become the embryo's contribution to the placenta and other supporting structures. The word "fate," however, might be too strong, because the cells of such very early embryos are resilient to perturbations—if one cell is removed, the remaining ones can compensate.

III. IMPLANTATION

After fertilization, the zygote proceeds immediately to the first cleavage and subsequent cell divisions follow rapidly. The zygote is a very large cell, but the first waves of rapid cell division occur without increase in cell volume. The result is a closely bound mass of cells each of more typical cell size. At this stage the cells are called *blastomeres*, ("parts of the blast," "blast" coming from the Greek for "bud" or "germ") and the organism as a whole is called a *morula* (from the Latin for mulberry, descriptive of its appearance) from the time it has 16 blastomeres to the next stage. The morula is still encased in the *zona pellucida*. As it is undergoing this very rapid cell division, the organism is also migrating down the uterine tube toward the uterus. After it arrives in the uterus, at about day five after the initiation of fertilization, the zona pellucida breaks up; the process is called "hatching" and is a necessary prelude to implantation.

Many zygotes do not survive this long. Estimates vary widely of the rate of natural embryo loss prior to implantation or after implantation but still early in gestation. One study of healthy women trying to conceive found 22 percent of pregnancies (identified by sensitive hormone measures) were lost prior to becoming detectable clinically. Even after implantation, there is a substantial rate of loss, still not known precisely but estimated at 25 to 40 percent.[4]

When the morula enters the uterus, fluid starts to accumulate between its blastomeres. The fluid-filled spaces run together, forming a relatively large fluid-filled cavity. At the point when the cavity becomes recognizable, the organism is called a *blastocyst* (Figure 6). The outer cells of the blastocyst, especially those around the blastocyst cavity, assume a flattened shape. The flattened cells of the exterior blastocyst are the *trophoblast*. They become the embryo's contribution to the placenta and other supporting structures. On one side of the blastocyst is a group of cells that project inside into the blastocyst cavity; this is the *inner cell mass*, or *embryoblast*, and its progeny form the body of the new offspring.

FORMATION OF THE BLASTOCYST

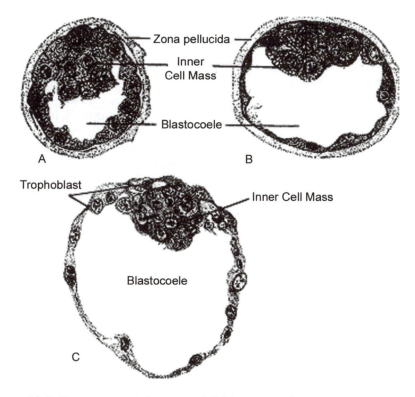

Figure 6A-C: Three stages of the mammalial (blastocyst) of the pig, drawn from sections to show the formation of the inner cell mass.
[Carlson, 4th Ed., Fig. 4-10, page 124.]

Figure 6D: Early blastocyst (see following page).

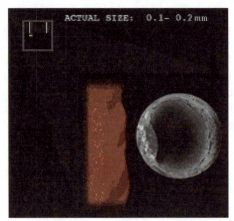

Figure 6D: Early Blastocyst [www.visembryo.com by Mouseworks.Inc.]

The cells of the inner cell mass can give rise to progeny differentiating into all the types of cells in the adult body, so they are called pluripotent. They have not usually been described as totipotent because, the inner cell mass having already differentiated from trophoblast, the cells of the inner cell mass were believed to be no longer able to give rise to the cells of the trophoblast. Recent work, however, describes culture conditions under which human embryonic stem cells can differentiate to trophoblast cells.[5] Although the new offspring itself develops only from the inner call mass, the trophoblast is not just passive padding. Its progeny are the essential and specialized connection between the embryonic and maternal systems. Embryonic stem cells can be isolated from the inner cell mass (see Chapter 4).

IV. TROPHOBLAST TO PLACENTA

After the embyo covering degenerates, the blastocyst, now in the uterus, enlarges and its trophoblast attaches to the endometrium (the uterine lining) at about six days after fertilization. This begins the process of implantation, during which the blastocyst becomes integrated with the endometrium through specialized membranes. The embryo is now beginning its second week of development. The process of implantation takes three to four days, but is generally completed by day twelve. The trophoblast area that binds to the endometrium first differentiates into an inner layer of cells and an exterior layer in which the membranes dividing the cells degenerate and the cells fuse. As the blastocyst become more deeply embedded in the endometrium, the layered area expands until finally the whole trophoblast surface has divided into one layer or the other.

Meanwhile, a sort of primitive circulation develops, supporting the embedded blastocyst while more complex structures continue to develop. The inner cell mass then separates itself from the overlying trophoblast. The resulting space is called the amniotic cavity and the layer of cells that forms its roof is called the amnion (Figure 7).

Another membrane called the chorionic sac develops from the trophoblast and nearby tissue. Finally, outgrowths of trophoblast from the chorion project into the endometrium and are called primary chorionic villi, later giving rise to the placenta. Although the blastocyst has become completely embedded in the endometrium and maternal blood bathes the chorionic villi, the maternal blood does not enter the blastocyst. Later, as the fetal circulation develops, the fetal and maternal blood systems still remain distinct and do not mingle. Nutrients, oxygen, and wastes diffuse in the appropriate direction across the placenta, but the two blood systems are individual and do not combine.

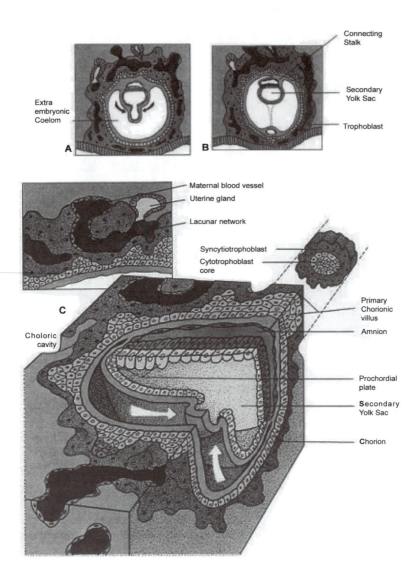

Figure 7A: Sections of completely implanted blastocysts at the end of the second week, illustrating how the secondary yolk sac forms. The presence of primary chorionic villi on the wall of the chorionic sac is characteristic of blastocysts at the end of the second week. A primitive uteroplacental circulation is now present. [Reprinted from Moore, *Essentials of Human Embryology*, 1988, Fig. 2-2, p. 13, with permission from Elsevier.]

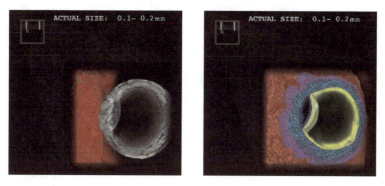

Figure 7B: Photo micrographs of implantation beginning and completed. [www.visembryo.com by Mouseworks, Inc.]

V. TWINNING

The usual case for human beings is for one ovum to be released, and if all goes well, fertilized and developed to term. Less commonly, more than one ovum may be released and fertilized so that more than one embryo develops. These embryos would be genetically distinct, sharing the uterus during the same gestation period. They will have a family resemblance but no more genetic commonality than any other set of siblings, and they may be of the same or different sexes. These are called dizygotic twins (because they came from two zygotes). More rarely, a single zygote may, during its early cleavages, separate completely into two groups of cells. As discussed above, the two cells resulting from the first cleavage may already have different probable fates, the progeny of one contributing to the body and the other to the supporting structures. Both, however, at this stage are still totipotent and can, if disrupted, go on to generate a full individual organism. If this separation occurs, then monozygotic twins may be born (Figure 8). Monozygotic twins, two offspring coming from one zygote, have the same genome and are always of the same sex. When the twinning occurs in the first cleavages and there are not yet any extraembryonic membranes (Figure 8A), the two develop separately as do dizygotic twins, with separate amnions, chorions and eventually placentae. If an embryo should divide into two later in its development, between about days four and eight, the twins will share the same chorion and therefore eventually the same placenta, but a separate amnion will form around each (Figure 8B). Should an embryo divide later than this, between about the ninth and thirteenth days, the resulting twins will share the same amnion, chorion, and placenta. It is very rare for embryos to divide still later than this, but occasionally they do divide after the fourteenth day. These divisions may not be complete, and then the twins remain conjoined and can only be surgically separated after birth (Figure 8C). The twin birth rate in the U.S. has increased markedly in recent years, and was 30.1 per 1,000 live births in 2001.[6] The rate of multiple births (most multiple births are twins; triplets and so on are more rare) is higher with assisted reproductive technologies and with higher maternal age. Dizygotic twins clearly can result in ART from transferring more than one embryo to the prospective mother. In addition, some assisted reproductive practices, like age of the embryo transferred, may be associated with more likelihood of monozygotic twinning,[7] though in general the causes of monozygotic twinning are not known.

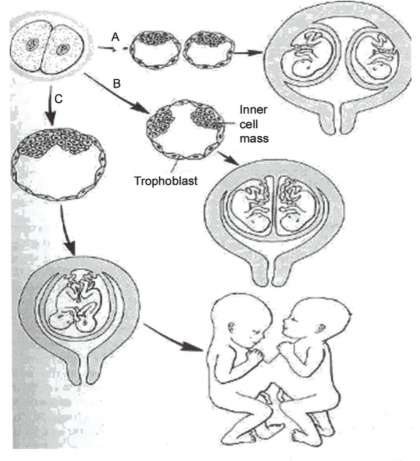

Figure 8A-C: Modes of monozygotic twinning. [Carlson, 4th Ed., Fig. 1-12, page 23]

VI. THE PRIMITIVE STREAK AND GASTRULATION

While implantation is occurring, the inner cell mass is also undergoing changes. First, the inner cell mass separates into two layers, the epiblast, which is next to the amniotic cavity, and the hypoblast, which is next to what was the blastocyst cavity but is by this stage called the primary yolk sac. The epiblast thus forms the floor of the amniotic cavity (as the amnion forms the roof) and is connected with the amnion around the edges. The hypoblast is connected around its edges with the exocoelomic membrane or primary yolk sac. Thus, the supporting structures, collectively called the extraembryonic membranes, are outside of the body that is starting to develop and that will eventually be born, but during embryonic development the membranes are also continuous with that body. By the end of the second week, the hypoblast has developed a thickened area, called the prochordal plate, that is located at what will be the cranial (head) end of the individual. In fact, the prochordal plate shows where the mouth will develop.

As the third week of development begins, dividing cells pile up in a line to form a thicker band in the epiblast. The line or band starts nearly directly across from the prochordal plate, and extends from the edge toward the center of the embryonic disc. The band is called the *primitive streak*. In many policy discussions, the appearance of the primitive streak is an important boundary. This summary will continue just a little longer, in order to discuss briefly the nature of the primitive streak.

The end of the primitive streak that is toward the middle of the disc (nearer the prochordal plate marking the mouth) is the cranial end, and this end thickens more as more cells divide. This especially thick end is called the primitive knot (formerly called Henson's node). The end of the primitive streak near the edge is the caudal (or tail-ward) end. As a model, think of the primitive streak as a zipper: the epiblast cells that made the thickness now start to migrate across the surface and into the zipper of the primitive streak. As the cells enter the primitive streak, they do a U-turn around the edge and continue to migrate back the way they came but underneath the surface, displacing the hypoblast cells. This movement results in three layers, all of epiblast origin: what was the epiblast on top, the cells that used to be part of the epiblast but are now underneath it, and the cells that remain in between (Figure 9).

These three layers get new names, and they also get newly specified fates for their progeny. In the same order as above, they are the ectoderm, mesoderm, and endoderm. The completion (during the third week after fertilization) of forming these three layers is called gastrulation. The ectodermal layer gives rise to progeny fated to become the skin, the nervous system, and sensory structures of the eye, ear, and nose; mesoderm gives rise to the skeletal and muscular systems, connective tissue and blood vessels, and endoderm gives rise to epithelial parts (e.g., the linings) of the digestive and respiratory systems.

Gastrulation is a crucial event in the development of the body plan of the individual, and it is a stage of development common to all vertebrates. Our understanding of the significance of establishing the three germ layers has grown more complex and subtle over the years. Once interpreted as three completely separate paths or compartments of development, we now know that the progeny of the three layers are not totally isolated in their fates. Cartilage, for example, was once thought to be entirely of mesodermal origin, but now we know that some cartilaginous structures of the head and neck come from ectoderm. Even more recently, work with certain adult stem cell populations in culture and under special conditions has suggested plasticity of cell progeny from one germ layer to develop characteristics of cells typically from another germ layer, long after gastrulation has assigned the cells of different germ layers their different fates. Gastrulation is not the first differentiating event: cells begin to acquire fates for different parts of the developing embryo before the inner cell mass separates into epiblast and hypoblast, indeed some results suggest even before the blastocyst develops an inner cell mass and trophoblast. Yet these findings in no way detract from the significance of gastrulation. They rather facilitate our understanding of gastrulation by placing it in the context of the entire process of differentiation, beginning from the very earliest stages.

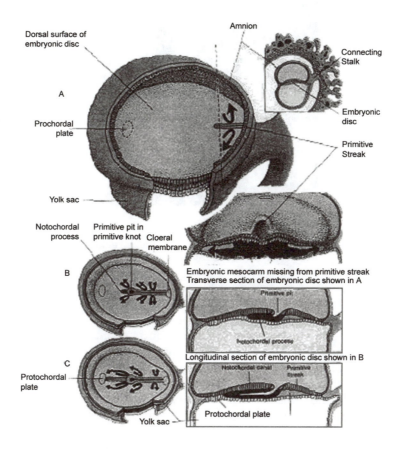

Figure 9A-C: Schematic drawings of the embryonic disc and its associated extraembryonic membranes during the third week. A: the amniotic cavity has been opened to show the primitive streak, a midline thickening of the epiblast. Part of the yolk sac has been cut away to show the bilaminar embryonic disc (epiblast and hypoblast). The transverse section (*lower right* of A) illustrates the proliferation and migration of cells from the primitive streak to form embryonic mesoderm. B and C: drawings illustrating early formation of the notochordal process from the primitive knot of the primitive streak. In the longitudinal sections on the right side, note that the notochordal process grows cranially in the median plane between the embryonic ectoderm and endoderm. [Reprinted from Moore, *Essentials of Human Embryology*, 1988, Fig. 3-1, page 17, with permission from Elsevier.]

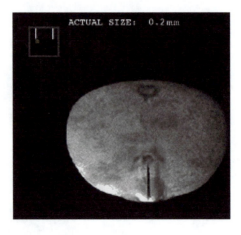

Figure 9D: Photo micrograph of Primitive Streak
[www.visembryo.com by Mouseworks, Inc.]

VII. NEURULATION

Neurulation is the series of developmental events that result in the beginnings of the central nervous system (Figure 10). From the cranial end of the primitive streak, a long stiff structure develops in the mesoderm, elongating still further in the cranial direction. This becomes the *notochord*, which marks the head/tail axis of the embryo. Later, the vertebral column develops around it. But at this time, the notochord and its adjacent tissue exert influence called *primary induction* on the ectoderm lying over them, such that the ectoderm thickens and becomes the neural plate.

The neural plate then actually pushes up to form folds (called the neural folds) along each side of the tissue over the notochord. The neural folds then meet and fuse to enclose the neural tube, beginning at the middle of the (future) tube, like a zipper closing from the middle toward each end. This process is completed by the end of the third week. Some cells along the crests of the folds migrate through the embryo. They are called neural crest cells, and they give rise to a variety of nerve cells including dorsal root (spinal) and autonomic nervous system ganglia, and some other nervous system and endocrine structures.

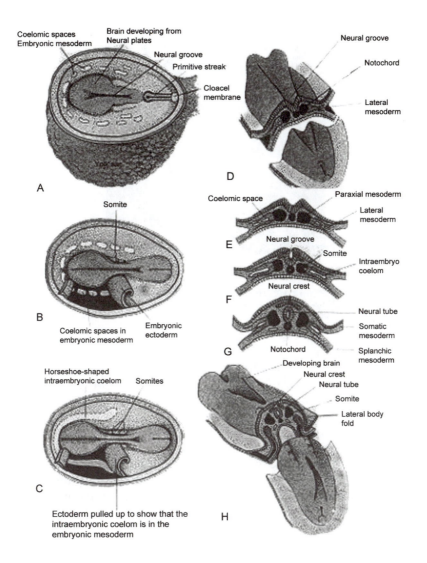

Figure 10A-H: Schematic drawings of the human embryo during the third and fourth weeks. *Left side*: Dorsal views of the developing embryo illustrating early formation of the brain, intraembryonic coelom, and somites. *Right side*: Schematic transverse sections illustrating formation of the neural crest, neural tube, intraembryonic coelom, and somites. [Reprinted from Moore, *Essentials of Human Embryology*, 1988, Fig. 3-3, page 20, with permission from Elsevier.]

Figure 10I: Neurulation and Notochordal Process
[www.visembryo.com by Mouseworks, Inc.]

The mesoderm still adjacent to the neural tube resolves into the form of paired blocks on either side of the tube, which are called *somites*. The first pair of somites appears at about the twentieth day after fertilization, at the cranial end of the neural tube. More pairs appear in the caudal direction, up until about the thirtieth day. Mesodermal cells from the somites give rise to most of the skeleton and skeletal muscle.

Blood cell and blood vessel formation actually start at the beginning of the third week after fertilization, first in the supportive structures of the yolk sac and chorion. Blood vessel formation begins in the embryo body about two days later, although blood is not formed in the embryo itself until the fifth week. The heart begins as a wide blood vessel, which later folds up to develop the chambers of the fully formed heart. But even as a tube, the membranes of its cells have the electrical and contractile capacity to begin beating in the third week, and thus to begin primitive circulatory function with blood. During this time the primary chorionic villi elaborate branches and form capillary networks and vessels connected with the embryonic heart. Oxygen and nutrients diffuse from the maternal blood to the embryonic blood through these capillaries, while carbon dioxide, urea, and other metabolic wastes diffuse from the embryonic blood into the maternal blood. Meanwhile, even firmer connections form between embryonic supporting membranes and the endometrium, finally completing the development of the placenta.

VIII. ORGANOGENESIS

The basic structures and relations of all the major organ systems of the body emerge during the fourth through the eighth weeks of embryonic development. First, the embryo folds in several ways so that the flat linear structure distinguished by neural tube flanked by somites become roughly C-shaped. The effect of this is to bring the regions of the brain, gut, and other internal organs into their familiar anatomical relations. During the fourth week the neural pores, the ends of the neural tube "zipper," close. First the one at the cranial or head end, which is called the anterior or rostral pore, closes, and later the caudal or tail-ward pore closes. Closure of the neural pores completes the closure of what will become the central nervous system. Also during the fourth week, limb buds become visible, first buds for arms and later for legs. Further, two accumulations of cells along the neural tube become distinguishable: the alar plate and the basal plate. Cells of the alar plate go on to become mostly sensory neurons, while basal plate cells give rise mostly to motor neurons. Already while the neural tube is closing, its walls along the cranial area are thickening to form early brain structure. Cranial nerves, for example the nerves for the eye and for the muscles of the face and jaw, also are beginning to develop at this time. The embryonic brain develops rapidly in both size and structure especially during the fifth week, and the optic cup that will form the retina of the eye becomes visible as well.

IX. CONCLUSION AND CONTINUATION

Embryonic development continues with the emergence and differentiation of organs, the skeleton, limbs, and digits, and with the development of the face and further differentiation and integration throughout the body. The development discussed above is summarized briefly in Table 1.[8] But development continues, and is a continuous process, past the eight-week mark, when the organism is no longer called an embryo and instead is called a fetus. Although the basic elements of the body plan have been established during embryogenesis, a great deal of development of that body plan, refinement and integration, continues in the fetal stage, also called phenogenesis (emergence of the normal appearance of the body). Development continues after birth as well.

Table1: Summary of Developmental Timecourse			
Stage	*Week after fertilization*	*Days after fertilization*	*Event*
Pregenesis: development of parents	4th week develop-ment of parents	24	Parents' primordial germ cells (PGCs) begin their migration to parents' gonads
Blastogenesis	1st week, embryo is unilaminar	1	Fertilization
		1.5-3	1st cleavages, move to uterus
		4	Free blastocyst in uterus
		5-6	Hatching, start implantation
	2nd week, embryo is bilaminar	7-12	Fully implanted
		13	Primary stem villi and primitive streak appear
	3rd week, embryo is trilaminar	16	Gastrulation begins, notochord forms
		18	Primitive pit, neural plate, neural groove
		20	First somites, primitive heart tube
	4th week	22	Neural folds fuse, pulmonary primordium,
		24	PGCs begin migration, Cranial neuropore closes, optic vesicles and pit form
		26	Caudal neuropore closes, arm limb buds
		28	Leg limb buds, more brain, eye/ear devel
Organogenesis	5th - 8th weeks	29-56	
Phenogenesis	9th - 38th weeks		

ENDNOTES

[1] There are many fine embryology texts, and the reader is urged to consult one or more for deeper, broader and more extended treatment of embryology. The following references are samples only, not a comprehensive bibliography, selected in part for accessibility to the general though committed reader, and in part for recent publication. A few examples concentrating on human embryology would include Larsen, W. J., *Essentials of Human Embryology.* New York: Churchill Livingstone, 1988; Sadler, T. W., *Langman's Medical Embryology* 8th Edition, Philadelphia: Lippincott Williams and Wilkens, 2002; or Sweeney, L. J., *Basic Concepts in Embryology: A Student's Survival Guide,* New York: McGraw-Hill, 1988. For a more comparative approach consider Carlson, B. M., *Patten's Foundations of Embryology.* 6th Edition, New York: McGraw-Hill, 1996; and for more comparison and inclusion of related topics, see Gilbert, S. J., *Developmental Biology,* 6th Edition, Sunderland, MA: Sinauer Associates Inc., 2000. In addition, there are many fine web-based resources, which the reader is encouraged to visit, for example http://anatomy.med. unsw.edu.au/cbl/embryo/Embryo.htm and http://www.visembryo.com/ and to accompany Gilbert's text, http://www.devbio.com/. These sites provide links to further resources as well.

[2] Shamblott M.J., et al., "Derivation of pluripotent stem cells from cultured human primordial germ cells" *Proceedings of the National Academy of Sciences.* 95(23): 13726-13731 (1998). [Erratum in: *Proc Natl Acad Sci* USA 96(3): 1162 (1999).]

[3] Pearson, H., "Your destiny, from day one." *Nature* 418(6893): 14-15 (2002).

[4] Wilcox, A. J., et al., "Incidence of early loss of pregnancy," *New England Journal of Medicine* 319(4): 189-194 (1988). See also this review article: Norwitz, E. R., et al., "Implantation and the survival of early pregnancy," *New England Journal of Medicine* 345(19): 1400-1408 (2001). Some estimates are indeed much higher (as high as 80 percent for embryo loss before and after implantation).

[5] Xu, R.H., et al., "BMP4 initiates human embryonic stem cell differentiation to trophoblast" *Nature Biotechnology* 20(12): 1261-1264 (2002).

[6] National Center for Health Statistics, "Births: Final Data for 2001." National Vital Statistics Reports 51(2) (2002), available at http://www.cdc.gov/nchs/data/nvsr/ nvsr51/nvsr51_02.pdf.

[7] Milki, A.A. et al., "Incidence of monozygotic twinning with blastocyst transfer compared to cleavage-stage transfer" *Fertility and Sterility* 79(3): 503-506 (2003).

[8] Table 1 follows closely the table of events shown in Larsen (1998), p. xi, and also the table presented by John M. Opitz, MD, at the January 16, 2003, meeting of the Council. Not all the events listed in Larsen's table were included in the Table 1 above, however.

Appendix B.

Remarks by President George W. Bush on Stem Cell Research

August 9, 2001

The Bush Ranch
Crawford, Texas

THE PRESIDENT: "Good evening. I appreciate you giving me a few minutes of your time tonight so I can discuss with you a complex and difficult issue, an issue that is one of the most profound of our time.

The issue of research involving stem cells derived from human embryos is increasingly the subject of a national debate and dinner table discussions. The issue is confronted every day in laboratories as scientists ponder the ethical ramifications of their work. It is agonized over by parents and many couples as they try to have children, or to save children already born.

The issue is debated within the church, with people of different faiths, even many of the same faith coming to different conclusions. Many people are finding that the more they know about stem cell research, the less certain they are about the right ethical and moral conclusions.

My administration must decide whether to allow federal funds, your tax dollars, to be used for scientific research on stem cells derived from human embryos. A large number of these embryos already exist. They are the product of a process called in vitro fertilization, which helps so many couples conceive children. When doctors match sperm and egg to create life outside the womb, they usually produce more embryos than are planted in the mother. Once a couple successfully has children, or if they are unsuccessful, the additional embryos remain frozen in laboratories.

Some will not survive during long storage; others are destroyed. A number have been donated to science and used to create privately funded stem cell lines. And a few have been implanted in an adoptive mother and born, and are today healthy children.

Based on preliminary work that has been privately funded, scientists believe further research using stem cells offers great promise that could help improve the lives of those who suffer from many terrible diseases — from juvenile diabetes to Alzheimer's, from Parkinson's to spinal cord injuries. And while scientists admit they are not yet certain, they believe stem cells derived from embryos have unique potential.

You should also know that stem cells can be derived from sources other than embryos — from adult cells, from umbilical cords that are discarded after babies are born, from human placenta. And many scientists feel research on these type of stem cells is also promising. Many patients suffering from a range of diseases are already being helped with treatments developed from adult stem cells.

However, most scientists, at least today, believe that research on embryonic stem cells offer the most promise because these cells have the potential to develop in all of the tissues in the body.

Scientists further believe that rapid progress in this research will come only with federal funds. Federal dollars help attract the best and brightest scientists. They ensure new discoveries are widely shared at the largest number of research facilities and that the research is directed toward the greatest public good.

The United States has a long and proud record of leading the world toward advances in science and medicine that improve human life. And the United States has a long and proud record of upholding the highest standards of ethics as we expand the limits of science and knowledge. Research on embryonic stem cells raises profound ethical questions, because extracting the stem cell destroys the embryo, and thus destroys its potential for life. Like a snowflake, each of these embryos is unique, with the unique genetic potential of an individual human being.

As I thought through this issue, I kept returning to two fundamental questions: First, are these frozen embryos human life, and therefore, something precious to be protected? And second, if they're going to be destroyed anyway, shouldn't they be used for a greater good, for research that has the potential to save and improve other lives?

I've asked those questions and others of scientists, scholars, bioethicists, religious leaders, doctors, researchers, members of Congress, my Cabinet, and my friends. I have read heartfelt letters from many Americans. I have given this issue a great deal of thought, prayer and considerable reflection. And I have found widespread disagreement.

On the first issue, are these embryos human life — well, one researcher told me he believes this five-day-old cluster of cells is not an embryo, not yet an individual, but a pre-embryo. He argued that it has the potential for life, but it is not a life because it cannot develop on its own.

An ethicist dismissed that as a callous attempt at rationalization. Make no mistake, he told me, that cluster of cells is the same way you and I, and all the rest of us, started our lives. One goes with a heavy heart if we use these, he said, because we are dealing with the seeds of the next generation.

And to the other crucial question, if these are going to be destroyed anyway, why not use them for good purpose — I also found different answers. Many argue these embryos are byproducts of a process that helps create life, and we should allow couples to donate them to science so they can be used for good purpose instead of wasting their potential. Others will argue there's no such thing as excess life, and the fact that a living being is going to die does not justify experimenting on it or exploiting it as a natural resource.

At its core, this issue forces us to confront fundamental questions about the beginnings of life and the ends of science. It lies at a difficult moral intersection, juxtaposing the need to protect life in all its phases with the prospect of saving and improving life in all its stages.

As the discoveries of modern science create tremendous hope, they also lay vast ethical mine fields. As the genius of science extends the horizons of what we can do, we increasingly confront complex questions about what we should do. We have arrived at that brave new world that seemed so distant in 1932, when Aldous Huxley wrote about human beings created in test tubes in what he called a "hatchery."

In recent weeks, we learned that scientists have created human embryos in test tubes solely to experiment on them. This is deeply troubling, and a warning sign that should prompt all of us to think through these issues very carefully.

Embryonic stem cell research is at the leading edge of a series of moral hazards. The initial stem cell researcher was at first reluctant to begin his research, fearing it might be used for human cloning. Scientists have already cloned a sheep. Researchers are telling us the next step could be to clone human beings to create individual designer stem cells, essentially to grow another you, to be available in case you need another heart or lung or liver.

I strongly oppose human cloning, as do most Americans. We recoil at the idea of growing human beings for spare body parts, or creating life for our convenience. And while we must devote enormous energy to conquering disease, it is equally important that we pay attention to the moral concerns raised by the new frontier of human embryo stem cell research. Even the most noble ends do not justify any means.

My position on these issues is shaped by deeply held beliefs. I'm a strong supporter of science and technology, and believe they have the potential for incredible good — to improve lives, to save life, to conquer disease. Research offers hope that millions of our loved ones may be cured of a disease and rid of their suffering. I have friends whose children suffer from juvenile diabetes. Nancy Reagan has written me about President Reagan's struggle with Alzheimer's. My own family has confronted the tragedy of childhood leukemia. And, like all Americans, I have great hope for cures.

I also believe human life is a sacred gift from our Creator. I worry about a culture that devalues life, and believe as your President I have an important obligation to foster and encourage respect for life in America and throughout the world. And while we're all hopeful about the potential of this research, no one can be certain that the science will live up to the hope it has generated.

Eight years ago, scientists believed fetal tissue research offered great hope for cures and treatments — yet, the progress to date has not lived up to its initial expectations. Embryonic stem cell research offers both great promise and great peril. So I have decided we must proceed with great care.

As a result of private research, more than 60 genetically diverse stem cell lines already exist. They were created from embryos that have already been destroyed, and they have the ability to regenerate themselves indefinitely, creating ongoing opportunities for research. I have concluded that we should allow federal funds to be used for research on these existing stem cell lines, where the life and death decision has already been made.

Leading scientists tell me research on these 60 lines has great promise that could lead to breakthrough therapies and cures. This allows us to explore the promise and potential of stem cell research without crossing a fundamental moral line, by providing taxpayer funding that would sanction or encourage further destruction of human embryos that have at least the potential for life.

I also believe that great scientific progress can be made through aggressive federal funding of research on umbilical cord placenta, adult and animal stem cells which do not involve the same moral dilemma. This year, your government will spend $250 million on this important research.

I will also name a President's council to monitor stem cell research, to recommend appropriate guidelines and regulations, and to consider all of the medical and ethical ramifications of biomedical innovation. This council will consist of leading scientists, doctors, ethicists, lawyers, theologians and others, and will be chaired by Dr. Leon Kass, a leading biomedical ethicist from the University of Chicago.

This council will keep us apprised of new developments and give our nation a forum to continue to discuss and evaluate these important issues. As we go forward, I hope we will always be guided by both intellect and heart, by both our capabilities and our conscience.

I have made this decision with great care, and I pray it is the right one.

Thank you for listening. Good night, and God bless America."

* * *

Appendix C.

Bush Administration NIH Guidelines for Embryonic Stem Cell Funding

Taken from the website of the National Institutes of Health on September 26, 2003.

The text follows.

NOTICE OF CRITERIA FOR FEDERAL FUNDING OF RESEARCH ON EXISTING HUMAN EMBRYONIC STEM CELLS AND ESTABLISHMENT OF NIH HUMAN EMBRYONIC STEM CELL REGISTRY

Release Date: November 7, 2001

NOTICE: NOT-OD-02-005

Office of the Director, NIH

On August 9, 2001, at 9:00 p.m. EDT, the President announced his decision to allow Federal funds to be used for research on existing human embryonic stem cell lines as long as prior to his announcement (1) the derivation process (which commences with the removal of the inner cell mass from the blastocyst) had already been initiated and (2) the embryo from which the stem cell line was derived no longer had the possibility of development as a human being.

In addition, the President established the following criteria that must be met:

o The stem cells must have been derived from an embryo that was created for reproductive purposes;
o The embryo was no longer needed for these purposes;
o Informed consent must have been obtained for the donation of the embryo;
o No financial inducements were provided for donation of the embryo.

189

In order to facilitate research using human embryonic stem cells, the NIH is creating a Human Embryonic Stem Cell Registry that will list the human embryonic stem cells that meet the eligibility criteria. Specifically, the laboratories or companies that provide the cells listed on the Registry will have submitted to the NIH a signed assurance. Each provider must retain for submission to the NIH, if necessary, written documentation to verify the statements in the signed assurance.

The Registry will be accessible to investigators on the NIH Home Page http://escr.nih.gov. Requests for Federal funding must cite a human embryonic stem cell line that is listed on the NIH Registry. Such requests will also need to meet existing scientific and technical merit criteria and be recommended for funding by the relevant National Advisory Council, as appropriate. Further guidance is accessible at http://grants.nih.gov/ grants/guide/ notice-files/NOT-OD-02-006.html.

Inquiries should be directed to the Deputy Director for Extramural Research DDER@nih.gov.

* * *

Appendix D.

Clinton Administration NIH Guidelines for Embryonic Stem Cell Funding

As printed in the Federal Register, August 25, 2000 ("National Institutes of Health Guidelines for Research Using Human Pluripotent Stem Cells," 65 Fed. Reg. 51,975, Aug. 25, 2000)

The text of the final Guidelines follows.

National Institutes of Health Guidelines for Research Using Human Pluripotent Stem Cells

I. Scope of Guidelines

These Guidelines apply to the expenditure of National Institutes of Health (NIH) funds for research using human pluripotent stem cells derived from human embryos (technically known as human embryonic stem cells) or human fetal tissue (technically known as human embryonic germ cells). For purposes of these Guidelines, "human pluripotent stem cells" are cells that are self-replicating, are derived from human embryos or human fetal tissue, and are known to develop into cells and tissues of the three primary germ layers. Although human pluripotent stem cells may be derived from embryos or fetal tissue, such stem cells are not themselves embryos. NIH research funded under these Guidelines will involve human pluripotent stem cells derived: (1) From human fetal tissue; or (2) from human embryos that are the result of in vitro fertilization, are in excess of clinical need, and have not reached the stage at which the mesoderm is formed.

In accordance with 42 Code of Federal Regulations (CFR) 52.4, these Guidelines prescribe the documentation and assurances that must accompany requests for NIH funding for research using human pluripotent stem cells from: (1) Awardees who want to use existing funds; (2) awardees requesting an administrative or competing supplement; and (3) applicants or intramural researchers submitting applications or proposals. NIH funds may be used to derive human

191

pluripotent stem cells from fetal tissue. NIH funds may not be used to derive human pluripotent stem cells from human embryos. These Guidelines also designate certain areas of human pluripotent stem cell research as ineligible for NIH funding.

II. Guidelines for Research Using Human Pluripotent Stem Cells That Is Eligible for NIH Funding

A. Utilization of Human Pluripotent Stem Cells Derived From Human Embryos

1. Submission to NIH

Intramural or extramural investigators who are intending to use existing funds, are requesting an administrative supplement, or are applying for new NIH funding for research using human pluripotent stem cells derived from human embryos must submit to NIH the following:

a. An assurance signed by the responsible institutional official that the pluripotent stem cells were derived from human embryos in accordance with the conditions set forth in section II.A.2 of these Guidelines and that the institution will maintain documentation in support of the assurance;

b. A sample informed consent document (with patient identifier information removed) and a description of the informed consent process that meet the criteria for informed consent set forth in section II.A.2.e of these Guidelines;

c. An abstract of the scientific protocol used to derive human pluripotent stem cells from an embryo;

d. Documentation of Institutional Review Board (IRB) approval of the derivation protocol;

e. An assurance that the stem cells to be used in the research were or will be obtained through a donation or through a payment that does not exceed the reasonable costs associated with the transportation, processing, preservation, quality control and storage of the stem cells;

f. The title of the research proposal or specific subproject that proposes the use of human pluripotent stem cells;

g. An assurance that the proposed research using human pluripotent stem cells is not a class of research that is ineligible for NIH funding as set forth in section III of these Guidelines; and

h. The Principal Investigator's written consent to the disclosure of all material submitted under Paragraph A.1 of this section, as necessary to carry out the public review and other oversight procedures set forth in section IV of these Guidelines.

2. Conditions for the Utilization of Human Pluripotent Stem Cells Derived From Human Embryos

Studies utilizing pluripotent stem cells derived from human embryos may be conducted using NIH funds only if the cells were derived (without Federal funds) from human embryos that were created for the purposes of fertility treatment and were in excess of the clinical need of the individuals seeking such treatment.

a. To ensure that the donation of human embryos in excess of the clinical need is voluntary, no inducements, monetary or otherwise, should have been offered for the donation of human embryos for research purposes. Fertility clinics and/or their affiliated laboratories should have implemented specific written policies and practices to ensure that no such inducements are made available.

b. There should have been a clear separation between the decision to create embryos for fertility treatment and the decision to donate human embryos in excess of clinical need for research purposes to derive pluripotent stem cells. Decisions related to the creation of embryos for fertility treatment should have been made free from the influence of researchers or investigators proposing to derive or utilize human pluripotent stem cells in research. To this end, the attending physician responsible for the fertility treatment and the researcher or investigator deriving and/or proposing to utilize human pluripotent stem cells should not have been one and the same person.

c. To ensure that human embryos donated for research were in excess of the clinical need of the individuals seeking fertility treatment and to allow potential donors time between the creation of the embryos for fertility treatment and the decision to donate for research purposes, only frozen human embryos should have been used to derive human pluripotent stem cells. In addition, individuals undergoing fertility treatment should have been approached about consent for donation of human embryos to derive pluripotent stem cells only at the time of deciding the disposition of embryos in excess of the clinical need.

d. Donation of human embryos should have been made without any restriction or direction regarding the individual(s) who may be the recipients of transplantation of the cells derived from the human pluripotent stem cells.

e. Informed Consent

Informed consent should have been obtained from individuals who have sought fertility treatment and who elect to donate human embryos in excess of clinical need for human pluripotent stem cell research purposes. The informed consent process should have included discussion of the following information with potential donors, pertinent to making the decision whether or not to donate their embryos for research purposes.

Informed consent should have included:

(i) A statement that the embryos will be used to derive human pluripotent stem cells for research that may include human transplantation research;

(ii) A statement that the donation is made without any restriction or direction regarding the individual(s) who may be the recipient(s) of transplantation of the cells derived from the embryo;

(iii) A statement as to whether or not information that could identify the donors of the embryos, directly or through identifiers linked to the donors, will be removed prior to the derivation or the use of human pluripotent stem cells;

(iv) A statement that derived cells and/or cell lines may be kept for many years;

(v) Disclosure of the possibility that the results of research on the human pluripotent stem cells may have commercial potential, and a statement that the donor will not receive financial or any other benefits from any such future commercial development;

(vi) A statement that the research is not intended to provide direct medical benefit to the donor; and

(vii) A statement that embryos donated will not be transferred to a woman's uterus and will not survive the human pluripotent stem cell derivation process.

f. Derivation protocols should have been approved by an IRB established in accord with 45 CFR 46.107 and 46.108 or FDA regulations at 21 CFR 56.107 and 56.108.

B. Utilization of Human Pluripotent Stem Cells Derived From Human Fetal Tissue

1. Submission to NIH

Intramural or extramural investigators who are intending to use existing funds, are requesting an administrative supplement, or are applying for new NIH funding for research using human pluripotent stem cells derived from fetal tissue must submit to NIH the following:

a. An assurance signed by the responsible institutional official that the pluripotent stem cells were derived from human fetal tissue in accordance with the conditions set forth in section II.A.2 of these Guidelines and that the institution will maintain documentation in support of the assurance;

b. A sample informed consent document (with patient identifier information removed) and a description of the informed consent process that meet the criteria for informed consent set forth in section II.B.2.b of these Guidelines;

c. An abstract of the scientific protocol used to derive human pluripotent stem cells from fetal tissue;

d. Documentation of IRB approval of the derivation protocol;

e. An assurance that the stem cells to be used in the research were or will be obtained through a donation or through a payment that does not exceed the reasonable costs associated with the transportation, processing, preservation, quality control and storage of the stem cells;

f. The title of the research proposal or specific subproject that proposes the use of human pluripotent stem cells;

g. An assurance that the proposed research using human pluripotent stem cells is not a class of research that is ineligible for NIH funding as set forth in section III of these Guidelines; and

h. The Principal Investigator's written consent to the disclosure of all material submitted under Paragraph B.1 of this section, as necessary to carry out the public review and other oversight procedures set forth in section IV of these Guidelines.

2. Conditions for the Utilization of Human Pluripotent Stem Cells Derived From Fetal Tissue.

a. Unlike pluripotent stem cells derived from human embryos, DHHS funds may be used to support research to derive pluripotent stem cells from fetal tissue, as well as for research utilizing such cells. Such research is governed by Federal statutory restrictions regarding fetal tissue research at 42 U.S.C. 289g-2(a) and the Federal regulations at 45 CFR 46.210. In addition, because cells derived from fetal tissue at the early stages of investigation may, at a later date, be used in human fetal tissue transplantation research, it is the policy of NIH to require that all NIH-funded research involving the derivation or utilization of pluripotent stem cells from human fetal tissue also comply with the fetal tissue transplantation research statute at 42 U.S.C. 289g-1.

b. Informed Consent

As a policy matter, NIH-funded research deriving or utilizing human pluripotent stem cells from fetal tissue should comply with the informed consent law applicable to fetal tissue transplantation research (42 U.S.C. 289g-1) and the following conditions. The informed consent process should have included discussion of the following information with potential donors, pertinent to making the decision whether to donate fetal tissue for research purposes.

Informed consent should have included:

(i) A statement that fetal tissue will be used to derive human pluripotent stem cells for research that may include human transplantation research;

(ii) A statement that the donation is made without any restriction or direction regarding the individual(s) who may be the recipient(s) of transplantation of the cells derived from the fetal tissue;

(iii) A statement as to whether or not information that could identify the donors of the fetal tissue, directly or through identifiers linked to the donors, will be removed prior to the derivation or the use of human pluripotent stem cells;

(iv) A statement that derived cells and/or cell lines may be kept for many years;

(v) Disclosure of the possibility that the results of research on the human pluripotent stem cells may have commercial potential, and a statement that the donor will not receive financial or any other benefits from any such future commercial development; and

(vi) A statement that the research is not intended to provide direct medical benefit to the donor.

c. Derivation protocols should have been approved by an IRB established in accord with 45 CFR 46.107 and 46.108 or FDA regulations at 21 CFR 56.107 and 56.108.

III. Areas of Research Involving Human Pluripotent Stem Cells That Are Ineligible for NIH Funding

Areas of research ineligible for NIH funding include:

A. The derivation of pluripotent stem cells from human embryos;

B. Research in which human pluripotent stem cells are utilized to create or contribute to a human embryo;

C. Research utilizing pluripotent stem cells that were derived from human embryos created for research purposes, rather than for fertility treatment;

D. Research in which human pluripotent stem cells are derived using somatic cell nuclear transfer, i.e., the transfer of a human somatic cell nucleus into a human or animal egg;

E. Research utilizing human pluripotent stem cells that were derived using somatic cell nuclear transfer, i.e., the transfer of a human somatic cell nucleus into a human or animal egg;

F. Research in which human pluripotent stem cells are combined with an animal embryo; and

G. Research in which human pluripotent stem cells are used in combination with somatic cell nuclear transfer for the purposes of reproductive cloning of a human.

IV. Oversight

A. The NIH Human Pluripotent Stem Cell Review Group (HPSCRG) will review documentation of compliance with the Guidelines for funding requests that propose the use of human pluripotent stem cells. This working group will hold public meetings when a funding request proposes the use of a line of human pluripotent stem cells that has not been previously reviewed and approved by the HPSCRG.

B. In the case of new or competing continuation (renewal) or competing supplement applications, all applications shall be reviewed by HPSCRG and for scientific merit by a Scientific Review Group. In the case of requests to use existing funds or applications for an administrative supplement or in the case of intramural proposals, Institute or Center staff should forward material to the HPSCRG for review and determination of compliance with the Guidelines prior to allowing the research to proceed.

C. The NIH will compile a yearly report that will include the number of applications and proposals reviewed and the titles of all awarded applications, supplements or administrative approvals for the use of existing funds, and intramural projects.

D. Members of the HPSCRG will also serve as a resource for recommendations to the NIH with regard to any revisions to the NIH Guidelines for Research Using Human Pluripotent Stem Cells and any need for human pluripotent stem cell policy conferences.

Dated: August 17, 2000.
Ruth L. Kirschstein,
Principal Deputy Director, NIH.

* * *

Appendix E - Appendix N

The following commissioned papers were prepared at the request of the President's Council on Bioethics; the Council has not itself verified the accuracy of the information contained therein, nor does it necessarily endorse any of the authors' conclusions or opinions. Additionally, the Council has not edited these papers either for style or content.

Appendix E.

Legislators as Lobbyists:
Proposed State Regulation of Embryonic Stem Cell Research, Therapeutic Cloning and Reproductive Cloning

LORI B. ANDREWS, J.D.
Distinguished Professor of Law, Chicago Kent College of Law, Illinois Institute of Technology

The issue of research involving embryos and fetuses has been a matter of serious social concern for the past three decades. As each new scientific and medical development is introduced that affects embryos and fetuses, new state laws are proposed to deal with them. In the wake of the 1973 U.S. Supreme Court abortion decision in Roe v. Wade,[1] for example, 24 states passed laws prohibiting research on human conceptuses.[2] When *in vitro* fertilization (IVF) was accomplished, at least one state passed a law attempting to discourage the procedure.[3] After Dolly the cloned sheep was born, nine states adopted laws restricting human reproductive cloning.[4] When researchers began creating and using human embryos for stem cell research, states began introducing laws to deal with that technology as well. In 2002 and 2003, dozens of bills were introduced in state legislatures on cloning and stem cell research. In total, 38 states considered such bills;[5] some states

---

[1] Roe v. Wade, 410 U.S. 113 (1973).

[2] See Lori B. Andrews, Medical Genetics: A Legal Frontier 70 (Chicago: American Bar Foundation, 1987).

[3] See description of the Illinois law in Lori B. Andrews, The Clone Age: Adventures in the New World of Reproductive Technology 24-28 (New York: Henry Holt, 1999).

[4] See Section IA, infra.

[5] Alabama, Arizona, Arkansas, California, Colorado, Connecticut, Delaware, Florida, Georgia, Illinois, Indiana, Iowa, Kansas, Kentucky, Louisiana, Maryland, Massachusetts, Michigan, Mississippi, Missouri, Nebraska, New

considered multiple and conflicting bills. In all, 133 separate bills were introduced on these subjects. (See overview in Chart II and specific provisions in Chart III).

Although state legislatures have addressed regulation of research on the unborn for 30 years, the more recent technologies are being debated within a new context. Pro-life and pro-choice advocates are aligning in unusual ways. Unlike the clear line of demarcation between the groups on the issue of abortion, some pro-choice advocates oppose all forms of human cloning[6] and some pro-life advocates support so-called "therapeutic" cloning.[7] It would seem that these unique alliances might allow more fruitful policy discussions. However, despite agreement across traditional boundaries on the substance of particular bills, some of the new alliances break down because the sponsors of the bills include preambles that evince either pro-life or pro-biotechnology sentiments that go far beyond the substance of the bills themselves. This is a new development in the policy considerations related to human conceptuses. The legislators themselves are acting like lobbyists by including within their bills language espousing a particular position that goes far beyond the bills' actual legal substance.[8] Some bills seem to be driven more by political posturing than by an attempt to actually get legislation enacted to deal with the issue at hand.

This paper surveys the state law landscape by analyzing 1) existing laws with implications for reproductive cloning, therapeutic cloning and embryo stem cell research, 2) proposed bills addressing those technologies, and 3) potential constitutional challenges to the laws.

I. The Existing Legal Framework

A.   Existing Bans on Cloning

1.   The Restrictions

Hampshire, New Jersey, New York, North Dakota, Ohio, Oklahoma, Oregon, Pennsylvania, Rhode Island, South Carolina, Tennessee, Texas, Vermont, Virginia, Washington, West Virginia, Wisconsin.

[6] See, for example, Nigel Cameron and Lori Andrews, "Cloning and the Debate on Abortion," Chicago Tribune, August 8, 2001, at 17.

[7] Orin Hatch, for example.

[8] See Section II, infra.

There are eight states that have enacted explicit laws to ban reproductive cloning,[9] and one state that prohibits only the use of state funds for reproductive cloning.[10] (See Chart I). Of these eight states, four also ban "therapeutic" cloning.[11] Some states also ban shipping, transferring, or receiving for any purpose an embryo produced by human cloning.[12]

Under the existing reproductive cloning bans, some states ban the purchase or sale of an ovum, zygote, embryo, or fetus for the purpose of cloning a human being.[13] Other states, however, have bans on payment for eggs, embryos, or human tissue that could also cover some of the transactions involved in reproductive or "therapeutic" cloning.[14]

California's initial anti-cloning bill, adopted in 1998, created an advisory committee that met frequently over a five year period, heard extensive testimony from scientists and members of the public, and issued a report.[15]

---

[9] Ark. Code § 20-16-1001 et seq. (2003) (formerly AR SB 185); Cal. Bus. & Prof. §§16004, 16105 (2003) and Calif. Health & Safety §§ 24185-24187 (2003) (formerly CA SB 1230); Iowa Code § 707B.1-.4 (2003) (formerly IA SB 2046, became IA SB 2118); La. Rev. Stat. Ann. § 1299.36.1-.6 (2003); Mich. Comp. Laws §§ 333.26401-06, 333.16274, 16275, 20197, 750.430a (2003); ND Cent. 16.4-1 to .4-4 (2003); Va. Code Ann. § 32.1-162.21 to .22 (2003). Two of these state laws could be evaded if a non-human egg was used instead of a human egg as the incubation for the somatic cell DNA. Ark. Code § 20-16-1001(5) (2003); La. Rev. Stat. § 1299.36.1 (2003).

[10] Mo. Rev. Stat. § 1.217 (2003).

[11] Arkansas, Iowa, Michigan, and North Dakota.

[12] Ark. Code § 20-16-1002 (A)(3); Iowa Code § 707B.4(1)(c); ND Cent. Code §§ 12.1-39-01 to 02; Va. Code Ann. § 32.1-162.22(A) (for purposes of implantation).

[13] See, for example, Cal. Health & Safety Code §§ 24185(b); La. Rev. Stat. § 1299.36.2(B).

[14] Lori B. Andrews, "State Regulation of Embryo Stem Cell Research," in National Bioethics Advisory Commission, Appendix II at pages A1 – A13.

[15] "Cloning Californians?" Report of the California Advisory Committee on Human Cloning, January 11, 2002, Sacramento, California, available at http://www.sfgate.com/chronicle/cloningreport/.

2.   Exceptions for Certain Research

In some states, the ban on reproductive cloning is accompanied by language supporting other types of research or medical practices. In Louisiana and Michigan, for example, the cloning bans say that they do not prohibit scientific research on a cell-based therapy.[16] Similarly, some states allow the use of nuclear transfer or other cloning techniques to produce molecules, DNA, cells other than human embryos, tissues, organs, plants, or animals other than humans.[17] The Arkansas and Rhode Island laws also specifically allow *in vitro* fertilization and fertility enhancing drugs, so long as they are not used in the context of human cloning.[18] The Iowa law more broadly allows *in vitro* fertilization and the use of fertility drugs.[19]

3.   Penalties

The penalties for violation of existing cloning bans range widely. In Louisiana and Michigan, for example, penalties can be up to 10 years imprisonment and fines of $10 million for an entity such as a clinic or corporation and $5 million for an individual.[20] In

---

[16] La. Rev. Stat. Ann. § 1299.36.2(c); Mich. Comp. Laws § 750.430a(2).

[17] See, for example, Ark. Code § 20-16-1003(A); Iowa 707B.4(2); R.I. Gen Laws § 23-16.4-1 (specifically allows animal cloning in another section, R.I. Gen Laws. § 23-16.4-2(c)(1)(iii)). Virginia allows gene therapy, cloning of non-human animals, and cloning molecules, DNA, cells or tissue. Va. Code Ann. § 32.1-162.22(B).

[18] Ark. Code § 20-16.1003(B); R.I. § 23-16.4-2(c)(2)(i). Rhode Island's bill goes overboard in its hype about scientific progress. Rhode Island's preamble to its reproductive cloning ban states, "recent medical and technological advances have had tremendous benefit to patients, and society as a whole, and biomedical research for the purpose of scientific investigation of disease or cure of a disease or illness should be preserved and protected and not be impeded by regulations involving the cloning of an entire human being; and . . . molecular biology, involving human cells, genes, tissues, and organs, has been used to meet medical needs globally for twenty (20) years, and has proved a powerful tool in the search for cures, leading to effective medicines to treat cystic fibrosis, diabetes, heart attack, stroke, hemophilia, and HIV/AIDS." R.I. Gen. Laws. § 23-16.4-1.

[19] Iowa Code § 707B.4(2).

[20] La. Rev. Stat. § 1299.36.3; Mich. Comp. Laws § 750.430(a)(3) (in Michigan even individuals too, can be fined up to $10 million.) Rhode Island has a penalty of $250,000 for individuals and $1 million for entities. R.I. Gen. Laws § 23-16.4-3.

contrast, in Arkansas the penalty is a mere $250,000, but there are also felony criminal penalties.[21] Moreover, in some states, cloning can result in the permanent revocation of a doctor's license[22] and the denial of any other type of license or permit from the state regarding any trade, occupation or profession.[23]

B.   Existing Bans on Embryo Research

Twelve states' laws apply to *in vitro* embryos. In New Hampshire, the regulation of research on embryos prior to implantation is minimal.[24] The research must take place before day 14 post-conception,[25] and the subject embryo must not be implanted in a woman.[26] These stipulations could readily be met by researchers wanting to use IVF embryos as a source of stem cells.

Nine states ban research on *in vitro* embryos altogether,[27] and two states ban destructive embryo research.[28] In Louisiana, an *in vitro* fertilized ovum may not be farmed or cultured for research purposes;[29] the sole purpose for which an IVF embryo may be used is for human *in utero* implantations.[30] In the other states, embryo research is banned as part of the broader ban on all research involving live conceptuses. These laws ban embryo stem cell

---

[21] Ark. Code § 20-16-1002 (B) to (D).

[22] See, for example, La. Rev. Stat. § 1299.36.4; See also, Iowa § 707B.4(5).

[23] La. Rev. Stat. § 1299.36.4; Iowa § 707B.4(6).

[24] N.H. Rev. Stat. Ann. § 168-B:15.

[25] N.H. Rev. Stat. Ann. § 168-B:15 at I.

[26] N.H. Rev. Stat. Ann. § 168-B:15 at II.

[27] Fla. Stat. Ann. § 390.0111(6); La. Rev. Stat. Ann. § 9:121 et. seq.; Me. Rev. Stat. Tit. 22 § 1593; Mass. Ann. Laws. Ch. 112 § 12J; Mich. Comp. Laws. §§ 333.2685 to 2692; Minn. Stat. Ann. § 145.421 (applies only until 265 days after fertilization); N.D. Cent. Code §§ 14-02.2-01 to –02; 18 Pa. Cons. Stat. Ann. § 3216; R.I. Gen. Laws. § 11-54-1.

[28] S.D. Codified Laws § 34-14-16; Iowa Code § 707B.1-4.

[29] La. Rev. Stat. Ann. § 9:122.

[30] La. Rev. Stat. Ann. § 9:122.

research. The penalties are high — in some states, the punishment includes imprisonment.[31]

In some instances, the woman or couple who donate an embryo or fetus for research purposes also face liability. Laws in Maine, Michigan, North Dakota, and Rhode Island prohibit the transfer, distribution, or giving away of any live embryo for research purposes.[32] In Maine, a person who does so is subject to a fine of up to $5,000 and up to five years' imprisonment.[33]

C.   Existing Laws Specifically Addressing Embryo Stem Cell Research

A California law declares that research shall be permitted on human embryonic or adult stem cells from any source, including somatic cell nuclear transfer.[34] Such research must be reviewed by an approved institutional review board and such research may not be undertaken without written informed consent of the embryo donor. Interestingly, the law does not say consent of the "donors," plural. So it would appear that the female patient of infertility services (and not the woman and her husband) is the sole source of consent. California also enacted a law urging Congress to ban reproductive cloning, while permitting therapeutic cloning and embryo stem cell research.[35]

---

[31] The Maine law, which applies both to research on embryos and research on fetuses, carries a maximum five year prison term. Me. Rev. Stat. tit. 22 § 1593. The Massachusetts and Michigan laws also carry with them a potential prison sentence of up to five years. Mass Ann. Laws 112 § 12J(a)(V); Mich. Comp. Laws § 333.2691.

[32] Me. Rev. Stat. Ann. tit. 22 § 1593; Mich. Comp. Laws Ann. § 333.2690; N.D. Cent. Code § 14-02.2-2(4); R.I. Gen. Laws § 11-54-1(f). In addition, S.D. Codified Law § 34-14-7 bans the sale or transfer of an embryo for nontherapeutic purposes.

[33] Me. Rev. Stat. Ann. tit. 22 § 1593.

[34] Cal. Health & Safety Code § 125115.

[35] California Senate Joint Resolution 38.

II. Proposed Bills[36]

A.   Preambles to the Bills – Hype versus Doom

Legislators are currently considering bills that govern the technologies of cloning or embryo stem cell research. But even when the provisions of those bills are narrowly tailored to deal with those specific technologies, lawmakers include extensive preambles that include sweeping pro-life or pro-biotechnology language. None of the bills include a pro-choice preamble.

A Kentucky bill attempts to fuel a biotechnology sector in the state. Its preamble mentions the "great potential" of nuclear transplantation to treat "diseases and disorders, including but not limited to Lou Gehrig's disease, Parkinson's disease, Alzheimer's disease, spinal-cord injury, cancer cardiovascular diseases, diabetes, rheumatoid arthritis, and many others."[37] Its provisions would require anyone engaging in therapeutic cloning to register with the health department and pay a $50 fee. The bill also implores public colleges in Kentucky to protect "the interest of the Commonwealth or the institution relating to all intellectual property and other rights to any research, experiments, or other activity related to human nuclear transplantation. . . ."[38]

On the other hand, some bills' preambles are designed to cast aspersions on the technologies. The preamble of a Virginia Senate resolution contains a hodge-podge of historical facts (from slavery in Virginia 1619 to 1865 to Advanced Cell Technologies cloning of an embryo), Presidential statements (from President Reagan through President George W. Bush), and concerns about

---

[36] This research was undertaken in Lexis current session bill text and bill text archive for 2002 and 2003, using the words "human cloning," "stem cell research," "embryo," "cloning," "twinning," "oocyte," and "somatic cell nuclear transfer" in order to locate bills.

[37] KY HB 265.

[38] KY HB 265.

[39] "WHEREAS, although the "commodification" of human of human beings existed in this Commonwealth and the United States from 1619 to 1865, the concept of human beings as property has been rejected by Americans in their constitution and in their deeply held belief in the value of human life; and

WHEREAS, Ronald Reagan, the 40th President of the United States, stated this country understood that the personhood of every American should be protected "from the moment of conception until natural death"; and

WHEREAS, the United States Patent and Trade Office rejected human commercialization in a ruling on April 7, 1987, which stated "A claim directed to or including within its scope a human being will not be considered to be patentable subject matter" under the federal patent law; and

WHEREAS, on August 25, 2000, the National Institutes of Health published guidelines relating to stem cell research and the funding thereof that called for the denial of funding for research involving stem cells derived from embryonic human beings created for research purposes and noted that President Clinton, many members of Congress and the NIH Human Embryo Research Panel and the National Bioethics Advisory Committee had all endorsed the "distinction between embryos created for research purposes and those created for reproductive purposes"; and

WHEREAS, the NIH guidelines also called for assurances that "there can be no incentives for donation" of human embryos and "any inducement for the donation of human embryos for research purposes" would be prohibited; and

WHEREAS, George W. Bush, the 43rd President of the United States noted on August 9, 2001, that he is "deeply troubled" by the creation of "human embryos in test tubes solely to experiment on them," and described this act as a "warning sign" to "all of us" as Americans; and

WHEREAS, Senator William Frist, distinguished physician representing the State of Tennessee in the United States Senate, has proposed as a first principle of ethical research that "the creation of human embryos solely for research should be strictly prohibited"; and

WHEREAS, recently a Massachusetts research company claimed that it had cloned the first human embryo, that " [T]his work sets the stage for human therapeutic cloning as a potentially limitless source of immune-compatible cells," and that this work provides "hope for people with spinal injuries, heart disease and other ailments"; and

WHEREAS, the Jones Institute of Norfolk recently published research involving stem cell research conducted through the creation of approximately 110 embryos developed with the purchased sperm and eggs from men and women; and

WHEREAS, this conduct established a trade in new human life that treats such lives as merchandise for manipulation and destruction; and

WHEREAS, reportedly, the Jones Institute screened and evaluated the fitness of new human life according to the absence of "cosmetic handicaps, and other eugenic formulations"; and

WHEREAS, the General Assembly of Virginia has, by way of HJR 607 of 2001, condemned past practices within the Commonwealth involving institutional involvement in eugenics and eugenic ideology; now, therefore, be it

RESOLVED by the House of Delegates, the Senate concurring, That a joint subcommittee be established to study the medical, ethical, and scientific issues relating to stem cell research conducted within the Commonwealth." VA HJR 148. See also VA HJR 573.

eugenics and commodification of humans.[39] The resolution would establish a legislative task force, but its agenda is directed in a one-sided way. It is supposed to "examine the medical, ethical and scientific policy implications of prohibiting the creation of embryos *in vitro* for any purpose other than bringing them to birth, and the criminalizing of the transfer of compensation, in cash or in-kind, to induce any person to donate sperm or eggs for any purpose other than procreation. The joint subcommittee shall also examine the efficacy of research using adult stem cells rather than embryonic stem cells."[40]

Some bills assert scientific or medical facts that are not necessarily accurate. For example, Louisiana's resolution says, "stem cells derived from adult or placental tissues may hold greater promise than embryonic stem cells for human medicine, due to a lesser danger of immune reactivity against a recipient."[41]

B. Proposed Bills on Reproductive Cloning

1. The Restrictions

In 2002 and 2003, 27 states introduced bills that would ban reproductive cloning.[42] One bill details a specific enforcement mechanism for the human cloning ban. A New Jersey bill provides that the Commissioner of Health and Senior Services may issue a cease and desist order, with a subsequent administrative hearing within 20 days, with judicial review available.[43]

---

[40] VA HJR 148. See also VA HJR 573.

[41] LA House Current Resolution (HCR) 29.

[42] AL HB 9; AL SB 314; AZ HB 2108; CA SB 133; CA AB 267; CA SB 1557; CO HB 1073; CT SB 407; CT HB 5639; CT SB 410; DE SB 55; DE SB 329; DE SB 344; FL HB 285; FL SB 1726; FL HB 805; FL SB 1164; IL HB 253; IN HB 1538; IN HB 1984; IN SB 151; IN SB 138; KS HB 2736; KY HB 138; KY HB 153; KY HB 265; LA HB 472; LA HB 1810; MA HB 1280; MA HB 3125; MA HB 2048; MA HB 2052; MA SB 1917; MO HB 163; MO HB 209; MO HB 1449; MO SB 191; NE LB 602; NE LB 1067; NH HB 1464; NJ AB 2040; NJ AB 2840; NJ AB 1379; NJ SB 542; NY AB 1819; NY AB 3295; NY AB 4533; NY AB 6249; NY SB 206; NY SB 7638; NY SB 612; NY SB 3013; OK HB 1130; OK HB 2011; OK HB 2036; OK HB 2142; OK SB 1552; OR HB 2538; OR HB 2504; SC HB 3819; SC HB 4408; SC SB 820; TN HB 1075; TN SB 1515; TN HB 2675; TN SB 2295; TX HB 1175; TX SB 1034; TX SB 156; VT HB 326; WA HB 1461; WA SB 5466; WA HB 2173; WA SB 5571; WV HB 2832; WV SB 402; WV SB 514; WI AB 104; WI AB 246; WI SB 45; WI SB 699; WI AB 736; WI SB 404.

[43] NJ AB 2040.

Many of these proposed laws suffer from drafting infirmities. For example, some of the states' proposals create a loophole by only prohibiting the reproduction of a "genetically identical" individual through cloning.[44] Since the current cloning technique uses a donated egg to create a clone, the resulting individual will have some additional mitochondrial DNA from the enucleated egg, so he or she will not be genetically identical to the original individual.[45] In contrast, other states' bills prohibit the reproduction of a "virtually genetically identical" individual through human cloning using a human or non-human egg,[46] which would apply even if a donated egg, containing additional mitochondria were used.

Certain states' bills would prohibit the transferring of a human cell into a human egg cell.[47] But, in January 1998, scientists at the University of Wisconsin revealed that cow eggs could serve as incubators for nucleic DNA of other mammalian species.[48] Thus, the proposed laws could be evaded by making a human clone via somatic cell DNA transfer into an enucleated cow egg. Other states' bills prohibit human reproductive cloning using a human or non-human egg.[49]

A New Jersey bill has the opposite problem. Rather than being too narrow, it is too broad, using vague language in places and even potentially banning the creation of children through sexual intercourse.[50] The focus of the bill is to ban human cloning and germline genetic engineering. But it defines human cloning as the replication of one or more human beings, through "sexual or asexual reproduction." When two people create a child sexually are they replicating themselves? The key word, "replication," is not defined.

---

[44] See, for example, IN HB 1538 § 1; IN SB 151 § 1.a; MO HB 163 § 565.305; TN HB 1075 § 1; TN SB 1515 § 1.

[45] Unless, of course, one used an egg from the to-be-cloned person's own mother.

[46] See, for example, CA SB 133 § 1; FL SB 1726 § 2.A; KY HB 153 § 1.A; NE LB 602 § 2.2; OK HB 1130 § C.1; SC SB 820 § 2.1; SC HB 3819 §2.1; WI SB 45 § 1.c; WI AB 104 § 1.c.

[47] See, for example, DE SB 55 § 3002; IL HB 253 § 10.c; NH HB 1464 § 141-J:2; NY SB 206 § 3230; NY SB 7638 § 3230; OR HB 2538 § 1.

[48] See Robert Lee Hotz, "Cow Eggs Used as Incubator in Cloning Boon," L.A. Times, Jan. 19, 1998, at A1.

[49] See, for example, CA SB 133 § 1.

[50] NJ AB 2040 (1). See also MA HB 3125 § 2A.

The bill prohibits the destruction of human or non-human embryos "for the purpose of human cloning."[51] There is a certain vagueness to that provision. If a researcher working on primate cloning generates data that might be useful to cloning human beings could he or she be prosecuted?

2. Public funds

Six states' bills would prohibit the use of public funds and facilities to participate in cloning or attempted cloning.[52]

3. Other Means to Discourage Cloning

Some states' bills attempt to discourage reproductive cloning by prohibiting either the sale or purchase[53] or the shipment or receival of an ovum, zygote, embryo, or fetus to clone a human.[54] Some states' laws attempt to discourage cloning through liability provisions. A Florida bill would create vast liabilities as another means to discourage cloning.[55] Participants in human reproductive cloning would be liable to "the individual, the individual's spouse, dependents, and blood relatives, and to any woman impregnated with the individual, her spouse, and dependents, for damages for all physical, emotional, economic, or other injuries suffered by such persons at any time as a result of the use of human cloning to produce the individual." Even more dramatically, participants in human reproductive cloning would be liable to the cloned individual and his or her legal guardian for support until the age of majority (or longer, if the clone has "any congenital defects or other disability" related to the cloning).

---

[51] NJ AB 2040 (2)(c).

[52] IN HB 1538; IN HB 1984; IN SB 138; IN SB 151; LA HB 472; MS SB 2747; MO HB 481; MO SB 191; MO HB 163; NY AB 4533; OK HB 1130.

[53] See, for example, CA SB 133.

[54] See, for example, CO HB 1073; FL SB 1726; KY HB 153; OK HB 1130; SC HB 820; SC HB 3819; WI SB 45; WI AB 104.

[55] FL SB 1726.

4.   Exceptions

Eighteen states' bills have an exception allowing cloning to create DNA, tissues, and organs.[56] Eleven states' bills would not restrict the use of *in vitro* fertilization or fertility enhancing drugs.[57]

C.   Proposed Bills on Therapeutic Cloning

There are 26 states that have bills about therapeutic cloning.[58] Two states have at least one bill that would ban therapeutic cloning and at least one that would allow therapeutic cloning.[59]

1.   Ban Therapeutic Cloning

Twenty-six states have bills that would prohibit all human cloning including therapeutic cloning.[60] Some states prohibit therapeutic cloning through a general cloning ban that defines

---

[56] AL HB 9; AZ HB 2108; CA SB 133; CA SB 1557; DE SB 55 § 3003; DE SB 329; DE SB 344; FL HB 285; FL HB 805; FL SB 1164; FL SB 1726 § 5; IL HB 253; KS HB 2736; KY HB 138; KY HB 153 § 3; LA HB 1810; MA HB 2048; MA HB 3125; NE LB 1067; NH HB 1464 § 1441-J:4; NY AB 3295; NY AB 6249 § 2; NY SB 206 § 3232; NY SB 7638 § 3232; NY SB 3013; OK HB 2142; SC HB 3819 § E; SC SB 820; TN HB 1075 § e; TN SB 1515 § e; TX HB 1175; WA HB 1461; WA HB 2173; WA SB 5466.

[57] AZ HB 2108; IN HB 1538; IN HB 1984; IN SB 138; IN SB 151; KS HB 2737; LA HB 1810; NH HB 1464 § 1441-J:4; NY AB 6249 § 2; NY SB 206 § 3232; NY SB 7638 § 3232; NY SB 612; NY SB 3013; OK HB 2142; OR HB 2538 § 2; VA HB 2366; WA HB 2173; WI AB 736; WI SB 404.

[58] AL HB 9; AZ HB 2108; CA SB 133; CA SB 1557; CO HB 1073; CT SB 407; CT HB 5639; DE SB 55; FL HB 805; FL SB 1164; FL SB 1726; IL HB 253; IN HB 1538; IN HB 1984; IN SB 151; IN SB 138; KS HB 2736; KY HB 138; KY HB 153; LA HB 1810; MA HB 3125; MO HB 1449; NE LB 602; NH HB 1464; NJ AB 2040; NY AB 4533; NY AB 6249; NY SB 206; NY SB 3013; NY SB 7638; OK HB 1130; OK HB 2011; OK HB 2142; OK SB 1552; OR HB 2538; SC HB 3819; SC SB 820; SC HB 4408; TN HB 1075; TN SB 1515; TX HB 1175; TX SB 156; WA HB 2173; WA SB 5571; WV HB 2832; WV SB 402; WV SB 514.; WI SB 699; WI AB 104; WI AB 736; WI SB 404; WI SB 45. (Therapeutic cloning, as currently undertaken, would also fall within the ban on destructive embryo research discussed <u>infra</u>.)

[59] <u>See</u>, for example, Indiana and New York.

[60] AL HB 9; AZ HB 2108; CA SB 133; CA SB 1557; CO HB 1073; CT SB 407; CT HB 5639; DE SB 55; FL HB 805; FL SB 1164; FL SB 1726; IL HB 253; IN HB

human cloning as the creation of any human organism through somatic cell nuclear transfer.[61] Others prohibit therapeutic cloning by banning not only the creation of the cloned embryo, but also the "derivation of any product from human cloning."[62]

2. Allow Therapeutic Cloning

Two states have bills that specifically allow therapeutic cloning.[63] In addition, twelve other states allow for the use of stem cells from any source, including those derived from somatic cell nuclear transplantation.[64]

One state bill would encourage researchers to do stem cell research using only human adult or placental tissues obtained with informed consent, and would discourage research involving human embryonic or fetal tissue.[65] One state's bill urges state universities to refrain from embryonic stem cell research.[66]

D. Proposed Bans on Embryo Stem Cell Research

1. Ban Embryo Stem Cell Research

Five states have bills that would prohibit acts where a human fetus or embryo is destroyed or subject to injury.[67] New Jersey has

---

1538; IN SB 151; IN SB 138; KS HB 2736; KY HB 138; KY HB 153; LA HB 1810; MA HB 3125; MO HB 1449; NE LB 602; NH HB 1464; NJ AB 2040; NY AB 4533; NY SB 206; NY SB 7638; OK HB 1130; OK HB 2011; OK HB 2142; OK SB 1552; OR HB 2538; SC HB 3819; SC SB 820; SC HB 4408; TN HB 1075; TN SB 1515; TX HB 1175; TX SB 156; WA HB 2173; WA SB 5571; WV HB 2832; WV SB 402; WV SB 514; WI AB 104; WI SB 699; WI AB 736; WI SB 404; WI SB 45.

[61] See, for example, FL SB 1726.

[62] See, for example, NE LB 602.

[63] IN HB 1984; NY AB 6249; NY SB 3013.

[64] MA HB 1280; MA HB 2052; NJ AB 2840; NJ SB 1909; NY AB 1819; NY SB 612; TX SB 1034; VT HB 326; WA HB 1461; WA SB 5466.

[65] LA HCR 29A.

[66] MI HR 189.

[67] KS HB 2737; MI HB 4507; NJ AB 2040; VA HB 2366; WI AB 736; WI SB 404.

these provisions in a cloning bill, while Wisconsin has them in an artificial insemination/*in vitro* fertilization bill.[68] The others (Kansas, Michigan, and Virginia) have these provisions in stem cell research bills.[69] One state's bill would make all contracts for payment for the extraction or use of embryonic stem cells against public policy, void, and unenforceable.[70] In addition, a Michigan bill, in its preamble, opposes the White House decision to allow certain embryonic stem cell research.[71] The substance of the bill is a provision urging the state's public universities to refrain from embryonic stem cell research.[72]

2.   Allow Embryo Stem Cell Research, Including Therapeutic Cloning

Twelve states have bills that would allow embryo stem cell research from any source.[73] While the bills literally say "from any source," some also have a list saying this includes human embryonic stem cells, human embryonic germ cells, and human adult stem cells, from any source, including somatic cell nuclear transplantation.[74] In the states where proposed bills would permit embryonic stem cell research, there are often provisions regarding how the research should be conducted. In eleven states, for example, physicians undertaking *in vitro* fertilization on patients must discuss options for the disposition of frozen embryos, including donation to

---

[68] NJ AB 2040; WI AB 736; WI SB 404.

[69] KS HB 2737; MI HB 4507; VA HB 2366.

[70] VA HB 1361.

[71] MI HR 189.

[72] MI HR 189.

[73] CA SB 771; CA SB 1272; IL HB 3589; MD HB 482; MA HB 2052; MA HB 1280; NJ AB 2840; NJ SB 1909; NY AB 1819; NY SB 612; PA HB 2984; PA HB 422; PA HB 945 (Limited to those from fertility clinics); RI SB 266; TN HB 945; TN SB 1654; TX SB 1034; VT HB 326; WA HB 1461; WA SB 5466.

[74] See, for example, IL HB 3589 § 10.2.

[75] CA SB 771; CA SB 1272; IL HB 3589 (does not require Dr. to inform patient of embryo disposition options, but requires written informed consent); MA HB 2052; MA HB 1280; NJ AB 2840; NJ SB 1909; NY AB 1819; PA HB 2984; PA HB 422; PA HB 945 (Does not require Dr. to inform patient of embryo disposition options, but requires written informed consent); RI SB 266; TN HB 945; TN SB 1654; TX SB 1034; VT HB 326; WA HB 1461; WA SB 5466.

research, and written informed consent for such donation would have to be obtained.[75] In at least one state, the proposed research must be assessed by a state department,[76] and in eleven states the research must be assessed by an institutional review board.[77] A Kentucky bill requires that individuals engaged in therapeutic cloning must register their name, address, and phone number, which information shall be public record.[78] Under one proposed California bill, the State Department of Health Services would have the responsibility to develop guidelines regarding the derivation or use of human embryonic stem cells.[79]

In Illinois, care was taken in the proposed bill favoring embryonic stem cell research not to create barriers for the research due to contract or tort liability. Consequently, the proposed Illinois bill states that "procuring, furnishing, donating, processing, distributing, or using embryonic or cadaveric fetal tissue for research purposes pursuant to this Act is declared for the purposes of liability in tort or contract to be the rendition of a service by every person, firm, or corporation participating therein, whether or not remuneration is paid, and is declared not to be a sale of any such items and no warranties of any kind or description nor strict tort liability shall be applicable thereto."[80]

3.   **Address Embryo Donation**

Twelve states' bills would expressly allow a person to donate human embryos for research.[81] Twelve states' bills would also prohibit the sale of embryonic or cadaveric fetal tissue for research purposes.[82]

---

[76] PA HB 2984; PA HB 422.

[77] CA SB 322; CA SB 771; CA SB 1272; IL HB 3589; MD HB 482; MA HB 1280; MA HB 2052; NJ AB 2840; NJ SB 1909; NY AB 1819; RI SB 266; TN HB 945; TN SB 1654; TX SB 1034; VT HB 326; VA HB 639.

[78] KY HB 265.

[79] CA SB 771.

[80] IL HB 3589.

[81] CA SB 771; CA SB 1272; IL HB 3589 § 15; MD HB 482 § 20-901.C; MA HB 1280 § 3; MA HB 2052; NJ AB 2840 § 2.b.3; NJ SB 1909 § 2.b.2; NY AB 1819 § 2452.2; PA HB 2984 § 5.b; PA HB 422 § 5.b; PA HB 945; RI SB 266 § 23-77-2; TN HB 945; TN SB 1654; TX SB 1034 § 168.003; VT HB 326 § 9341.B; WA HB 1461§ 4; WA SB 5466 § 4.1.

4.   Review Committees

Six states' bills would establish a subcommittee or task force to review and/or study issues related to stem cell research.[83] Some of these states have a bill that only establishes a task force.[84] Some states establish this task force, subcommittee, or review board within a bill relating to stem cell research.[85] Maryland has one of each.[86]

It has often been true in the bioethical area that when controversial new technologies are introduced, states propose commissions to address the issue.[87] Often, the creation of the commission is a stalling tactic to avoid having to confront the issue head on or to postpone having to limit a technology, or to facilitate a moratorium on the technology (while the Commission evaluates the technology). Sometimes, though, state commissions have admirably overseen a technology's development, application and evaluation. Such was the case in the 1970's and 1980's with Maryland's newborn screening commission. So it is no surprise that of the bills proposing a state task force or commission, the Maryland bill provides the best guidance about what the commission would do. The Maryland task force would:

---

[82] CA SB 771; CA SB 1272; IL HB 3589 § 15a; IN HB 1538; IN SB 138; IN SB 151; MD HB 482 § 20-901; MA HB 1280; MA HB 2052; NJ AB 2840 § c.1; NJ SB 1909 § c.1; NY AB 1819 § 2453; PA HB 2984 § 6; PA HB 422 § 6; PA HB 945; TN HB 945; TN SB 1654; VT HB 326 § 9342; VA HB 2366 (If used in destructive research); WA HB 1461 § 5.2; WA SB 5466 § 5.2.

[83] CA SB 322; CA SB 771 (State department to oversee ESC research); MD HB 72 (Task force would cover bioscience in general, including ESC research); MD HB 1171; MA HB 1280 § 220; MA HB 2052; NY AB 6249 § 4372; NY SB 3013; PA HB 422 § 4; PA HB 2984 § 4; VA HJR 148; VA HJR 573; VA HB 639 (Task force would cover all human research in general, including ESC).

[84] See, for example, California, Maryland, and Virginia.

[85] See, for example, Illinois, Maryland, Massachusetts, New Jersey, New York, Pennsylvania, Rhode Island, Texas, and Vermont.

[86] MD HB 72; MD HB 1171.

[87] This was the case, for example, with newborn screening and surrogate motherhood.

"(1) Make recommendations for state policy regarding stem cells derived from embryos created solely for research purposes;

(2) evaluate whether there should be written informed consent requirements for prospective donors of embryonic stem cells derived from embryos created during infertility treatment;

(3) study constitutional, ethical, and policy issues with respect to whether the state should regulate or prohibit commerce in embryos or stem cells;

(4) (i) determine what stem cell research is already being done in the state by private companies and academic institutions; and

(ii) evaluate how these companies and institutions are regulating themselves when doing stem cell research;

(5) determine the effect on Maryland businesses and institutions of Federal restrictions that limit federally funded research and development to existing cell lines on stem cell research;

(6) analyze the roles of, and interrelationship between, federal and state oversight of stem cell research; and

(7) review any other matter relating to stem cell research that the task force considers necessary and proper."[88]

5.   Public Funding or Tax Credits for Stem Cell Research

States have also proposed bills allowing for or prohibiting the granting of state funds for stem cell research. California's proposed "Biomedical Research and Development Act of 2004" would award grants or loans to public or private institutions conducting research in stem cell biology.[89] To be eligible for funding, all research projects must be reviewed and approved by an institutional review board.

The states of Indiana, Michigan, Mississippi, and Nebraska have gone the opposite direction and introduced bills specifically

---

[88] MD HB 1171.

[89] CA SB 778.

prohibiting the use of state money or tax credits for stem cell research.[90] The Indiana bill proposes an amendment to the state tax code denying tax exemptions and credits for research involving human cloning or stem cell research with embryonic tissue.

A Mississippi House Bill and a Mississippi Senate Bill provide that no public funds shall be used to assist in or to provide facilities for stem cell research that uses cells from human embryos and human cloning.

Finally, although not establishing public funds for embryonic stem cell research, six Florida bills seek to establish a center for "Universal Research to Eradicate Disease" which will facilitate funding opportunities for stem cell research.[91]

E.   Proposed Bills on Health Care Providers' Right of Conscience

In addition to proposing bills on the prohibition of or allowance of embryonic stem cell research and cloning, a number of states have proposed bills protecting a health care provider, payer, or institution's right to refuse certain services based on moral or ethical convictions. Kansas, Louisiana, and Rhode Island have all introduced bills entitled "Health Care Providers Rights of Conscience." [92]

The Kansas bill introduced on January 28, 2002, specifically allows all persons to refuse to participate in the provision of, or pay for, a health care service related to human cloning and embryonic stem cell and fetal experimentation. The Louisiana bill and the identical Rhode Island House and Senate bills allow all health care providers, institutions and payers to decline to counsel, advise, pay for, provide, perform, assist, or participate in providing or performing health care services that violate their consciences, including, human cloning and embryonic stem cell research.

F.   Constitutionality of State Law

Not all regulations affecting research are constitutional. Laws restricting research on conceptuses may be struck down as

---

[90] IN HB 1001; MI SB 1485; MI HB 4454; MS HB 361; MS SB 2747.

[91] FL SB 2212; FL SB 572; FL SB 2390; FL HB 107A; FL HB 845; FL SB 2142.

[92] KS HB 2711; LA SB 850; RI SB 906; RI HB 5846.

too vague or as violating the right to privacy to make reproductive decisions. Such a challenge was successful in a federal district court case, Lifchez v. Hartigan,[93] which held that a ban on research on conceptuses was unconstitutional because it was too vague in that it failed to define the terms "experimentation" and "therapeutic."[94] The court pointed out that there are multiple meanings of the term "experimentation."[95] It could mean pure research, with no direct benefit to the subject. It could mean a procedure that is not sufficiently tested so that the outcome is predictable, or a procedure that departs from present-day practice. It could mean a procedure performed by a practitioner or clinic for the first time. Or it could mean routine treatment on a new patient. Since the statute did not define the term, it was unconstitutionally vague. It violated researchers' and clinicians' due process rights under the fifth amendment since it forced them to guess whether their conduct was unlawful.[96]

A similar result was reached by a federal appellate court assessing the constitutionality of a Louisiana law prohibiting nontherapeutic experimentation on fetuses in Margaret S. v. Edwards.[97] The appeals court declared the law unconstitutional because the term "experimentation" was so vague it did not give researchers adequate notice about what type of conduct was banned.[98] The court said that the term "experimentation" was impermissibly vague[99] since physicians do not and cannot distinguish clearly between medical experimentation and medical

---

[93] 735 F. Supp. 1361 (N.D. Ill. 1990).

[94] Id. at 1364. The court also held that the statute violated couples' right to privacy to make reproductive decisions to undertake preimplantation genetic screening, or procreate with a donated embryo. Id. at 1377. With embryo stem cell research, the progenitors' reproductive freedom is not an issue, however.

[95] Id. at 1364-65.

[96] Id. at 1364.

[97] Margaret S. v. Edwards, 794 F.2d 994 (5th Cir. 1986).

[98] Id. at 999.

[99] Id.

[100] Id. A concurring judge found this analysis to be contrived and opined that the provision was not unconstitutionally vague. Id. at 1000 (Williams,

tests.[100] The court noted that "even medical treatment can be reasonably described as both a test and an experiment."[101] This is the case, for example, "whenever the results of the treatment are observed, recorded, and introduced into the data base that one or more physicians use in seeking better therapeutic methods."[102]

A third case struck down as vague the Utah statute that provided that "live unborn children may not be used for experimentation, but when advisable, in the best medical judgment of the physician, may be tested for genetic defects."[103] The Tenth Circuit held "[b]ecause there are several competing and equally viable definitions, the term 'experimentation' does not place health care providers on adequate notice of the legality of their conduct."[104]

In a fourth case, Forbes v. Napolitano,[105] a number of plaintiffs[106] challenged the constitutionality of an Arizona statute that criminalized medical experimentation or investigation involving fetal tissue from induced abortions.[107] The statute at issue in that litigation, A.R.S. § 36-2302(A), provides: A person shall not knowingly

---

J., concurring). Instead, he suggested that the prohibition was unconstitutional because "under the guise of police regulation the state has actually undertaken to discourage constitutionally privileged induced abortions." Id. at 1002, citing Thornburgh v. American College of Obstetricians and Gynecologists, 106 S. Ct. 2169, 2178 (1986). The concurring judge pointed out that the state had "failed to establish that tissue derived from an induced abortion presents a greater threat to public health or other public concerns than the tissue of human corpses [upon which experimentation is allowed]." Id. Moreover, the state had not shown a rational justification for prohibiting experimentation on fetal tissue from an induced abortion, rather than a spontaneous one. Id.

[101] Margaret S. v. Edwards, 794 F.2d 994, 999 (5th Cir. 1986).

[102] Id.

[103] Utah Code Ann. § 76-7.3-310.

[104] Jane L. v. Bangerter, 61 F.3d 1493, 1501 (10th Cir.), rev'd on other grounds sub nom., Leavitt v. Jane L., 518 U.S. 137 (1996).

[105] 2000 U.S. App. LEXIS 38596 (9th Cir. 2000).

[106] Plaintiffs include numerous doctors and individuals suffering from Parkinson's disease who cannot receive transplants of fetal brain tissue because of the statute.

[107] 2000 U.S. App. LEXIS 38596 (9th Cir. 2000).

use any human fetus or embryo, living or dead, or any parts, organs, or fluids of any such fetus or embryo resulting from an induced abortion in any manner for any medical experimentation or scientific or medical investigation purposes except as is strictly necessary to diagnose a disease or condition in the mother of the fetus or embryo and only if the abortion was performed because of such disease or condition.[108]

An exception to this is found in A.R.S. § 36-2302(C), which reads:

> This section shall not prohibit any routine pathological examinations conducted by a medical examiner or hospital laboratory provided such pathological examination is not part of or in any way related to any medical or scientific experimentation.[109]

The penalty for violation of A.R.S. § 36-2302, a class 5 felony, is one and a half years in prison and fines up to $150,000.[110]

Plaintiffs argued that the statute prevented fetal tissue transplantation treatment of Parkinson's disease, certain fertility treatments, and development of treatments for illness.[111] The court ultimately affirmed the decision of the lower court[112] in holding that the statute was unconstitutionally vague.[113] Specifically, the "distinction between experiment and treatment in the use of fetal tissue is indeterminate, regardless of whether the tissue is obtained after an induced abortion."[114] Furthermore, the court determined that a criminal statute that serves to prohibit medical experimentation but provides no guidance as to where to draw the line between experiment and treatment does not provide doctors

---

[108] A.R.S. § 36-2302(A)(2000).

[109] A.R.S. § 36-2302(C)(2000).

[110] See A.R.S. § 36-2303(2000).

[111] 2000 U.S. App. LEXIS 38596, *5 (2000).

[112] Forbes v. Woods, 71 F. Supp.2d 1015, 1999 U.S. Dist. LEXIS 17025 (1999).

[113] 2000 U.S. App. LEXIS 38596, *12 (9th Cir. 2000).

[114] 2000 U.S. App. LEXIS 38596, *11 (9th Cir. 2000). The court determined that the "knowingly" scienter requirement within the statute did not serve as a clarification for the distinction.

[115] 2000 U.S. App. LEXIS 38596, *12 (2000).

with constructive notice, nor does it provide police, prosecutors, juries, and judges with standards for application.[115]

Specific bans on embryo stem cell research are unlikely to raise the same constitutional concerns. Part of the constitutional deficiency in the Arizona, Illinois, Louisiana, and Utah cases was that physicians offering health care services for their patients (such as embryo donation from an infertile woman, preimplantation genetic screening on an embryo, or treatment of a pregnant woman for diabetes) would not know if the activity would be considered by prosecutors to be experimental with respect to the embryo or fetus. Moreover, the female patient's reproductive freedom was implicated statute. Stem cell researchers do not have those potential legal arguments. There is a slight possibility, however, a reproductive liberty challenge could be raised against a law banning reproductive cloning.[116]

Conclusion

Bills on cloning and stem cell research are the subjects of vast media attention and public debate. As technologies develop that focus increasingly on the human embryo, lawmakers are having difficulty coming to agreement on legislative policy in the absence of societal accord about the moral and legal status of the embryo.

[Note from the President's Council Staff: Please see the accompanying Charts I and II detailing state laws and bills relevant to this area. In the interest of space, Chart III, referenced by the author, has not been included in this report. Additional information is available on the Council's website at www.bioethics.gov in the transcript of the author's presentation to the Council at its July 24, 2003, meeting.]

---

[116] Andrews, Lori B., "Is There A Right to Clone? Constitutional Challenges to Bans on Human Cloning," 11(3) Harvard Journal of Law and Technology 643 (1998).

Chart I **State Cloning Legislation**					
State	Reproductive cloning ban	"Therapeutic" cloning ban	Prohibition of State Funds	Permanent ban	Definition of "cloning"
Arkansas Ark. Code §20-16-1001 et. seq.	X	X		X	Asexual human reproduction, accomplished by introducing the genetic material from one or more human somatic cells into a fertilized or unfertilized oocyte whose nuclear material has been removed or inactivated so as to produce a living organism, at any stage of development, that is genetically virtually identical to an existing or previously existing human organism
California Cal. Bus. & Prof. §16004, 16105; Cal. Health & Safety §§24185-24187 (2003)	X			X	Nucleus transfer from a human cell from whatever source into a human or nonhuman egg cell from which the nucleus has been removed for the purpose of, or to implant, the resulting product to initiate a pregnancy that could result in the birth of a human being.
Iowa Iowa Code §707B.1-.4 (2003)	X	X		X	Human asexual reproduction, accomplished by introducing the genetic material of a human somatic cell into a fertilized or unfertilized oocyte whose nucleus has been or will be removed or inactivated, to produce a living organism with a human or predominantly human genetic constitution
Louisiana La. Rev. Stat. Ann. §1299.36.1-.6 (2003) Sunset Provision: July 1, 2003	X	Does not prohibit use of state funds for scientific research or cell-based therapies that do not involve human cloning	X		Identical to California
Michigan Mich. Comp. Laws §§333.26401-06, 333.16274, 16275, 20197, 750.430a (2003)	X	X	X	X	Use of human somatic cell nuclear transfer technology (transferring the nucleus of a human somatic cell into an egg from which the nucleus has been removed or rendered inert) to produce a human embryo

Chart I State Cloning Legislation (Cont.'d)					
Missouri Mo. Rev. Stat. §1.217 (2003)			X	X	Replication of a human person by taking a cell with genetic material and cultivating such cell through the egg, embryo, fetal and newborn stages of development into a new human person
North Dakota N.D. Cent. Code §§12.1-39-01-02 (2003)	X	X		X	Human asexual reproduction, accomplished by introducing the genetic material of a human somatic cell into a fertilized or unfertilized oocyte, the nucleus of which has been or will be removed or inactivated, to produce a living organism with a human or predominantly human genetic constitution
Rhode Island R.I. Gen. Laws §23-16.4-1 -.4-4 (2003) Sunset provision: July 7, 2010	X	Cell transfer and other cloning technologies not included in ban; mitochondrial, cytoplasmic and gene therapy not prohibited for research or animal creation			Use of somatic cell nuclear transfer for pregnancy prohibited (transferring the nucleus of a human somatic cell into an oocyte from which the nucleus has been removed)
Virginia Va. Code Ann. §32.1-162.21-.22 (2003)	X			X	Transferring the nucleus from a human cell from whatever source into an oocyte from which the nucleus has been removed or rendered inert in order to create a human being

	CHART II LAWS AND 2002-2003 BILLS				
State	Prohibit All Human Cloning (including therapeutic)	Prohibit Reproductive Cloning	Allow Therapeutic Cloning Specifically	Ban Destructive Embryo Research	Allow Embryo Stem Cell Research
AL	X	X			
AZ	X	X			
AR	Law	Law			
CA	X	X Law			X Law
CO	X	X			
CT	X	X			
DE	X	X			
FL	X	X			
GA					
HI					
ID					
IL	X	X			X
IN	X	X	X		
IA	Law	Law		Law	
KS	X	X		X	
KY	X	X			
LA	X	X Law			
ME					
MD					X
MA	X	X			X
MI	Law	Law		X	
MN					
MS					
MO	X	X			

	CHART II LAWS AND 2002-2003 BILLS (Cont.'d)				
State	Prohibit All Human Cloning (including therapeutic)	Prohibit Reproductive Cloning	Allow Therapeutic Cloning Specifically	Ban Destructive Embryo Research	Allow Embryo Stem Cell Research
MT					
NE	X	X			
NV					
NH	X	X			
NJ	X	X		X	X
NM					
NY	X	X	X		X
NC					
ND	Law	Law			
OH					
OK	X	X			
OR	X	X			
PA					X
RI		Law			X
SC	X	X			
SD					
TN	X	X			X
TX	X	X			X
UT					
VT		X			X
VA		Law		X	
WA	X	X			X
WV	X	X			
WI	X	X		X	
WY					

* The Laws that have already been enacted are indicated by the word "Law"
*The Bills are indicated by an "X"

The Meaning of Federal Funding

PETER BERKOWITZ, J.D., PH.D.[1]

I. INTRODUCTION

How should the government approach the question of public funding of activities that are deemed controversial by the American people? Is it appropriate to make such decisions on moral grounds? Can moral grounds for such decisions be avoided? If they can't, whose moral views, and of which sort, should govern, and with what consequences for those in the minority?

Questions of this sort have been frequently discussed in the wake of President Bush's 2001 decision regarding federal funding of embryonic stem cell research. In that decision, the President permitted federal funds to be used, for the first time, to support research on embryonic stem cells, but only those already in existence. At the same time, he made it clear that there would be no federal support for any research that involved or depended on any future destruction of human embryos. In so doing, he was upholding both the letter and the spirit of a Congressional enactment, the 1996 Dickey Amendment, which prohibited the creation of embryos for use in experiments, or the use of embryos in research that led to their destruction.

President Bush's decision has generated a great deal of controversy. Most scientists and patient advocacy groups believe that he made the wrong decision, and that the Dickey Amendment is a terrible mistake. Among the objections one commonly hears to the President's policy are several that concern the meaning of federal funding:

(1) By withholding federal funding for research that involved the creation of *new* embryos or the future destruction of embryos in existence, the President has effectively banned embryonic stem cell research.

---

[1]Peter Berkowitz teaches at George Mason University School of Law and is a research fellow at Stanford's Hoover Institution. He has served as a senior consultant to the President's Council on Bioethics.

(2) The decision was wrong because the President allowed his personal moral views to govern federal policy. Or, along the same lines, the congressional ban is wrong because it represents the imposition of moral views—religiously based moral views at that— to frustrate sound and beneficial public policy.

(3) The decision is morally incoherent, for if an act is so immoral as to deserve the governmental disapproval implicit in withholding funding, it should be accompanied by efforts to prohibit the activity altogether.

Whatever the merits of the current law, or the President's 2001 stem cell decision, these objections, once closely examined, cannot pass muster. The first confuses a limitation on funding with the imposition of a ban or prohibition. The second wrongly supposes that legislating morals through federal budget decisions is always or generally wrong. And the third incorrectly assumes that government has an obligation to bring an end to all conduct it believes immoral.

Explaining these errors requires an exploration of the meaning of all government funding decisions. Such an exploration can not decide the difficult question of the merits of the President's stem cell policy. It can, however, put to rest the objections built around the claim that the policy somehow violates the letter or the spirit of sound constitutional government.

II. FEDERAL FUNDING

A. Basic Considerations

The common objections to the President's policy fail to come to grips with what government funding in a liberal democracy really means. Several fundamental features of our constitutional system need to be emphasized.

First, no one and no activity has a constitutional right to federal funding. There is no governmental obligation to fund most activities, not even the most worthy, save for such matters as the Constitution explicitly proclaims to be the responsibility of government, such as national defense, the maintenance of federal courts, the holding of elections, and so on. And even concerning these constitutional essentials, it is an open question, to be resolved by our elected representatives, of how government will choose to allocate taxpayer dollars.

Second, no individual or cause has a right to sit at the government trough. Resources are scarce, and insufficient to support all worthy activities. People with different causes and interests compete to obtain them, and in order to succeed they are forced to bring their case to members of Congress. Funds are distributed only through the political process, within limits set by the Constitution, as the result of deliberation, lobbying, deal-making, and the like.

Third, in a healthy democracy people will always have disagreements about what activities should receive government funding. Sometimes the disagreements will be intense, and sometimes not. Sometimes the disagreements will include moral disagreements, and sometimes not. Sometimes the political process will generate a stable compromise on the issue, and sometimes one side or the other decisively prevails.

Fourth, while the Constitution prohibits the government from establishing religion, it does not interfere with the right of citizens to form moral judgments based on their deeply held religious beliefs, and to persuade a majority of their fellow citizens to enact legislation informed by those moral judgments, provided of course that the legislation does not interfere with other constitutionally guaranteed rights.

Fifth, those who fail in the democratic process to obtain federal funding will always feel that they did not get what they need or want, but in the absence of a clear legal entitlement to such funding, they cannot properly complain that the government has thereby denied their rights or interfered with their liberty to exercise them.

Sixth, those who lose have several alternatives built into the democratic process. They can try to persuade their representatives in Congress to reconsider, they can vote in others more sympathetic to their cause, they can seek to influence public opinion, or they can seek non-government funding for their activities.

All of this is straight-forward and uncontroversial. It suggests the legitimacy, indeed the routine character, of the President's policy. It might be regarded as the end of the story.

B. Are There Special Cases?

Although the framework laid out above may correctly describe the situation for most or even all federal funding decisions, there are moral and political reasons why people might regard, for example, withholding of support for selected aspects of biomedical research as a special case, an exception that demands a different approach.

The nation strongly and overwhelmingly backs biomedical research. And we generally leave the mapping of research strategies to scientists and those who administer the institutions in which they work. The entire biomedical enterprise in the US, including also the training of the next generation of scientific researchers, has come to depend heavily on government support. The public generally favors this arrangement, and relies on government-funded research for the treatment and for the cure of all still untreatable diseases, such as cancer and Alzheimer's disease.

Consequently, the decision to withhold public funds from any particular piece of the biomedical research portfolio looks and feels, both to scientists and to the public, like an intrusion of government into a place where it does not belong, and it prompts harsh accusations that government is engaging in censorship or even outright prohibition of medically necessary scientific research. To be sure, the FDA regularly imposes restrictions on research, but mainly on grounds of safety. When, however, government's objection to research is moral in nature, it strikes scientists as a deprivation: a restriction of freedom to inquire, a thwarting of worthy community goals, an intrusion of morals into a sphere where they do not belong. At the same time, it appears to those members of the public who disagree with the decision as a failure by the government to abide by its putative moral obligation to use its resources to explore *all* fruitful areas of research in search of cures for dread diseases.

Moreover, there is reason to single out for special attention those decisions about federal funding where powerful moral principles are at loggerheads, and the nation is deeply and passionately divided. This is the case of stem cell research. It poses a confrontation between genuine and conflicting goods: on the one side, respect for nascent human life and, on the other side, commitment to unfettered scientific inquiry and to the fight against disabling and deadly disease. The clash between those who hold that the moral status of the embryo is no different from that of a fully developed human being, and those who believe that the embryo is a clump of cells, utterly devoid of moral worth is not resolved by appeal to shared moral premises. This is because what the debate over stem cell policy calls into question is how to *apply* our shared belief in the rights and dignity of the individual.

Despite the powerful presumption in favor of federal funding of biomedical research, and the moral stakes which both sides see as exceedingly high, the controversy over federal funding of stem cell research does not present a special case. In politics, though, how a policy appears is important. For this reason, and despite the fact that the common arguments condemning the president's policy rest on false assumptions or unreasonable expectations, and though they may be mere rationalizations for the failure to win the policy battle, it is worth examining their flaws fully.

III. MORALS, FEDERAL FUNDING, AND LEGISLATION

A. Federal Funding

In the first place, federal funding is about resource distribution—who and what will get how much of the nation's scarce taxpayer dollars. It is usually not about restricting basic rights. For example, there is no constitutional right to the funding of biomedical research.

But often the question of whether government will or will not fund an activity is about more than mere distribution. It is about government shaping choices among various and competing goods or undertakings. It is a statement of approval and encouragement by government, a declaration by the nation that an activity or undertaking is meritorious and has priority. Or, in the decision to withhold funds, government policy can be a statement of disapproval and discouragement, a declaration by the nation that a permitted activity or undertaking lacks merit or has low priority.

Policy decisions about funding resemble policy decisions about taxing. Both sorts of decisions create incentives and disincentives by making activities more or less costly. The child tax credit, for example, reduces the financial cost of child rearing. In so doing, it strengthens families in two ways: it enables families to save money, and it conveys an important message about the political importance of the well-being of the family. Similarly, government funding of research into disease and its prevention and treatment increases the supply of these goods, and reflects our nation's considered judgment that the relief of physical suffering is a high national priority.

While all law either requires, forbids, or permits, the provision or withholding of funding and the use of tax incentives and disincentives allows government to express a range of attitudes toward that which it permits. In the United States, through such decisions government strongly endorses charity and higher education. It looks favorably on national service and the arts. It shows a preference for marriage to cohabitation. It frowns upon smoking. It is the distinction between permitting or tolerating an activity (which is the case with embryo research and destruction) and actively promoting it through governmental funding (which the president's policy on stem cells prohibits), that is crucial to understanding the president's stem cell research policy, but not only to the stem cell controversy.

The question of federal funding routinely implicates questions about the nation's *moral* priorities among permissible activities. And the question of moral priorities in politics is not so simple as a question of good or bad; rather it is a question of better or worse. One consequence of this is that in sorting out funding decisions it is not a matter of one side introducing moral considerations. Most of the time, both sides in disputes over policy are of necessity engaged in making moral arguments.

This is true, and to an extraordinary degree, in the stem cell controversy. *Both* sides–those who wish to defend the rights of nascent human life, and those who wish to defend unfettered scientific research directed at the relief of human suffering through the cure of deadly disease–defend moral principles. To make matters more difficult, *both* sides tend to defend those principles in their absolute form. While the president's position attempts to give weight to *both* sides' principles, *both* sides, because of the passion with which each holds its principle, are dissatisfied. However, insofar as the president's approach reflects that adopted by the Dickey Amendment, it follows a determination made by a majority of the people's representatives, and with that *both* sides should be satisfied. But they aren't.

But the dissatisfaction of both sides takes a recognizable form. In general, because moral principles are so frequently at stake in the fight for federal taxpayer dollars, funding decisions create bitterness. This truer still, as in the stem cell debate, when the moral principles are wielded in their absolute form, so that *both* sides can claim defeat. If funding is withheld, those who believe the activity is worthy can claim that their tax dollars, which they contribute in the hope that they will serve the good of the country, are being held back from what they deem a deserving or even overriding moral purpose. This is the position taken by many scientists and progressives with regard to the limitations imposed by the President's policy. If funding is provided, those who believe the activity is immoral can claim that their tax dollars are being used to advance a cause they believe is unworthy, or even abhorrent. This is the position taken my some social conservatives who believe that the limitations imposed by the President's policy did not go nearly far enough and pave the way, sooner rather than later to the routine creation and destruction of human embryos for biomedical research. Typically, both sides make moral claims, and one or the other–and in the really tough cases such as stem cell policy, *both* sides–will have to live with the fact that their moral principles are being rejected (if not assaulted) by the government, in their own name and with their own tax dollars.

Why should those who lose the political struggle put up with this? For the simple reason that living in a liberal democracy means sometimes being in the minority, even on questions of the utmost importance, but so long as the laws which one opposes are consistent with the Constitution and enacted according to legally appropriate procedures, one has an obligation to obey them.

B. Legislation

Is it really a legitimate aim of a liberal democracy to adopt laws and take actions to shape the moral beliefs of its citizens? Perhaps federal funding is the exception, and to the extent possible the moral dimension should be eliminated from policy formation. Doesn't government in a liberal democracy have an obligation to remain neutral toward competing conceptions of a good life, and so refrain from enacting morals into law? Otherwise, doesn't it impermissibly infringe on people's right to choose how to live their lives.

According to a common and sound criticism of this common view of the liberal state, such neutrality is a chimera: it is impossible for any government to remain neutral about morality and the nature of a well-lived life, since the resolution of controversies over public policy–for what purposes is the state permitted to classify citizens by race? what is the meaning of marriage? to what extent may the public schools engage in civic education?– always draw upon, reinforce, or suppress a view about what is deserving, proper and good. It is possible, as a matter of policy, to tolerate a wide variety of choices and forms of life, but toleration itself is a moral principle based on a certain interpretation of how to secure human freedom and respect the dignity of the individual.

Some then object that because of its very foundational commitments, our liberal democracy privileges the autonomous or freely choosing life. And so in sense it does. But it need not and should not do this unwittingly or surreptitiously. The mistake is to think that liberal state stands or falls with the commitment to neutrality. It doesn't. It stands or falls with the commitment to creating the conditions under which individuals can exercise political freedom.

Law and public policy in a liberal democracy properly seek to create conditions in which citizens can make informed and responsible choices. It does this in a variety of ways. The first and most taken for granted is through the establishment of public order. It also does this through establishing a system of public schools, promoting research in the arts and sciences, and enacting a wide variety of social and economic legislation, all with a view to forming a citizenry that is at home in, and capable of taking advantage of, freedom. Legislation designed to encourage biomedical research can be seen as creating circumstances in which we are better able to enjoy the blessings of freedom. The same can be said of laws designed to respect, and encourage respect, for nascent human life, which can reasonably be understood as contributing to the conditions under which individuals learn to respect humanity in others and in themselves.

To be sure, even within the limits provided by law, government's encouragement of informed and responsible choice can easily become a tool for the ill-conceived circumscribing of choice. Even well meaning government efforts to help prepare citizens for liberty and toleration can undermine both. Government funded education can be dogmatic and ideological; government funded research may be biased and unaccountable; government supported arts may disseminate tawdry or jingoistic sentiments and images; government funded programs directed at the family may fail to adapt to changing times. Of course, these familiar abuses are not arguments against government promoting the conditions that enable citizens to take advantage of freedom. Rather, they are reasons for proceeding with care, and with an appreciation of the complexities of contemporary moral and political life.

IV. AMERICAN DILEMMAS

The president's policy on stem cells is not the only funding decision in contemporary American politics that has generated controversy. Brief discussion of others sheds light on what is common to all and what is distinctive to the stem cell debate.

Consider first the battle over abortion, which involves a long standing struggle over the question of government funding for lawful conduct. Shortly after entering office, President Bush ordered the withholding of funding from international organizations that performed abortions, a decision that was neither required of him nor forbidden to him but within his discretion. The principle behind this policy is common to his position on stem cell research: government funds should not be used to destroy nascent human life.

At home, a line of Supreme Court decisions stretching from 1977 to 1991dealing with abortion and government funding established the principle that the constitution does not require government to fund activities that the Constitution protects. In *Maher v. Roe* 432 U.S.464 (1977), the Court held 6-3 that Connecticut could provide Medicaid benefits for childbirth while withholding benefits from women who wished to have non-medically necessary abortions. Justice Powell, writing for the majority, maintained that the right to abortion announced in *Roe v. Wade*, "protects the woman from unduly burdensome interference with her freedom to decide whether to terminate her pregnancy. It implies no limitation on the authority of a State to make a value judgment favoring childbirth over abortion, and to implement that judgment by the allocation of public funds." Powell's analysis emphasized the "basic difference between state interference with a protected activity and state encouragement of an alternative." In dissent, Justice Brennan disagreed vociferously. He argued that the denial of funds unconstitutionally interfered with the right of women to choose an abortion. Also in dissent, Justice Marshall argued that by withholding funds for non-medically necessary abortions, Connecticut was seeking "to impose a moral viewpoint that no State may constitutionally enforce."

In *Harris v. McRae* 448 U.S. 297 (1980), by a 5-4 margin, the Court upheld the Hyde Amendment, which banned federal funding of medically necessary abortions. The majority's argument was much the same as in *Maher*: "although government may not place obstacles in the path of a woman's exercise of her freedom of choice, it need not remove those not of its own creation." The government, the majority reasoned, does not have an obligation to provide taxpayer dollars so that individuals can exercise their individual rights to the maximum. To this, Justice Brennan replied in dissent that Hyde Amendment actually left poor women in a worse off position. By refusing to provide poor women with funding for even medically necessary abortions while subsidizing childbirth, the government demonstrated profound disapproval for abortion and thereby burdened the exercise of the right to privacy declared in *Roe*.

In *Rust v. Sullivan* 500 U.S. 173 (1991), again by 5-4 margin, the Court upheld federal regulations that barred health care professionals who received federal funding from offering counseling about abortion. Chief Justice Rehnquist, writing for the majority, reiterated the *Maher* and *Harris* principle: "The Government has no constitutional duty to subsidize an activity merely because the activity is constitutionally protected." In dissent, Justice Blackmun, joined by Justices Stevens and Marshall, echoing the dissenters in *Maher* and *Harris,* insisted that by conditioning federal funding on the withholding of counseling about abortion, the government was actually placing "formidable obstacles" in the path of women's exercise of their privacy rights.

While these cases appear closely analogous to the stem cell controversy, they differ in a crucial respect. Although some of the same forces are politically engaged, and although the moral issue concerns the question of the inviolability of nascent human life, the abortion cases would have been unlikely to come to the Court for adjudication had the Court not declared in *Roe v. Wade* a constitutional right to an abortion. For the argument made by those who seek federally funded abortions is that by withholding funding, the government is seeking to frustrate the exercise of a constitutionally protected right. Absent such a right, there could be no plausible legal claim. Indeed, absent an entitlement to government sponsored health care benefits, there is no valid legal claim that Medicare *must* pay for cosmetic surgery, sex-change operations, contraceptive benefits, heart transplants, or any other procedure one wants for oneself and can find a doctor to do. Only if there were a constitutionally protected right not to be poor, not to be without resources to fully take advantage of all the things that we are legally entitled to pursue, could such a claim prevail as a matter of law. While there is no such right, this is just the kind of claim that dissenters in the Court's cases concerning federal funding and abortion defend. In short, because the Constitution provides no special protection to biomedical research, the argument for legal entitlement to funding of stem cell research proceeds on dramatically weaker grounds than the rejected arguments in the abortion funding cases.

Title VI of the Civil Rights Act of 1964 furnishes another example of how government may withhold funds from practices it does not outlaw. It provides that, "No person in the United States shall, on the ground of race, color, or national origin, be excluded from participation in, be denied the benefits of, or be subjected to discrimination under any program or activity receiving Federal financial assistance." It is this provision that requires private universities to avoid those racial classification in admissions and hiring that would violate the prohibitions imposed on *state* action by the equal protection clause of the 14th Amendment. Title VI is far reaching, because most private universities rely heavily on government funding for the support of basic research. And it provides a way for the federal government to shape the moral contours of what is largely private conduct, and bring that conduct in line with fundamental constitutional principles. Of course private institutions are free to continue to practice activities that disqualify them for federal funding. All they have to do is refuse to take federal funds.

Close in form to federal policy on stem cell research are social security regulations regarding marriage and survivor benefits. For example, although cohabitation without matrimony is not illegal, indeed it is quite common, the federal government refuses to pay social security survivor benefits to all but legal spouses. This is a way for government to provide financial incentives for marriage. And for government to take sides on the value of marriage, proclaiming the union marked by it as good for individuals and good for the polity. This is a policy decision that does not engender bitterness or controversy. It must be acknowledged that the withholding of a reward could, under imaginable circumstances, stigmatize those who choose to live together as a loving couple but not to marry. But just as it can not plausibly be claimed today that the child tax credit confers social disapprobation on married couples without children, so too it cannot be plausibly claimed that unmarried couples suffer social disapprobation because of government policy that restricts the paying of social security survivor benefits to legal spouses.

Or, from a different angle, consider the question of elementary level and high school education. In 1923, in a landmark decision, *Meyer v. Nebraska* 262 U.S. 390 (1923), the Supreme Court ruled that parents have a right to educate their children in a foreign language. In 1925, in a related case, *Pierce v. Society of Sisters* 268 U.S. 510 (1925), the Supreme Court ruled that parents have a right to educate their children in private schools. But nobody concludes that the rights that these cases protect *prohibit* states from policy decisions encouraging public education. And nobody claims that the right of parents to privately educate their children creates an *entitlement* to have that private education funded by the government.

As these examples illustrate, the controversy over stem cells should be seen as one among many political battles over the allocation of limited federal funds. It is distinguished not by the presence of moral principles, or the presence of moral principles on both sides, but by the particular moral principles at stake, the absoluteness with which they are wielded, and the intensity of the passions their defense provokes.

V. CONCLUSION

When the question of federal funding is placed in perspective, it can be seen that the common objections to the President's policy on stem cell research are misplaced.

First, by withholding federal funding for research that involved the creation of new embryos or the future destruction of embryos, the President did not effectively *ban* embryonic stem cell research. His August 2001 decision for the first time provided federal funding for stem cell research, and it permitted private individuals and companies to pursue it.

Second, by basing his policy in part on moral considerations, the President did not violate an obligation to keep morals out of politics, because funding decisions, whichever way they go, typically and unavoidably contain a moral component. Indeed, the moral component often lies at the heart of the dispute and at the core of the decision.

Third, by refusing to seek a blanket prohibition on an activity from which he withheld funding on moral grounds, the President did not make an incoherent decision. The complexities of a free society frequently create situations in which it makes sense for government to express doubt, anxiety or disapproval for an activity it is unwilling or unable to outlaw.

None of this is to deny that the president's policy on stem cell research is open to criticism on the merits. It is only to claim that the policy reflects a perfectly appropriate exercise of governmental powers in a liberal democracy.

Appendix G.

Report on the Ethics of Stem Cell Research

PAUL LAURITZEN, PH.D.

Director, Program in Applied Ethics, Professor, Department of Religious Studies, John Carroll University, University Heights, Ohio

> ...the final stage is come when man by eugenics, by prenatal conditioning, and by an education and propaganda based on perfect applied psychology, has obtained full control over himself. Human nature will be the last part of nature to surrender to man. (Lewis, 1947)

> This sudden shift from a belief in Nurture, in the form of social conditioning, to Nature, in the form of genetics and brain physiology is the great intellectual event, to borrow Nietzsche's term, of the late twentieth century. (Wolfe, 2001)

I begin with passages from an unlikely pair of authors because although C. S. Lewis and Tom Wolfe are somewhat distant in time, certainly different in temperament, and extravagantly different in personal style, they share an imaginative capacity to envision the possible consequences of modern technology. The technology that occasioned Lewis's reflections, "the aeroplane, the wireless, and the contraceptive" may now seem quaint, but his warning about turning humans into artifacts, that accompanied the passage quoted above, is eerily prescient. Similarly, although he does not directly take up stem cell research, Tom Wolfe's reflections on brain imaging technology, neuropharmacology, and genomics are worth noting in relation to the future of stem cell research. In his inimitable way, Wolfe summarizes his view of the implications of this technology in the title of the essay from which the above passage comes. "Sorry," he says, "but your soul just died."

The point of beginning with Lewis and Wolfe, then, is not that I share their dire predictions about the fate to which they believe technology propels us; instead, I begin with these writers because they invite us to take an expansive view of technology. I believe that such a perspective is needed and is in fact emerging in recent work on stem cell research. This is not to say that the sort of traditional analysis that has framed much of the debate on stem cells, analysis that involves issues of embryo status, autonomy, and informed consent, for example, is unhelpful; far from it. Nevertheless, traditional moral analysis of stem cell research is nicely complemented by a consideration of the "big picture" questions that Lewis and Wolfe both wish to press. This report will therefore seek to draw attention to the literature on stem cell research that attends both to the narrow and to the expansive bioethical issues raised by this research.

<u>The Moral Status of the Embryo</u>

There is little doubt that public reflection on stem cell research in the United States has been affected by the extraordinarily volatile cross-currents of the abortion debate. Although I will indicate below several reasons why framing the stem cell debate as a subset of that on abortion is problematic, nevertheless, in its current form, stem cell research is debated in terms dictated by the abortion controversy, and that has meant that questions about the status of the embryo have been particularly prominent.[1] For example, the National Bioethics Advisory Commission (NBAC) described the ethical issues raised by stem cell research as "principally related to the current sources and/or methods of deriving these cells" (NBAC, 1999, 45). A policy brief from the American Association for the Advancement of Science (AAAS) begins its discussion of the ethical dispute over stem cell research by citing the disagreement over the status of the embryo as the decisive variable leading to fundamentally different views on this research (AAAS, 1999, 11). The National Academy of Sciences's report on stem cell research claims that "the most basic [ethical] objection to embryonic stem cell research is rooted in the fact that such research deprives a human embryo of any further potential to develop into a complete human being" (National Academy of Sciences, 2002, 44). The Ethics Advisory Board of the Geron Corporation lists the moral status question as the first moral consideration relevant to deciding the acceptability of stem cell research (Geron Corporation Ethics Advisory Board, 1999, 32). The list could go on.

Despite the fact that these statements all insist on the importance of the status question, they also recognize that the debate about the status of the early embryo is not new and that the controversy over stem cell research does not, strictly speaking, raise novel issues in this regard. Indeed, it is probably best to place the initial skirmishes over stem cell research in the context of moral debates about human embryo research generally. In fact, it is worth noting that the report of the NIH Human Embryo Research Panel (HERP), published in 1994, explicitly identified the isolation of human embryonic stem cells as one of thirteen areas of research with preimplantation embryos that might yield significant scientific benefit and that should be considered for federal funding (See NIH HERP, 1994, ch. 2).

Although the recommendations of the HERP were never implemented, the fact that a high-profile panel reviewed ex utero preimplantation human embryo research and explicitly endorsed stem cell research, meant that the panel report would affect the policy debate about stem cell research, even though its recommendation that the derivation and use of stem cells be federally funded was not adopted. For one thing, the panel's anticipatory support for stem cell research assured that when human stem cells were actually derived several years later, the debate that ensued would be tied to the abortion controversy. As members of the HERP panel have made clear, from the start, the work of the panel was embroiled in controversy. For example, shortly after the HERP was impaneled, thirty-two members of Congress wrote to Harold Varmus, the director of NIH, to complain about the composition of the panel. A lawsuit was filed in an attempt to prevent the panel from meeting, and members of the panel received threatening letters and phone calls (Green, 1994; Tauer, 1995; Hall, 2003).

Given the pro-life opposition to the HERP panel and its recommendations, it is no real surprise that initial reactions to the prospect of human stem cell research fell out along the fault lines of abortion politics in the country. By and large, individuals and groups opposed to abortion tended to be opposed to stem cell research, and individuals and groups supportive of legalized abortion tended to support stem cell research.[2] For example, the testimony that Richard Doerflinger, the principal spokesperson for the U.S. Catholic bishops on pro-life matters, offered before the Senate Appropriation Subcommittee on Labor, Health, and Education in 1998 was substantially the same as that he offered before the HERP in 1994 on stem cell research (Doerflinger, 1998, 1994). In both cases, the fundamental issue was the status of the embryo. Given Catholic teaching that the embryo must be treated as a person from conception, no experimentation on the embryo can be allowed that would not also be allowed on infants or children. Hence, the Catholic church treats stem cell research as it has treated previous issues involving the destruction of human embryos; it is condemned as morally abhorrent.

In similar fashion, the arguments reviewed by the HERP panel that supported embryo research generally in 1994, were mobilized again four years later when stem cell research was the specific point of contention; and again the focal point was embryo status. Consequently, just as the HERP report opted for a "pluralistic" view of the embryo that emphasized its developmental potential, so, too, did the NBAC endorse the idea that the early embryo deserves respect, but is not to be treated fully as a person.[3]

Moreover, the fact that the HERP defended its support of stem cell research by stressing the developmental capacity of the embryo also shaped the trajectory of much subsequent support for this work, because insisting on respect for the embryo but denying its personhood meant explaining how one could respect the embryo while nevertheless destroying it. Daniel Callahan, for example, posed this problem very strongly in response to the HERP report. If "profound respect" for the embryo is compatible with destroying it, he asked, "What in the world can that kind of respect mean?" It is, he says, "an odd form of esteem, at once high-minded and altogether lethal" (Callahan, 1995). Callahan was not alone in raising this issue and attempts to answer his question continue to appear in the literature (See Lebacqz, 2001; Meyer and Nelson, 2001; Ryan, "Creating Embryos," 2001; Steinbock, 2001, 2000).

In retrospect, then, it seems that the HERP report served almost as choreography for the initial debates about stem cell research, and, as a result, the steps in the debate closely followed those that are familiar from the abortion controversy (On this point, see Hall 2003). The upshot, in my view, is that much of the debate has been too narrowly focused and has a kind of repetitive and rigid quality to it. As I noted above, for example, the Catholic church has repeatedly claimed that the central issue raised by stem cell research is that it involves the destruction of human embryos, embryos it believes should be treated as persons.[4] For that reason, the rhetoric with which the Catholic church condemns embryonic stem cell research closely parallels that used to condemn abortion. Yet, because the American bishops do not want to be perceived as anti-science, they have also repeatedly and uncritically praised adult stem cell research, even though there are good reasons, given Catholic concerns about social justice, to be concerned about the pursuit of adult stem cell research. I will return to this point below, but for now I wish simply to note that much of the opposition to embryonic stem cell work has resembled Catholic opposition in being circumscribed by questions of embryo status, narrowly construed.

A similar constriction, however, is also apparent in the preoccupations of supporters of stem cell research. Just as opponents of this research have ritualistically condemned the destruction of early embryos but uncritically celebrated adult stem cell work, supporters of embryonic stem cell research have typically insisted on using embryos left over from IVF procedures, while repudiating the use of embryos created solely for research. Indeed, insisting on the distinction between so-called "spare" embryos and "research" embryos and endorsing only the use of spare embryos has been one way that supporters of embryo research have tried to demonstrate their "respect" for the embryo. Yet, it is worth asking whether the spare embryo/research embryo distinction does not, to borrow Daniel Callahan's image, provide a kind of "wafting incense" to mask what supporters still find a disquieting smell (Callahan, 1995).[5]

Although the debate about stem cell research might have been framed in terms of the abortion controversy in any event, the HERP report insured that the initial debate over stem cell work that followed in aftermath of the public announcement of the work of John Gearhart (Shamblott et al., 1998) and James Thomson in 1998 (Thomson et al.) would be navigated in the wake of the conflict over abortion. As I indicated, the upshot is that the discussion about stem cell research has been more cramped than it might otherwise have been. The discussion has been too focused on the details of embryological development; too focused on the differences between those who view the early embryo as a person and those who do not; and far too individualistically oriented. Before turning to ways that the debate might be become less cramped, let me focus more concretely on these difficulties.

The point about the debate being framed too individualistically is nicely illustrated in an article on abortion by Lisa Sowle Cahill entitled "Abortion, Autonomy, and Community" (Cahill, 1996). Cahill begins this article by claiming that, in discussing the morality of abortion, there is no way to avoid the question of the status of the fetus. Nevertheless, she says, the debate about fetal status is almost always conducted with the goal of determining the rights involved, where rights are understood very individualistically. To the degree that the fetus is acknowledged to have rights, those rights are pitted against the rights of the pregnant women. Although Cahill doubts that we can jettison the use of rights language altogether, if we are going to use rights language, she says, we must "remove that language from the context of moral and political liberalism" (361). If we do so, we might be able to see that we have duties and obligations to which we do not explicitly consent. As Cahill puts it, "such obligations originate simply in the sorts of reciprocal relatedness that constitutes being a human" (361).[6] For example, moving away from an individualistic liberal view of the pregnant woman as primarily or exclusively an autonomous moral agent might lead us to recognize the obligations that individuals and communities have to support her during and after a burdensome pregnancy (363).

We do not need to accept Cahill's commitment to the Catholic common good tradition to recognize the truth in her conclusion that pitting the rights of the fetus against the rights of the pregnant woman individualistically construed leads us to overlook important social dimensions of the problem of abortion. It seems to me that much the same dynamic is evident in the stem cell research debate.

Consider again the central argument that the Catholic church has made against stem cell research. The Pontifical Academy for Life suggests that the fundamental ethical issue is whether it is morally licit to produce or use human embryos to derive embryonic stem cells. The reasoning the Academy provides for concluding it is not licit is worth reproducing in full. The Academy lists five points:

> 1. On the basis of a complete biological analysis, the living human embryo is—from the moment of the union of the gametes—a <u>human subject</u> with a well defined identity, which from that point begins its own <u>coordinated, continuous and gradual development</u>, such that at no later stage can it be considered as a simple mass of cells.

> 2. From this it follows that as a "<u>human individual</u>" it has the <u>right</u> to its own life; and therefore every intervention which is not in favor of the embryo is an act which violates that right. ...

> 3. Therefore, the ablation of the inner cell mass (ICM) of the blastocyst, which critically and irremediably damages the human embryo, curtailing its development, is a <u>gravely immoral</u> act and consequently is <u>gravely illicit</u>.

> 4. <u>No end believed to be good</u>, such as the use of stem cells for the preparation of other differentiated cells to be used in what look to be promising therapeutic procedures, <u>can justify an intervention of this kind</u>. A good end does not make right an action which in itself is wrong.

> 5. For Catholics, this position is explicitly confirmed by the Magisterium of the Church which, in the Encyclical <u>Evangelium Vitae</u>, with reference to the Instruction <u>Donum Vitae</u> of the Congregation for the

Doctrine of the Faith, affirms: "The Church has always taught and continues to teach that the result of human procreation, from the first moment of its existence, must be guaranteed that unconditional respect which is morally due to the human being in his or her totality and unity in body and spirit: The human being is to be respected and treated as a person from the moment of conception; and therefore from that same moment his rights as a person must be recognized, among which in the first place is the inviolable right of every innocent human being to life'"(No. 60). (Pontifical Academy for Life, 2000; emphasis in original)

Notice that the core of the argument, namely points one and two, is framed in terms of the rights of the individual embryo. We have seen this emphasis already in noting Richard Doerflinger's various statements on stem cell research. Yet, notice also the claim that we know the embryo to be an individual with rights on the basis of "a complete biological analysis." This is not, of course, the first time that the Catholic church has made this claim. In the Declaration on Procured Abortion, the Congregation for the Doctrine of the Faith claimed that "modern genetic science" confirms the view that "from the first instant, the programme is fixed as to what this living being will be: a man, this individual-man with his characteristic aspects already well determined" (Congregation, "Declaration on Procured Abortion,"1974, 13). The Instruction on reproductive technology, Donum Vitae, also makes this claim. "The conditions of science regarding the human embryo provide a valuable indication for discerning by the use of reason a personal presence at the moment of the first appearance of a human life: how could a human individual not be a human person?" (Donum Vitae, 13).

One reason the Catholic church has played such a major role in framing the stem cell debate is that, in defending its position, it combines the two claims we have just noted, neither of which is explicitly religious. First, the early embryo is an individual person with rights and, second, the fact that the embryo is an individual person is confirmed by modern science. Indeed, a fair amount of the literature that supports embryo research generally can be read as an attempt to answer the question posed in Donum Vitae: How can a human individual not be a human person?

Certainly Catholic writers who reject the church's teaching on the status of the embryo have responded directly to that question

(See Cahill, 1993; Farley, 2001; McCormick, 1994; Shannon, 2001; Shannon and Walter, 1990), but so too have non-Catholics. For example, in a statement issued by their ethics committee, what was then called the American Fertility Society rejected the claim in Donum Vitae that science supports the personhood of the embryo. According to the ethics committee "… it remains fundamentally inconsistent to assign the status of human individual to the human zygote or early pre-embryo when compelling biological evidence demonstrates that individuation, even in a primitive biologic sense, is not yet established. Thus, homologues (identical) twins may result from spontaneous cleavage of the pre-embryo at some point after fertilization but prior to the completion of implantation. Furthermore, during very early development, an embryo is not clearly established and awaits the differentiation between the trophoblast and the embryoblast" (American Fertility Society, 1988, 3S).

Arguably, writers like Mary Anne Warren and Bonnie Steinbock, who distinguish between biological or genetic humanity and moral humanity, are also at least indirectly answering the question posed in Donum Vitae (Warren, 1997; Steinbock 2001, 1992). Yet, whether writers are responding more or less directly to Catholic discourse, or not at all, the important point is that the stem cell debate has been remarkably preoccupied with the question of whether the early embryo is an individual person and whether and how the minute details of embryological development help us to answer this question. This is one reason why a fair amount of the ethics literature on the topic reads like a textbook on embryology.

I want to be clear here: I am not suggesting that the details of embryological development are unimportant. The maxim from the field of research ethics applies here as well: bad science is bad ethics. My point is rather that the preoccupation with the details of early embryogenesis may lock us even more rigidly into an individualistic human rights framework than we are in debates about abortion. It also leads us to frame the debate as fundamentally about one question, and, indeed, it tempts us to treat the question as if there is one and only one answer. In this frame of mind, once we have that answer, there is not a lot more to talk about. Either the early embryo is a person with the right to life, in which case embryonic stem cell research is wrong, or the early embryo is not a person with rights, and then there is no moral reason to object to stem cell work. Gene Outka has made a similar point in his assessment of stem cell literature. As Outka puts it, in its starkest form, the crystallizing question is whether it is cogent to claim that embryonic stem cell research is morally indistinguishable from murder (Outka,

184). The problem with framing the question this way, he says, is that it "encourages an unfortunate tendency to restrict evaluative possibilities to a single either/or. Either one judges abortion and the destruction of embryos to be transparent instances of treating fetuses and embryos as mere means to other's ends, or one judges abortion and embryonic stem cell research to be, in themselves, morally indifferent actions that should be evaluated solely in terms of the benefit they bring to others." (Outka, 2002, 184).

The frame of human rights reinforces this either/or because, as I noted, a being is either a rights-bearing entity or it is not. I have argued elsewhere, that this either/or tends to drive people to the extremes. Either the embryo is a person or it is essentially a kind of property (Lauritzen, 2001). Although I will not rehearse the argument for rejecting the two extremes here, it is worth noting that the rhetoric associated with each extreme does not appear to match the practice of those who adopt the rhetoric or in fact to match the considered moral judgments of most Americans on these issues.

I can illustrate my point in relation to the view that the early embryo is a person with the right to life by describing a cartoon that hangs on my office door (See Figure A). The cartoon depicts protestors in front of a stem cell research lab condemning those who work there as being anti-life. Down the street at the abortion clinic, the workers are noting how quiet things have gotten at their facility since the stem cell lab opened. The point of the cartoon, of course, is that we may soon see protests and demonstrations of the sort that are common at abortion clinics at facilities that conduct stem cell research and that there is an irony in the fact that pro-life advocates would be demonstrating against research being done to find treatments for Alzheimer's, Parkinson's, and other devastating illnesses. This is not entirely fair to the pro-life community, but it makes a point.

In fact, I do not have trouble imaging protestors picketing stem cell research facilities for, as we just noted, when stem cell research and abortion are evaluated together and when the evaluative option is a single either/or, then abortion and stem cell research may appear indistinguishable from murder. Certainly the rhetoric of someone like Richard Doerflinger has been consistent in condemning both abortion and stem cell research as equivalent to murder. The cartoon draws attention to this consistency, even while it questions the commitment of pro-life advocates to scientific research designed to promote the quality of life.

In one sense, then, the cartoon probes whether there is an inconsistency between being pro-life and opposed to research an Alzheimer's disease, Parkinson's disease, and other devastating illnesses. I do not myself think that there is any inconsistency in being pro-life and opposed to stem cell research, but the cartoon does point in the direction of a fairly significant disconnect between the rhetoric and the reality of those opposed to stem cell research because they believe the early embryo is a person. To see this point, imagine that, instead of a stem cell clinic, the cartoon depicted on IVF clinic down the block from the abortion clinic and that the workers at the abortion clinic are noting how quiet things have gotten since the IVF clinic opened. The dramatic tension that made the original cartoon funny would be missing from our revised cartoon precisely because it is hard to imagine protestors disrupting the work at IVF clinics. To be sure, the Catholic church and others have argued that IVF is morally wrong, but the rhetoric condemning IVF is exceptionally muted compared to that condemning abortion or stem cell research. Nor has there been a concerted effort to put an end to IVF practice in this country as there has been in the case of abortion and stem cell research. Yet, if the embryo is a person from conception, then participating in IVF as it is practiced in this country, when early embryos are routinely frozen or discarded or both, is to be complicit with murder. Why, then, are there no organized efforts to shut down IVF clinics in this country?[7]

Indeed, opponents of stem cell research and cloning often write as if these technologies raise the haunting specter of human embryo research for the first time. The reality, of course, is that the existence of in vitro fertilization depended entirely on embryo research and that every variation or innovation in IVF protocols involves experimentation on human embryos. Carol Tauer is one of the few scholars who has pressed this point. As Tauer sees it:

> ...the entire history of the research leading to the first successful IVF is the history of attempts to fertilize oocytes in the laboratory. Eventually these attempts succeeded, and the first IVF baby was born, followed by thousands of others in the ensuing decades.
>
> The ethics literature contains scholarly discussions as to whether it is ethically permissible to make use of medical advances that result from unethical research. This discussion sometimes focuses on medical research conducted by the Nazis

in concentration camps and institutions for retarded, mentally ill, and handicapped persons. Yet I have never seen reference to reproductive technologies in this context. If the fertilization of embryos in research is a practice that is abhorrent to many or most people, then would it not be logical to question the continuing use of the results of such research? (Even the Catholic Church, which opposes the use of IVF and most other forms of assisted reproduction, does not invoke this argument to support its opposition.) (Tauer, 2001, 153)

If the embryo research associated with IVF points to a problem of consistency for those who oppose stem cell research because it involves destroying persons, it is no less problematic for those who support stem cell research but insist on respecting the embryo and embrace the distinction between "research" and "spare" embryos. For as Tauer points out, Robert Edwards, the scientist involved in the first successful IVF procedure, began studying fertilization nearly thirty years before Louise Brown was born in 1978, and the first successful laboratory fertilization of human eggs took place a full ten years before she was born. Tauer quotes Edwards' report on this work: "We fertilized many more eggs and were able to make detailed examinations of the successive stages of fertilization. We also took care to photograph everything because we would have to persuade colleagues of the truth of our discoveries" (Tauer, 154).[8] Nor was the creation of these "research" embryos done secretly: Edwards and Steptoe published their work in the journal, Nature in 1970 (Edwards, Steptoe, and Purdy, 1970).

At the very least, then, there is something of an irony in the fact that so much attention has been devoted to developing and defending the distinctions between embryos created solely for research and embryos left over from IVF procedures, because there would be no embryos left over from IVF procedures had there not been embryos created solely for research purposes to develop IVF in the first place. Given this fact, and given that this fact is no great secret even though it has not been discussed very much it appears disingenuous to endorse the distinction between "research" and "spare" embryos as a way of demonstrating respect for the early embryo while nevertheless encouraging its destruction.

I have suggested that the fact that so much of the stem cell debate has been framed in terms of whether the embryo is a person with rights has been unfortunate because it has cast the debate in sharply individualistic terms and has led to a preoccupation with embryological development narrowly construed. In addition, however, framing the debate in terms of embryo status and embryo rights tends to exaggerate the differences among commentators in contrast to their similarities. Consider, for example, the response of conservative Judaism in the United States to this issue. Rabbi Elliot Dorff has prepared a responsum on stem cell research for the Rabbinical Assembly Committee on Jewish Law and Standards, and his responsum is instructive.[9]

As responsa are, it is structured in terms of relevant questions: in this case, two questions frame Dorff's discussion. First, "may embryonic stem cells from frozen embryos originally created for purposes of procreation or embryonic germ cells from aborted fetuses be used for research?" (Dorff, 2002, 1). Second, "may embryonic stem cells from embryos created specifically for research, either by combining donated sperm and eggs in a petri dish or by cloning be used for research?" (1) I think it is noteworthy that the very questions that frame Dorff's analysis both reflect and perpetuate a certain construction of the issue, but at this juncture, my point is different: given the way the debate has been framed, what most (non-Jewish) readers of Dorff's analysis are likely to focus on is the difference between his treatment of the early embryo and that of others in the literature. Indeed, even where you might expect to find and do in fact find on closer inspection similarities between this Jewish analysis and Catholic reflection on this issue, the first impression will be that of difference. The reason, of course, is that our attention is drawn to Dorff's analysis of the early embryo, and Jewish views are sharply different about embryo status than those of the Catholic church and other pro-life opponents of stem cell work.

For example, Dorff points out that, according to the Talmud, during the first forty days of gestation, the embryo and the fetus are considered as simply water. From the forty-first day until birth, Jewish tradition considers the fetus as "the thigh of its mother." Moreover, says Dorff:

> As it happens, modern science provides good evidence to support the Rabbis' understanding. As Rabbi Immanuel Jakobovits noted long ago, the Rabbis' "forty days" is, by our obstetrical count, approximately fifty-six days, for the Rabbis counted from the woman's first missed menstrual flow, while doctors today count from the point of conception, which is usually about two weeks earlier. By 56 days

of gestation by obstetrical count the basic organs have already appeared in the fetus. Moreover, we now know that it is exactly at eight weeks of gestation that the fetus begins to get bone structure and therefore looks like something other than liquid. Indeed, the Rabbis probably came to their conclusion about the stages of development of the fetus because early miscarriages indeed looked like "merely water," while those from 56 days on looks like a thigh with flesh and bones. (16)

The contrast with Catholic teaching could hardly be more striking. Not only are Jewish views of the status of the early embryo notably different, but Jewish tradition claims scientific validation for its view of the embryo, just as Catholic tradition does. Not surprisingly, therefore, where Dorff answers both questions posed by the responsum in the affirmative, Catholic tradition would answer both negatively.

These differences are significant and must be attended to, but it is worth asking whether focusing on these differences does not obscure important similarities. Consider some of the similarities. In sketching the Jewish view of stem cell research, Dorff notes that certain theological commitments are central. He lists at least three that would be strikingly similar to Catholic and other Christian theological commitments.

- Our bodies are not ours; they belong to God and God commands that we seek to preserve life and health.

- All human beings, regardless of ability or disability are created in the image of God and are therefore to be valued as such.

- Humans are not God. We are finite and fallible and this fact ought to promote humility and urge caution.

Now if we focused merely on questions of embryo status, we would miss entirely these similarities between Catholic and Jewish views. More importantly, we would miss the fact that these similarities may underwrite significant moral reflection on stem cell research that is not rooted in concerns about the early embryo.

For example, Dorff notes that, given Jewish theological and legal commitments, the provision of health care must be understood as a communal responsibility. Thus, access to therapies developed through stem cell research is a crucial issue of justice for the Jewish community. This theme is echoed in Laurie Zoloth's work on stem cell research. As Zoloth puts it:

> Research done always will mean research foregone. Will this research help or avoid the problem of access to health, given that poverty and poor health are so desperately intertwined in this country?...How can difficult issues of global justice and fair distribution be handled in research involving private enterprise? (Zoloth, 2001, 238)

Surely these are questions that any Catholic moral theologian would gladly press.

Indeed, attending to the similarities between Zoloth's work and Catholic reflection on stem cell research brings us back to Lisa Cahill's observation about debates on abortion: they tend to be too focused on questions of rights individually construed. When one shifts the frame of analysis, new and different issues and new and different ways to approach the same issues come into view. Notice, for example, how close Zoloth and Cahill are on the issue of rights. According to Zoloth, Jewish tradition foregrounds questions of "obligations, duties, and just relationships to the other, rather than the protection of rights, privacy, or ownership of the autonomous self" (96). This leads Zoloth to ask: "Can the interests of the vulnerable be heard in our debate?" (105). To be sure, the American bishops have wanted to emphasize the vulnerability of the early embryo when they have asked this question, but Catholic tradition, like Jewish tradition, requires that we ask this question in a way that is not captured when moral emphasis is merely about individual rights and personal autonomy.

Or consider another shared sensibility that emerges if we move away from questions of embryo status, namely a wariness about the human tendency to hubris and overreaching. Zoloth put this point eloquently in relation to the biblical story of the Tower of Babel. She notes a rabbinic midrash on this text: "when a worker was killed, no one wept, but when a brick fell, all wept." Zoloth comments on this midrash as follows:

> It was this decentering of the human and reification of the thing that was the catastrophe that felled the enterprise... It is not just that they breached a limit between what is appropriate to create and what is

not, the process of the creation must be carefully mediated, with deep respect for persons over the temptations of the enterprise. Such a text elaborates on the tension between repairing the world ... and acts that claim that the world is ours to control utterly (Zoloth, 2001, 106-107).

Beyond Questions of Embryo Status

This passage from Zoloth helps to illustrate the point I wish to make in arguing that the stem cell debate has been too focused on questions of embryo status and that we must move beyond status questions if we are fully to do justice to the moral questions raised by technological developments associated with stem cell work. For concern about human efforts utterly to control the world is not a moral worry narrowly tied to status questions.

Let me put this point in the form of a question that has not typically been asked in the stem cell debate: Is adult stem cell work as unproblematic as it is often assumed to be? That this is a productive question is suggested by testimony of Francis Collins before the President's Council on Bioethics in December 2002. Collins was asked to speak about the topic, "genetic enhancements: current and future prospects" and he specifically addressed the issue of pre-implantation genetic diagnosis (PGD). Although PGD is usually understood to involve screening IVF embryos and discarding unwanted ones, it is also possible to screen gametes. Because gamete screening may not have broad utility, Collins did not discuss the issue at length. He did, however, offer an interesting observation about gamete selection. Focusing on gametes, he says, is useful because it "isolates you away from some of the other compelling arguments about moral status of the embryo and allows a sort of cleaner discussion about what are the social goods or evils associated with broad alterations in the sex ratio and inequities in access to that technology" (Collins, 2002, 7). In other words, if in the future we could screen gametes in the same way that we can now screen embryos, most of the moral issues raised by PGD would apply to gamete screening, even though gametes are not embryos. Might we not make a similar claim about embryonic and adult stem cell research? Do not many of the most pressing issues raised by embryonic stem cell technology remain when our focus is adult stem cell work rather than embryonic stem cell research?

The fact that we do not immediately answer yes to this question, is testament to how decisively the debate about abortion has structured the stem cell debate. Nevertheless, we need to see that the answer to this question is yes and we need to see why.

Although I will not try to address all of the issues raised jointly by embryonic and adult stem cell research, it is worth highlighting several that I think require fuller discussion particularly with respect to adult stem cell work than they have yet received.

Commodification Issues

Moral concerns about the commodification of gametes and embryos have been discussed extensively in the bioethics literature both in relation to reproductive technology and in connection with embryonic stem cell work (See Andrews and Nelkin, 2001; Annas, 1998; Corea, 1986; Radin, 1996; Resnik, 2002; Ryan, Ethics and Economics, 2001). Suzanne Holland, for example, has discussed the growing commodification of the human body in the biotech age. She cites a series of articles published in the Orange County Register that documents a vast for-profit market in human body tissue (Holland, 2001, 266). The Register's investigative reporters documented that most nonprofit tissue banks obtain tissue from cadavers donated by family members of the deceased for altruistic reasons. Most relatives are not told, and in fact have no idea, that donated body parts will be sold for profit. As Holland puts it:

> The "gift of life" is big business in America. For a nonprofit tissue bank, one typical donation can yield between $14,000 and $34,000 in downstream sales, sometimes far more than that. "Skins, tendons, heart valves, veins, and corneas are listed at about $110,000. Add bone from the same body, and one cadaver can be worth about $220,000." Four of the largest nonprofit tissue banks told the Orange County Register that together they expected to produce sales totaling $261 million in 2000. (226)

Nor is the issue of downstream commodification restricted to the sale of donated cadaveric tissue; it arises in relation to IVF embryos donated for research. As Dorothy Nelkin and Lori Andrews point out in their book, Body Bazaar, IVF patients are not generally told what the research involving their donated embryos will include. Many will be unaware that their embryos will be used to develop commercial stem cell lines (Nelkin and Andrews, 2001, 35).

It is significant that even the most vocal advocates of procreative liberty and laissez-faire arrangements in reproductive matters recoil from the prospect of selling human embryos. Yet, although the commodification of tissue may be particularly troubling when it involves embryos, if there is a problem with commodifying and commercializing human tissue, it is a problem we confront with adult stem cell research as well as with hES cell work.[10] Lori Knowles has made a similar point about being consistent in our moral judgments about commodifying embryos. She notes that fears about commodifying reproduction have led many to oppose the sale of embryos and to reject the idea that couples who donate embryos have any proprietary interest in the result of the research done with their embryos. As Knowles puts it, "if it is wrong to commercialize embryos because of their nature, then it is wrong for everyone. It is simply inconsistent to argue that couples should act altruistically because commercializing embryos is wrong, while permitting corporations and scientists to profit financially from cells derived by destroying those embryos" (Knowles, 1999, 40).

Knowles draws attention here to the fact that there is a tension between our moral and legal traditions as they apply to developments brought about by cloning, stem cell research, and the existence of <u>ex utero</u> human embryos, among other technological breakthroughs. For example, we patent embryonic stem cell lines, thereby insuring massive profits for patent holders, while decrying the commodification of embryonic life. Knowles is correct that there is a tension between our profit-based medical research model and our commitment to altruism and the access to health care that a commitment to justice demands. Adult stem cell research, of course, raises very similar issues, because the same tension exists between the need for proprietary control of technology and the need for affordable access. For example, patents have been sought and granted for human adult stem cells work as well as for embryonic stem cell technologies. According to a study on the patenting of inventions related to stem cell research commissioned by the European Group on Ethics in Science and New Technologies, as of October 2001, two thousand twenty-nine patents were applied or granted for stem cells and 512 patents were applied or granted for embryonic stem cell work (Van Overwalle, 2002, 23; see also McGee and Banger, 2002). As a result, access to therapies developed from adult stem cell research is likely to be as serious an issue as access to embryonic stem cell therapies.

Indeed, issues arising from commodification of both adult and embryonic stem cells are likely to dominate the next phase of the debate, if only because corporate interest in this work, both nationally and internationally, is so strong (Hill, 2003). Noting that, as of 2002, "a dozen biotech companies have entered the stem cell industry and have invested millions of dollars," David Resnik suggests that the next stage of the stem cell debate will involve a battle over property rights relating to stem cells (Resnik, 2002, 130-31).

To be sure, the battle over property rights raises important legal and policy question, but it also raises ethical questions as well. For example, Resnik provides the following table of possible ethical objections to patenting stem cells.

Deontological Objections to Property Rights in ES cells	Consequentialist Objections to Property Rights in ES cells
ES cells (and their products) should not be treated as property; they should be viewed as having inherent dignity or respect.	Treating ES cells as property will have dire social consequences, such as exploitation, destruction of altruism, and loss of respect for the value of human life.
ES cells should not be treated as private property; they are *res communis* or common property.	Treating ES cells as property will undermine scientific discovery and technological innovation in the field of regenerative medicine. (Resnik, 2002, 139)

Although the ethical objections to commodifying stem cell work can not be sorted out as neatly as this table suggests, Resnik correctly identifies a variety of such objections. I will not try to defend it here, but my own view is that the problem is not just that stem cells and their products may be commodified, but that market rhetoric may come to dominate the discussion (and practice) of regenerative medicine in a way that is dehumanizing. That is, market-rhetoric may lead us ultimately to think of humans as artifacts. In short, I take very seriously Margaret Radin's argument that the rhetoric in which we conceive our world affects who and what we are (Radin, 1996, 82).[11] At the same time, however, stem cell research is most likely to bear therapeutic fruit, if there is a market in stem cells and their products. The pressing moral question, then, is how do we promote the benefits that stem cell research may yield without succumbing to a market rhetoric that reduces humans to commodities?

Several writers have suggested that one answer may be to promote greater governmental regulation. For example, Holland argues for moving beyond a policy of restricting federal funding of stem cell research but allowing an unregulated private market in this field to active regulation to curb the private sector's work on stem cells. Lori Knowles suggests that the United States might adopt a body like Canada's Patented Medicine Prices Review Board as a way to allow a market to function, but with oversight that would provide access to potential cells for further research and price controls of products to insure widespread access (Knowles, 1990, 40). George Annas has suggested that we need to establish a federal Human Experimentation Agency to regulate in the area of human experimentation. (Annas, 1998, 18). As Annas puts it:

> Virtually all those who have studied the matter have concluded that a broad-based public panel is needed to oversee human experimentation in the areas of genetic engineering, human reproduction, xenografts, artificial organs, and other boundary-crossing experiments. (19)

Francis Fukuyama has argued that a new agency with a mandate to regulate biotechnology on broad grounds and in both the public and private sector may be needed (Fukuyama, 2002, 215). Vanessa Kuhn argues that "it is time to put in place legislation that will deter stakeholders from licensing their technology to one exclusive distributor and thus creating a monopoly market, which would set artificially-high prices and lead to less access for the sick especially for the uninsured, the poor, and the elderly" (Kuhn, 2002). To that end, Kuhn identifies four possibilities:

- Development of a new kind of patent.

- Set limits on exclusive licensing through compulsory licensing.

- Lower the lifespan of hES cell patents

- Set stricter guidelines for hES patent utility (2)

I do not have the expertise to make policy recommendations, but let me stress two points. First, the policy issues with regard to commodifiying adult stem cell work will be as vexing as those confronting regulation of embryonic stem cells. Second, although these questions may at first appear to be strictly legal or largely political matters, they involve serious value judgments about the common good that are every bit as morally vital as questions about the status of the embryo. I thus agree with Gene Outka, that not to confront directly questions about how stem cell research will be organized, financed, and overseen is a kind of ethical failure (Outka, 2002, 177). Obviously, for example, the institutional arrangements for conducting stem cell research have implications for the questions of justice we previously noted. The fact that so much stem cell research is being done by private corporations insures future conflict. On the one hand, corporations have fiduciary obligations to their shareholders and will therefore seek to control access to stem cell lines or therapies developed from those lines through patent protection and licensing agreements. On the other hand, such a system is likely to further widen the gap between the health care haves and have-nots (See Lebacqz 2001; McLean 2001).

Moreover, as Karen Lebacqz notes, if justice is an important consideration in deliberations about stem cell work, then it ought to shape the research agenda. The example she gives to make this point is worth noting. Just as with organ transplants, tissue rejection may be a major problem for stem cell therapies. This is one reason that the prospect of combining stem cell work with somatic cell nuclear transfer (SCNT) has been so enticing. With this combination, you could in theory develop tissue that would be completely histocompatible. Nevertheless, according to Lebacqz, developing stem cell therapies with SCNT is "highly questionable," if justice is a primary consideration. The reason is that unique cell lines would need to be created for each patient, and that is likely to be very expensive and thus unaffordable for many. Although it would certainly be less expensive, the same would likely be true for adult stem work. For that reason, rather than pursuing an individualized approach to stem cell research, concerns about affordable access to new therapies might urge the pursuit of universal donor cell lines.

Embodiment, Boundary Issues, and Human Nature

In discussing the debate about embryo status, I focused primarily on the contested question of whether the early embryo is a person with the right to life. We saw that this question tends to lead to the mobilization of minute details of embryological development to support one's view of the embryo. Yet, if attention to embryo status tends to focus us on the microscopic, viewing stem cell research through the lens of embryological development can also have a kind of telescopic function through which larger issues come into view.[12] For example, Catherine Waldby and Susan Squier argue in a forthcoming issue of the journal Configurations that focusing on stem cells and embryonic life leads us fundamentally to question what it means to be human. According to Waldby and Squier:

> Stem cell technologies have profound temporal implications for the human life course, because they can potentially utilize the earliest moments of ontogenesis to produce therapeutic tissues to augment deficiencies in aging bodies. Hence they may effect a major redistribution of tissue vitality from the first moments of life to the end of life. In doing so however they demonstrate the perfect contingency of any relationship between embryo and person, the non-teleological nature of the embryo's developmental pathways. They show that the embryo's life is not proto-human, and that the biology and biography of human life cannot be read backwards into its moment of origin. (Waldby and Squier, forthcoming)

The claim that there is a perfect contingency in the relationship between embryo and person may at first appear to be just another "microscopic" claim about embryo status, but it is clear that Waldby and Squier mean to imply much more in asserting that the embryo's development is non-teleological. In effect, they reject the notion that there is a meaningful trajectory to human life. What was killed, they say, when stem cells were first derived from the inner cell mass of a blastocyte was not a person, but a "biographical idea of human life, where the narrative arc that describes identity across time has been extended to include the earliest moments of ontogeny" (Waldby and Squier, forthcoming).

That much more is at stake here than the question of whether the embryo is a person is clear if we attend to the notion of a trajectory of a human life. Gilbert Meilaender, for example, has argued that our attitudes toward death and dying are importantly shaped by our conception of what it means to have a life (Meilaender 1993). Indeed, according to Meilaender, two views of what it means to have a life of what it means to be a person have been at war with each other within the field of bioethics over the past thirty years and these views underwrite sharply different views not just about the issues of abortion or euthanasia that are implicated here but with regard to practically every moral issue we might confront in the field of bioethics.

On Meilaender's view, having a life means precisely that there is a trajectory that traces a "natural pattern" in embodied life that "moves through youth and adulthood toward old age and, finally, decline and death" (29). As he puts it elsewhere in this essay, "to have a life is to be terra animata, a living body whose natural history has a trajectory" (31). Although Meilaender develops the notion of a natural trajectory of bodily life primarily to address the issue of euthanasia and not stem cell research, his talk of "natural history," "natural pattern," and "natural trajectory" draws attention to one of the most significant issues raised by stem cell research and related technologies. Does stem cell research undermine the very notion of a human life constrained by natural bodily existence? The example on which Meilaender focuses here is instructive for thinking about the broad implications of stem cell research in this regard. If stem cell therapies fundamentally alter our sense of a natural pattern to aging would they not also fundamentally alter our sense of what it means to be human? Meilaender's answer is that such a change would fundamentally affect what it means to be human precisely because we are embodied creatures and for that reason our identity is tied to the body and the body's history.

Leon Kass has made a similar point recently in reflecting on the prospect that regenerative medicine might significantly lengthen the human life span (Kass, 2003). He, too, invokes the notion of a natural trajectory, one that stem cell research may undermine. Although it is possible to approach the prospect of extending the human life-span in an abstract way, he says, to think of what such a change would mean experientially is to recognize that "the 'lived time' of our natural lives has a trajectory and a shape, its meaning derived in part from the fact that we live as links in the chain of generations" (13). Indeed, says Kass, without something like the natural trajectory of bodily life that currently exists, the relationship between the generations would be decidedly different, and probably not better. "A world of longevity," writes Kass, "is increasingly a world hostile to children" (13). Walter Glannon has argued that, at the very least, increased longevity would increase competition for scarce resources between older and younger generations. According to Glannon, "it is at least intuitively plausible that an over populated world with substantially extended human lives and scarce resources could adversely affect the survival and reproductive prospects of the young and harm them by thwarting their interest in being healthy enough so that they could survive and procreate" (Glannon, "Extending," 347).[13]

Francis Fukuyama has also suggested some of the reasons why increased longevity may imperil children, but he also notes that our relationship to death may change as well (See 2002, ch. 4). "Death," he says, "may come to be seen not as a natural and inevitable aspect of life, but a preventable evil like polio or measles. If so, then accepting death will appear to be a foolish choice, not something to be faced with dignity or nobility" (Fukuyama, 2002, 71).

Sometimes the question of the transformative possibilities that come with stem cell research is raised even more starkly when the question asked is not how may stem cell work affect what it means to be human, but instead: Does stem cell research open the door to a post human future? This is a point Waldby and Squier raise explicitly when they discuss the combination of genetic engineering and stem cell therapy. They suggest, for example, that xenotransplantation forces us to confront the prospect of transgressing species boundaries.[14] The conclusion of their paper is worth quoting in full:

> Thus the ontological status of the embryo is not the only thing in question. The ontological status of the graft recipient must be negotiated, when the graft involves genetically-engineered stem cells from another species. And the ontological status of the illnesses to which biomedical technology responds is equally challenged, in an endless regression, as the division between veterinary and human medicine, or between zoonoses (diseases humans can catch from animals) and what has recently been dubbed humanooses, is called into question. This increasingly permeable, increasingly constructed barrier between human and animal presents us with another form of life to negotiate, whose boundary

lies not between silicon and carbon, but rather between steps in the evolutionary ladder or the branching development tree of phylogenetic lifeforms. Stem cell technologies thus challenge both the temporal and spatial boundaries of human life, both our biography and our biological niche, giving a much broader meaning to the questioning of embryonic personhood. (Waldby and Squier, forthcoming)

Regrettably, with some notable exceptions, the ethical debate about stem cell research has not taken up in a sustained way what it would mean to pursue stem cell therapies that might significantly undermine the notion of a natural human life or erode the boundary between human and non-human species. When the issue is framed in terms of the status of the embryo, the question tends to be whether the research should be conducted at all. By contrast, when the issue is framed in terms of adult stem cell work, the question is not whether, but how and with what consequences. Yet, that is a question we have not systematically answered. Given the potential for good embedded in the prospects of adult stem cell research, it is not surprising that there appears to be widespread and largely uncritical acceptance of adult stem cell research. But, if the promise of stem research is as revolutionary as is often claimed, we are going to need a much more expansive discussion of stem cell research both embryonic and adult than we have had heretofore. Obviously, I cannot explore this more expansive horizon in any detail in this report, but let me in closing suggest one direction we need to explore.

To signal the decisive break that I think we may need from the usual bioethics frame, I want to draw attention to Martha Nussbaum's recent article in the journal Daedalus entitled, "Compassion & Terror" (Nussbaum, 2003). Discussing Euripides' play, Trojan Women, Nussbaum reflects on the fact that the Greek poets returned obsessively to the sacking of Troy and the acts of the "rapacious and murderous Greeks." She explores the poets' compassionate imagining of the fate of Trojan women and children to reflect on the conditions and limits of a compassionate vision. Although Nussbaum is ultimately concerned about engendering a compassionate vision for Americans in the face of terror and particularly compassion for innocent women and children far from our shores, her analysis of compassion is thought-provoking in relation to stem cell research.

Nussbaum notes that compassion is a complex emotion requiring a series of judgments involving another person's suffering or lack of well-being. We must judge that someone has been harmed, that the harm is serious, and that it was not deserved. Moreover, says Nussbaum, Western tradition has stressed what could be called the "judgment of similar possibilities." In other words, "we have compassion only insofar or we believe that the suffering person shares vulnerabilities and possibilities with us" (Nussbaum, 2003, 15).

Now surely in just about everyone's catalogue of human vulnerabilities are illness, old age, and death. Yet, as we have just seen, stem cell research might significantly transform the "human" experience of illness and death, at least for some. If stem cell therapies were to erode the notion of human nature or species membership, might they not also erode some basic moral sensibilities? Mary Midgley, for example, has argued that both the notion of human nature and that of human rights are importantly tied to membership in our species because rights are "supposed to guarantee the kind of life that all specimens of Homo sapiens need" (Midgley, 2000, 9).

Although Nussbaum avoids the language of human nature, it is precisely this sort of point that she highlights when she argues that compassion requires the belief that others share vulnerabilities and possibilities with us. Indeed, like Midgley, Nussbaum ties the notion of universal human rights to important human functions and capabilities. The basic idea, she says, is to ask what constitutes the characteristic activities of human beings. In other words:

> 'What does the human being do, characteristically, as suchand not, say, as a member of a particular group, or a particular local community?' To put it another way, what are the forms of activity, of doing and being, that constitute the human form of life and distinguish it from other actual or imaginable forms of life, such as the lives of animals and plants, or, on the other hand, of immortal gods as imagined in myths and legends (which frequently have precisely the function of delimiting the human)? (Nussbaum, 1995, 72)

Nussbaum notes that this inquiry proceeds by examining a wide variety of self-interpretations and that comparing characteristic human activities with non-human activities and, through myths and stories, comparing humans and the gods is particularly helpful. For one thing, such an inquiry helps us to define limits that derive from membership in the world of nature.

Indeed, although Nussbaum is particularly attentive to the wide variety of cultural interpretations of what it means to be human, she insists that to ground any essentialist or universal notion of human rights, one must attend to human biology. Although her account of the human is neither ahistorical nor a priori, it is linked to an "empirical study of a species-specific form of life" (1995, 75). When she develops her account of central human capabilities, she begins with the body. She writes:

> We live all our lives in bodies of a certain sort, whose possibilities and vulnerabilities do not as such belong to one human society rather than another. These bodies, similar far more than dissimilar (given the enormous range of possibilities) are our homes, so to speak, opening certain options and denying others, giving us certain needs and also certain possibilities for excellence. The fact that any given human being might have lived anywhere and belonged to any culture is a great part of what grounds our mutual recognitions; this fact, in turn, has a great deal to do with the general humanness of the body, its great distinctness from other bodies. The experience of the body is culturally shaped, to be sure; the importance we ascribe to its various functions is also culturally shaped. But the body itself, not culturally variant in its nutritional and other related requirements, sets limits on what can be experienced and valued, ensuring a great deal of overlap. (Nussbaum, 1995, 76)

Nussbaum's work both in identifying the judgments that underwrite compassion and in tying an account of rights to human function and capabilities that are presumably universal highlights what is at stake, not merely with stem cell research but with a growing list of biotechnological developments which appear to destabilize the concept of human nature and which require that we think carefully and hard about what it might mean for some humans to have access to these technologies while other humans do not.[15] At the very least, the combination of what Rabinow describes as the biologicalization of identity around genetics rather than gender and race with the possibility of manipulating that genetic identity for those with the money or power to do so does not bode well for securing wide-spread compassion across economic or technological divides. Even more important, however, is the recognition that the very notion of human rights may ultimately rest on the idea (and what, until recently has always been the reality) of a natural human condition that is relatively stable.

I believe that Nussbaum is correct when she claims that inquiring into characteristic human activities and comparing these to non-human activities helps us to define limits and thereby to promote human flourishing. Unfortunately, what Susan Squier calls the "pluripotent rhetoric" of stem cell research is that of limitless possibilities (Squier, Liminal Lives). The ultimate limit, of course, is death and yet even this limit appears illusory in some visions of our biotech future.

It is worth noting in closing that William Safire's New Year's Day column at the dawn of the twenty-first century, in January 2000, was entitled "Why Die?" The longing behind this question is neither new nor unfamiliar. What is new is that this longing to escape the vulnerabilities and limitations of the body is united with a technology that holds out the prospect of fundamentally changing that body. Yet, I agree with Gerald McKenny that we need to ask whether we wish to accept and promote a view of bodily vulnerability as merely an obstacle to human flourishing, which ought to be overcome at any cost (McKenny, 1998, 223).

Although a longing for invulnerability is perhaps a quintessentially human trait, and although the quest to reduce the human suffering wrought by illness and disease is morally admirable, there is no mistaking the hubris behind the question, Why Die? Opponents of stem cell research have, from the start, argued that there is a kind of idolatry in a science that would reduce the human embryo to just so much biological material. What I have tried to show in this report is that such concerns need not be limited to those who think that the early embryo is fully a person. Nor should this kind of concern be limited to those who oppose stem cell research. I do not think the early embryo is a person and I believe that both embryonic and adult stem cell research should go forward under a system of strict regulation. Nevertheless, I confess to being haunted by the passages with which I began and I believe that future debates on stem cell research must take very seriously the worries about commodification and the possibility of fundamentally changing the trajectory of a human life.[16]

FIGURE A

Jimmy Margulies, Editorial Cartoonist
The Record, Hackensack, New Jersey

Cartoon used with permission of the artist.

Endnotes

[1] Given the current status of technology, deriving human embryonic stem cells requires destroying embryos. If the cells could be derived without the destruction of embryos or if parthenogenetically stimulated eggs produced stem cells, issue of status would almost certainly fade. Nevertheless, serious ethical issues would still remain. This is one reason I believe it is a mistake to focus narrowly on embryo status.

[2] Gene Outka has argued that there is an "internal coherence" to views of the embryo, issues of complicity, and views on adult stem cell research (Outka, 2002).

[3] Although the HERP report claimed that "it is not the role of those who help form public policy to decide which of these views [of the embryo] is correct," there is little doubt that the panel adopted the pluralistic view. For that reason, most commentators found the above claim disingenuous.

[4] Compare, for example, the various statements that Richard M. Doerflinger has made on the U. S. bishops' behalf. See Doerflinger, 1988, 1998, 1999, 2001. Margaret Farley has argued that the Catholic preoccupation with abortion has eroded its credibility on other important social issues, including stem cell research. (See Farley, 2000).

[5] Another debate that is at least partly shaped by focusing on embryo status revolves around the question of complicity. For example, supporters of stem cell research may harbor a residual uneasiness about endorsing the destruction of human embryos, at least if the number of articles in the literature explaining the concept of complicity with wrongdoing is any indication. John Robertson, Ronald Green, and Thomas Shannon, have all written on the issue of cooperation with evil in relation to stem cell research. (Robertson, 1999; Green, 2002; Shannon, 2001; see also Kaveny, 2000; and Gilliam, 1997). To be sure, the issue of complicity or cooperation with wrongdoing is a very traditional one in moral philosophy and theological ethics. Still, if the early embryo does not deserve the respect accorded persons and if destroying the embryo is compatible with respecting it, then deriving stem cells is not an act of wrongdoing and issues of complicity do not arise.

[6] Cahill also emphasizes the way a liberal individualist view of the person discounts the significance of embodiment. I will return to this point below.

[7] That both those who view the embryo as a person and those who do not but who insist on respect for the embryo, have been remarkably cavalier with regard to the use of embryos in IVF programs can be seen by the fact that there are currently over 400,000 embryos frozen in the United States, a number we did not even know until quite recently (Hoffman et al. in association with The Society for Assisted Reproductive Technology and RAND, 2003).

[8] In a commentary published in <u>Nature</u>, in September 2001, Edwards writes: "On the verge of clinical application, stem cells offer a startlingly fundamental approach to alleviating severe incurable human maladies. Fondly believed to be a recent development, they have in fact been part and parcel of human <u>in-vitro</u> fertilization (IVF) from as long ago as 1962."

[9] Dorff's responsum was accepted by the Committee on Law and Standards by a vote of twenty-two to one in March 2002. On the basis of Dorff's responsum the Rabbinical Assembly passed a resolution in April 2003 supporting stem cell research for therapeutic purposes. (Resolution in Support of Stem Cell Research and Education, April 2003: available at www.rabassembly.org.)

[10] Alpers and Lo draw the distinction between commodification and commercialization as follows: "The issue of commodification involves treating either human beings or symbols of human life as merchandise or vendible goods. ... Commercialization refers to the practice of realizing large profits from the development and sale of techniques or products that involve distinctive human material, such as embryos, eggs, or tissue" (Alpers and Lo, 1995).

[11] Radin quotes Georg Lukács on the reification of commodities and the effects on human consciousness. Lukács writes: "The transformation of the commodity relation into a thing of 'ghostly objectivity' cannot therefore content itself with reduction of all objects for the gratification of human needs to commodities. It stamps its imprint upon the whole consciousness of man; his qualities and abilities are no longer an organic part of his personality, they are things which he can 'own' or 'dispose of' like the various objects of the external world" (Radin, 1996, 82). When we think about genes for enhancing memory or muscle mass, it is worth keeping in mind Lukács's claim that human qualities and abilities may come to be thought of as objects for sale in the external world.

[12] Erik Parens has noted the importance of attending to the big picture raised by stem cell work and how the politics of abortion has obscured that picture. See Parens, 2000.

[13] In another essay, Glannon argues that substantially increasing the human life span would profoundly affect issues of personal identity and thus a sense of personal responsibility for one's action. He ties his argument in interesting ways to the biology of memory function (Glannon, "Identity"). For a classic philosophical discussion of the problems associated with immortality, see Williams, 1973.

[14] Although he is not discussing stem cell research explicitly, Paul Rabinow's discussion of technological change wrought during the last two decades is worth noting. He writes: "In the United States, for example, in the last two decades, while the most passionate value conflicts have raged around abortion, a general reshaping of the sites of production of knowledge has been occurring. To cite the biotechnology industry, the growing stock of genomic information, and the simple but versatile and potent manipulative tools (exemplified by the polymerase chain reaction) is to name a few key elements; a more complete list would include the reshaping of American universities, the incessant acceleration in the computer domains, and the rise of 'biosociality' as a prime locus of identity a biologicalization of identity different from the older biological categories of the West (gender, age, race) in that it is understood as inherently manipulable and re-formable" (Rabinow, 1999, 13). A couple of pages later, he writes: "My analysis points to the fact that the basic understanding and practices of 'bare life' have been altered. The genome projects (human, plant, animal, microorganismic) are demonstrating a powerful approach to life's constituent matter. It is now known that DNA is universal among living beings. It is now known that DNA is extremely manipulable. One consequence among many others is that the boundaries between species need to be rethought; transgenic animals made neither by God nor by the long-term processes of evolution now exist (16).

[15] For a science fiction exploration of this theme of selected genetic enhancement, species boundary crossing see Octavia Butler, <u>Dawn</u>.

[16] A number of people either helped with the preparation of this report or provided feedback on an earlier draft. Thanks to Christa Adams, Diana Fritz Cates, William FitzPatrick, James L. Lissemore, Charlie Ponyik, Mary Jane Ponyik, Kristie Varga, Lisa Wells, and the ethics writers group at John Carroll University.

References

Alpers, Ann, and Bernard Lo. "Commodification and Commercialization in Human Embryo Research." <u>Stanford Law and Policy Review</u>, vol. 6, no. 2 (1995): 40.

American Association for the Advancement of Science and Institute for Civil Society. <u>Stem Cell Research and Applications Monitoring the Frontiers of Biomedical Research</u>. (November 1999).

American Fertility Society Ethics Committee. "Ethical Considerations of the New Reproductive Technologies in Light of Instruction on the Respect for Human Life in Its Origin and on the Dignity of Procreation." <u>Fertility and Sterility,</u> vol. 49, (1988): Supplement.

Andrews, Lori, and Dorothy Nelkin. <u>Body Bazaar: The Market for Human Tissue in the Biotechnology Age</u>. New York, NY: Crown Publishers, 2001.

Annas, George J. <u>Some Choice: Law, Medicine, and the Market</u>. New York: Oxford University Press, 1998.

Butler, Octavia E. <u>Dawn</u>. Boston, MA: Warner Books, 1997.

Cahill, Lisa Sowle. "Abortion, Autonomy, and Community." In <u>Abortion: A Reader</u>. Ed. Lloyd Steffen (Cleveland: Pilgrim Press, 1996), 359-72.

——. "The Embryo and the Fetus: New Moral Contexts." <u>Theological Studies</u>, vol. 54 (1993): 124-42.

Callahan, Daniel. <u>Hastings Center Report</u>, vol. 25, no. 1 (1995): 39.

Collins, Francis S. "Testimony before the Presidents Council on Bioethics." (December 13, 2002). Available at: http://www.bioethics.gov.

Congregation for the Doctrine of the Faith, "Declaration on Procured Abortion," 1974.

——. "<u>Donum vitae</u>," 1987.

Corea, Gena. <u>The Mother Machine.</u> New York: Harper and Row, 1985.

Doerflinger, Richard M. "Hearing on Stem Cell Research," Testimony of Richard M. Doerflinger on behalf of the Committee for Pro-Life Activities United States Conference of Catholic Bishops before the Subcommittee on Labor, Health, and Human Services, and Education Senate Appropriations Committee (July 18, 2001): available at <u>http://www.usccb.org/prolife/issues/bioethic/ stemcelltest71801.htm</u>.

——. "Public Comment before the National Bioethics Advisory Commission," National Conference of Catholic Bishops (April 16, 1999): available at http://www.usccb.org/prolife/issues/bioethic/nbac.htm.

——. "Hearing on Legal Status of Embryonic Stem Cell Research," Testimony of Richard M. Doerflinger on behalf of the Committee for Pro-Life Activities, National Conference of Catholic Bishops before the Senate Appropriations Subcommittee on Labor, Health, and Education (January 26, 1999): available at http://www.usccb.org/p;rolife/issues/bioethic/test99.htm.

——. "Testimony of Richard M. Doerflinger on behalf of the Committee for Pro-Life Activities National Conference of Catholic Bishops before the Senate Appropriations Subcommittee on Labor, Health and Education," Hearing on Embryonic Cell Research (December 2, 1998): available at http://www.usccb.org/prolife/issues/bioethics/1202.htm.

——. "Public Comment: NIH Human Embryo Research Panel." United States Conference of Catholic Bishops (February 2, 1994).

——. "Ethical Problems in the Use of Tissues from Abortion Victims," Public Comment, Meeting of the Human Fetal Tissue Transplantation Research Panel, (Bethesda, Maryland: National Institutes of Health, September 14, 1988): available at http://www.usccb.org/prolife/issues/ fetalresearch/rd9 1488.htm.

Dorff, Elliot N. "Stem Cell Research." Final Draft (August 2002).

Edwards, R. G. "IVF and the History of Stem Cells." Nature, vol. 413 (September 27, 2001): 349-51.

Edwards, Steptoe, and Purdy. "Fertilization and Cloning In Vitro of Preovulation Human Oocytes." Nature 277 (1970): 1307-1309.

Farley, Margaret A. "Roman Catholic Views on Research Involving Human Embryonic Stem Cells." The Human Embryonic Stem Cell Debate: Science, Ethics, and Public Policy. Cambridge, MA: The MIT Press, 2001. 113-18.

——. "The Church in the Public Forum: Scandal or Prophetic Witness?" The Catholic Theological Society of America, Proceedings of the Fifty-fifth Annual Convention, Vol. 55 (June 8-11, 2000): Presidential Address, 87-101.

FitzPatrick, William. "Surplus Embryos, Nonreproductive Cloning, and the Intend/ Foresee Distinction." Hastings Center Report (May-June 2003): 29-36.

Fukuyama, Francis. Our Posthuman Future. New York: Farrar, Straus and Giroux, 2002.

Gearhart, John. "New Potential for Human Embryonic Stem Cells." Science, vol. 282 (November 6, 1998): available at: http://www.sciencemag.org/cgi/content/full/282/5391/1061.

Geron Ethics Advisory Board. "Research with Human Embryonic Stem Cells: Ethical Considerations." Hastings Center Report, vol. 29, no. 2 (March-April 1999).

Gilliam, Lynn. "Arguing by Analogy in the Fetal Tissue Debate." Bioethics, vol. 11, no. 5 (1997): 397-412.

Glannon, Walter. "Extending the Human Life Span." Journal of Medicine and Philosophy, vol. 27, no. 3 (2002): 339-54.

——. "Identity, Prudential Concern, and Extended Lives." Bioethics, vol. 16, no. 3 (2002): 266-83.

Green, Ronald M. "Benefiting from 'Evil': An Incipient Moral Problem in Human Stem Cell Research." Bioethics, vol. 16, no. 6 (2002): 544-56.

——. The Human Embryo Research Debates: Bioethics in the Vortex of Controversy. New York: Oxford University Press, 2001.

——. "At the Vortex of Controversy." Kennedy Institute of Ethics Journal vol. 4, no. 4 (1994): 345-56.

Hall, Stephen S. Merchants of Immortality: Chasing the Dream of Human Life Extension. New York: Houghton Mifflin Company, 2003.

Hoffman, David I. et al. "Cryopreserved Embryos in the United States and Their Availability for Research." Fertility and Sterility, vol. 79, no. 5 (May 2003): 1063-69.

Holland, Suzanne. "Contested Commodities at Both Ends of Life: Buying and Selling Gametes, Embryos, and Body Tissues," Kennedy Institute of Ethics Journal, vol. 11, no. 3 (2001): 266.

Kass, Leon R. "Ageless Bodies, Happy Souls: Biotechnology and the Pursuit of Perfection." The New Atlantis, no. 1 (spring 2003).
——. Life, Liberty and the Defense of Dignity: The Challenge for Bioethics. San Francisco, CA: Encounter Books, 2002.

Kaveny, M. Cathleen. "Appropriation of Evil: Cooperation's Mirror Image," Theological Studies, vol. 61, no. 2 (June 2000): 280-313.

Knowles, Lori P. "Property, Progeny, and Patents." Hastings Center Report, vol 29. no. 2 (1999): 38.

Kuhn, Vanessa. "Stem Cells: Equity or Ownership?" The American Journal of Bioethics, vol. 2, no. 1 (winter 2002): 2.

Lauritzen, Paul. "Neither Person nor Property: Embryo Research and the Status of the Early Embryo," America (March 26, 2001).

——. Cloning and the Future of Human Embryo Research. New York: Oxford University Press, 2001.

——. "Expanding the Debate over Stem Cell Research," Dialog: A Journal of Theology, vol. 41, no. 3, (fall 2002): 238-39.

Lebacqz, Karen. "Stem Cells and Justice," Dialog, vol. 41, no. 3 (fall 2002).
——. "On the Elusive Nature of Respect." The Human Embryonic Stem Cell Debate: Science, Ethics, and Public Policy. Cambridge, MA: The MIT Press, 2001. 149-162.

Lewis, C. S. The Abolition of Man. New York: The Macmillan Company, 1947.

Macklin, Ph.D., Ruth. "Ethics, Politics, and Human Embryo Stem Cell Research." Women's Health Issues, vol. 10, no. 3 (May/June 2000): 111-15.

Magnus, David, Arthur Caplan, and Glenn McGee, ed. In Who Owns Life? Amherst, New York: Prometheus Books, 2002.

McCormick, R. A. "Who or What is the Preembryo?" In Corrective Vision: Explorations in Moral Theology. Kansas City: Sheed & Ward, 1994.

McGee, Glenn, and Elizabeth Banger. "Ethical Issues in the Patenting and Control of Stem Cell Research." In Who Owns Life? Ed. David Magnus, Arthur Caplan, and Glenn McGee. Amherst, New York: Prometheus Books, 2002. 243-264.

McKenny, Gerald P. "Enhancements and the Ethical Significance of Vulnerability." In Enhancing Human Traits: Ethical and Social Implications. Washington, D.C.: Georgetown University Press, 1998. 222-37.

McKibben, Bill. Enough: Staying Human in an Engineered Age. New York: Henry Holt and Company, 2003.

McLean, Margaret. "Stem Cells: Justice at the Gate." Dialog, vol. 41, no. 3 (fall 2002).

Meilaender, Gilbert. "Terra es animata: On Having a Life." Hastings Center Report, vol. 23, no. 4 (July-August 1993): 25-32.

Meyer, Michael J., and Lawrence J. Nelson. "Respecting What We Destroy: Reflections on Human Embryo Research." Hastings Center Report, vol. 31, no. 1 (January/February 2001).

Midgley, Mary. "Biotechnology and Monstrosity: Why We Should Pay Attention to the 'Yuk Factor.'" Hastings Center Report, vol. 30, no. 5 (2000): 7-15.

National Academy of Sciences. Stem Cells and the Future of Regenerative Medicine. Washington D.C.: National Academies Press, 2002.

National Bioethics Advisory Commission. Ethical Issues in Human Stem Cell Research. vol. 1. Rockville, Maryland: NBAC, 1999.

Nelkin, Dorothy, and Lori B. Andrews. Body Bazaar: The Market for Human Tissue in the Biotechnology Age. New York, NY: Crown Publishing, 2001.

National Institutes of Health, "Report of the Human Embryo Research Panel." (September 1994).

Nussbaum, Martha C. "Compassion and Terror." Daedalus, vol. 128, no. 4 (winter 2003): 10-26.

------. "Human Capabilities, Female Human Beings." Women, Culture, and Development: A Study of Human Capabilities. Ed. Martha C. Nussbaum and Jonathan Glover (Oxford: Clarendon Press, 1995), 61-104.

Outka, Gene. "The Ethics of Stem Cell Research." Kennedy Institute of Ethics Journal vol. 12, no. 2 (2002): 175-213.

Parens, Ph.D., Erik. "Embryonic Stem Cells and the Bigger Reprogenetic Picture." Women's Health Issues, vol. 10, no. 3 (May/June 2000): 116-20.

Pontifical Academy for Life. "On the Production and the Scientific and Therapeutic Use of Human Embryonic Stem Cells." Vatican (August 25, 2000): available at Http://www.petersnet.net/browse/ 3021.htm.

Rabinow, Paul. French DNA: Trouble in Purgatory. Chicago, IL: The University of Chicago Press, 1999.

Radin, Margaret Jane. Contested Commodities. Cambridge, MA: Harvard University Press, 1996.

Resnik, David. "The Commercialization of Human Stem Cells: Ethical and Policy Issues." Health Care Analysis, vol. 10 (2002):127-54.

Robertson, John A. "Ethics and Policy in Embryonic Stem Cell Research." Kennedy Institute of Ethics Journal, vol. 9, no. 2 (1999): 109-36.

Roche, J.D., Patricia A. and Michael A. Grodin, M.D., F.A.A.P. "The Ethical Challenge of Stem Cell Research." Women's Health Issues, vol. 10, no. 3 (May/June 2000): 136-39.

Ryan, Maura. "Creating Embryos for Research: On Weighing Symbolic Costs." Cloning and the Future of Human Embryo Research. Ed. Paul Lauritzen. New York: Oxford University Press, 2001.

——. The Ethics and Economics of Assisted Reproduction. Washington, D.C.: Georgetown University Press, 2001.

——. "The Politics and Ethics of Human Embryo and Stem Cell Research." Women's Health Issues, vol. 10, no. 3 (May/June 2000): 105-10.

Safire, William. "Why Die?" The New York Times (January 1, 2000).

Shamblott, M. J. et al. "Derivation of Pluripotent Stem Cells from Cultured Human Primordial Germ Cells." Proceedings of the National Academy of Sciences, vol. 95 (1998): 13726-31.

Shannon, T. A., and A. B. Walter. "Reflections on the Moral Status of the Preembryo." Theological Studies, vol. 51 (1990): 603-26.

Shannon, Thomas. "Human Embryonic Stem Cell Therapy." Theological Studies, vol. 62, (2001): 811-24.

Squier, Susan. Liminal Lives, unpublished manuscript.

Steinbock, Bonnie. "Respect for Human Embryos." In Cloning and the Future of Human Embryo Research. Ed. Paul Lauritzen. New York: Oxford University Press, 2001.

——. "What Does 'Respect for Embryos' Mean in the Context of Stem Cell Research?" Women's Health Issues, vol. 10, no. 3 (May/June 2000): 127-30.

——. Life before Birth: The Moral and Legal Status of Embryos and Fetuses. New York, NY: Oxford University Press, 1992.

Tauer, Carol. "Responsibility and Regulation: Reproductive Technologies, Cloning, and Embryo Research." In Cloning and the Future of Human Embryo Research. Ed. Paul Lauritzen. Oxford University Press: New York, 2001. 153.

——. "Preimplantation Embryos, Research Ethics, and Public Policy." Bioethics Forum, vol. 11, no. 3 (1995): 30-37.

Thomson, James A. et al. "Embryonic Stem Cell Lines Derived from Human Blastocysts." Science, vol. 282 (November 6, 1998).

Van Overwalle, Geertrui. Study on the Patenting of Inventions Related to Human Stem Cell Research. Luxembourg: Office for Official Publications of the European Communities, 2002. 23.

Waldby, Ph.D., Catherine, and Susan Squier, Ph.D. "Ontogeny, Ontology, and Phylogeny: Embryonic Life and Stem Cell Technologies." Configurations (Forthcoming).

Warren, Mary Anne. Moral Status: Obligations to Persons and Other Living Things. New York, NY: Oxford University Press, 1997.

Williams, Bernard. Problems of the Self: Philosophical Papers 1956-1972. Cambridge, England: Cambridge University Press, 1973.

Wolfe, Tom. Hooking Up. New York, NY: Farrar, Straus and Giroux, 2001.

Zoloth, Laurie. "Jordon's Banks: A View from the First Years of Human Embryonic Stem Cell Research." In The Human Embryonic Stem Cell Debate. Ed. by Suzanne Holland, Karen Lebacqz, and Laurie Zoloth. Cambridge, MA: The MIT Press, 2001. 238.

Appendix H.

**Human Embryonic Germ Cells:
June 2001 – July 2003
The Published Record**

JOHN GEARHART, PH.D.
*C. Michael Armstrong Professor,
Institute for Cell Engineering,
Johns Hopkins University School of Medicine*

There have been but two original research articles published on human embryonic germ cell in the period covered by this report. It is appropriate that only peer-reviewed articles be considered in this report, as the field of stem cell research is rife with undocumented or unsubstantiated claims. There have been several publications on EG cells derived in mice and chick and comment will be made on these reports as they may impact on the eventual use or study of the human cells.

There is a concern, given the imprinting-based developmental abnormalities observed in humans, and those that have been produced experimentally in animals, that dysregulated gene expression in stem cell derived tissues could pose a serious problem in the use of such tissues for cell-based therapies. Genomic imprinting is defined as an epigenetic modification of the DNA (other than sequence) in the germ line that leads to the preferential expression of a specific allele (monoallelic expression) of some genes in the somatic cells of the offspring in a parental dependent manner. Because a maternal gene allele may be a paternal allele in the next generation, the imprinting must be reprogrammed in the germ line, that is, the epigenetic 'marks' must be erased (during germ cell development) and established in the newly formed embryo. The timing of the erasure in humans is not known. Imprinting involves methylation at specific sites in the DNA and most likely, other changes as well.

There have been several reports from experiments with the mouse that indicate that imprinting is abnormal in pluripotent stem cells

273

derived from mouse embryos. These studies include abnormal or variable imprints in mouse EG cells, abnormal imprints in ES cells derived from interspecific crosses, and abnormal gene expression in mice derived from ES cell nuclear transfer.

The results of the study by Onyango *et al.* utilizing EG cell lines derived at John Hopkins, clearly demonstrate that general dysregulation of imprinted genes will not be a barrier to their use in transplantation therapies. The report has determined that the EG cells are not imprinted, that is, imprinting has been erased in the primordial germ cells that gave rise to the EG cells, that the erasure is maintained in the EG cells, but in all informative cases, they observed the transcription of only a single allele in differentiated cells derived from the human EG cells. These results, although on a limited number of lines and only a few imprinted genes, would indicate that these human EG cell lines will serve as reliable and safe sources for the study of EG cell differentiation and, perhaps, cells for cell-based interventions.

Another area of interest in genetic regulation within EG cells is that of X inactivation, the mammalian method for equalization of the dosage of X-linked genes in males and females. This equalization is accomplished by the down regulation of the transcriptional output of the X chromosomes in females, so that only one X is active in diploid somatic cells of both sexes. Inactivation is initiated in female blastocysts. Both X chromosomes in female primordial germ cells are active. Migeon *et al.* (2001) have demonstrated that in the very early stages of differentiation of cells from human EGs, only one X chromosome is active, indicating normal genetic regulation has occurred. In a report by Nesterova *et al.* (2002) on the use of mouse female EG cells in the study of X chromosome inactivation/ reactivation during primordial germ cell migration and EG cell formation. Both X chromosomes appear to be active in XX EG cells, and presumably, one becomes inactive when cells differentiate from the EG cells.

One of the goals of stem cell research is to provide sources of cells for cell-based therapies. As a step in this direction, proof of concept or proof of principle studies involve the use of human cells in animal models of human disease or injury. Although few, if any, animal models are true models for the human diseases, they are the closest approximation that can be made. The first report on the use of cells derived from stem cells of human embryonic sources was recently published: Kerr *et al.* Human Embryonic Germ Cell Derivatives Facilitate Motor Recovery of Rats with Diffuse Motor Neuron Injury,

J. Neuroscience 23, June 15, 2003. This is the first demonstration that a human pluripotent stem cell derived form embryonic or fetal tissue can ameliorate a disease process in an animal model.

Neural progenitor cells, derived from human EG cells, were introduced into the cerebrospinal spinal fluid of rats that had been paralyzed as a result of infection with a neuroadapted Sindbis virus that specifically targets motor neurons in the spinal cord. All animals in which human cells were found had some degree of hindlimb recovery. It was clear from the histology of the animals that the human cells had differentiated into appropriate neural cell types within the ventral horns, including motor neurons, the results indicated that the major effect of the human cells was to protect host neurons from death and to facilitate reafferentiation of motor neuron cell bodies. Growth factors responsible for this recovery, produced by the human cells, were identified as brain-derived neurotrophic factor and transforming growth factor.

The significance of this experiment appears to be that for this disease model, the human cells supply factors that facilitate motor neuron recovery following viral damage. However, the cells did migrate to the site of injury in the spinal cord, which differentiates them from other cellular grafts, and then delivered the factors. Also, there was considerable differentiation of the human cells into various neural cells types in the cord.

In a review article on deriving glucose-responsive insulin-producing cells from stem cells (Kaczorowski et al. 2002), mention is made on the isolation of such cells from mouse and human ES cells and human EG cells, but no new data was presented, only a reference to a paper published in 2001.

EG cell studies in other species

Several studies on EG cells from other species have been published. EG cells from the chick have been demonstrated to yield germ line chimeras when transferred to early embryos (Park et al. 2003). EG cell lines of the mouse were found to colonize not only the epiblast but also the primary endoderm of the gastrulating embryo following aggregation with 8-cell embryos (Durcova-Hills et al. 2003). This observation was permitted as a result of using a transgenic construct effect as a lineage marker for cells of the early embryo. Horii et al. (2002) report the serum-free culture of mouse EG cells, a system which inhibits the 'spontaneous' differentiation due to the presence of various growth factors in the serum. The cells are, however, are cultured on a feeder layer.

An area of great interest in EG cell derivation is the basis of the underlying mechanism for the conversion, derivation or transformation (terms reflecting our ignorance of the process) of primordial germ cells to EG cells. Kimura *et al.* (2003) report that the loss of a tumor suppressor gene in mice, PTEN (phosphatase and tensin homology deleted form chromosome ten, aka MMAC1 and TEP1), leads to a high incidence of testicular teratomas and enhances the production of EG cells from PGCs of the mutant mice. While it is highly unlikely that the loss of this gene in the culture of normal PGCs leads to the derivation of EG cells, it may implicate downstream signaling pathways that are involved in cell cycle progression, cell survival and cell migration as important players in the process.

References

Durcova-Hills, G., Wianny, F., Merriman, J., Zernicka-Goetz, M., and McLaren, A. 2003. Developmental fate of embryonic stem cells (EGCs), *in vivo* and *in vitro*. Differentiation 71, 135-141.

Horii, T., Nagao, Y., Tokunaga, T., and Imai, H. 20003. Serum-free culture of murine primordial germ cells and embryonic germ cells. Theriogen. 59, 1257-1264.

Kaczorowski, D.J., Patterson, E.S., Jastromb, W.E., Shamblott, M.J. 2002. Review Article. Glucose-responsive insulin-producing cells from stem cells. Diabetes Metab. Res. Rev. 18, 442-450.

Kerr, D.A., Lladó, J., Shamblott, M.J., Maragakis, N.J., Irani, D.N., Crawford, T.O, Krishnan, C., Dike, S., Gearhart, J.D., and Rothstein, J.D. 2003. Human embryonic germ cell derivatives facilitate motor recovery of rats with diffuse motor neuron injury. J. Neurosci. 23, 5131-5140.

Kimura, T., Suzuki, A., Fujita, Y., Yomogida, K., Lomeli, H., Asada, N., Ikeuchi, M., Nagy, A., Mak, T.W., and Nakano, T. 2003. Development and Disease. Conditional loss of PTEN leads to testicular teratoma and enhances embryonic germ cell production. Development 130, 1691-1700.

Migeon, B.R., Chowdry, A.K., Dunston, J.A., and McIntosh, I. 2001. Identification of *TSIX*, encoding an RNA antisense to human *XIST*, reveals differences from its murine counterpart: Implications for X inactivation. Am. J. Human Genet. 69, 951-960.

Onyango, P., Jiang, S., Uejima, H., Shamblott, M.J., Gearhart, J.D., Cui, H., and Feinberg, A.P. 2002. Monoallelic expression and methylation of imprinted genes in human and mouse embryonic germ cell lineages. Proc. Natl. Acad. Sci. USA 99, 10599-10604.

Park, T.S., Hong, Y.H., Kwon, S.C., Lim, J.M., and Han, J.Y. 2003. Birth of germline chimeras by transfer of chicken embryonic germ (EG) cells into recipient embryos. Molec. Reprod, Develop. 65, 389-395.

Sapienza, C. 2002. Commentary. Imprinted gene expression, transplantation medicine, and the "other" human embryonic stem cell. Proc. Natl. Acad. Sci. USA 99, 10243-10245.

Appendix I.

Current Progress in
Human Embryonic Stem Cell Research

TENNEILLE E. LUDWIG, PH.D., and
JAMES A. THOMSON, PH.D
*Wisconsin National Primate Research Center, University of
Wisconsin-Madison*

The immortality and potentially unlimited developmental capacity of human embryonic stem (ES) [1] cells ignite the imagination. After months or years of growth in culture dishes, these cells retain the ability to form cell types ranging from heart muscle to nerve to blood—possibly any cell in the body. Because of their unique developmental potential, human ES cells have widespread implications for human developmental biology, drug discovery, drug testing, and transplantation medicine. Indeed, human ES cells promise an essentially unlimited supply of specific cell types for *in vitro* experimental studies and for transplantation therapies for diseases such as heart disease, Parkinson's disease, leukemia, and diabetes.

The derivation of a human ES cell line destroys a human embryo. Thus, the derivation of human ES cells resurrected a fierce controversy over human embryo research in the United States, a controversy originally created by the development of human *in vitro* fertilization decades ago, but never completely resolved. In particular, the derivation of human ES cells led to a re-examination of the role of federal funding of human embryo research. In response to the intense public interest, President George W. Bush reviewed the potential of human ES cell research to improve the health of Americans. In his national address on August 9, 2001, he stated "Federal dollars help attract the best and brightest scientists. They ensure new discoveries are widely shared at the largest number of research facilities, and the research is directed toward the greatest public good." On that basis, he directed federal funding to "explore the promise and potential of stem cell research," including, for the first time, human ES cell research. However, the President went on to restrict federal funding to research that used only those human

279

ES cell lines derived prior to his address on August 9, 2001. This paper reviews progress in human ES cell research in the wake of that decision.

Human ES Cell Publication Summary

Since their initial derivation, only a limited number of independent (i.e., derived from different embryos) human ES cell lines that meet President Bush's criteria have been used in published research. Just nine human ES cell lines meeting President Bush's criteria are currently listed by the National Institutes of Health as readily available for distribution to investigators. Despite limited availability to date, research with human ES cells is proceeding. Forty human ES cell primary research papers have been published in peer-reviewed journals since the initial publication of human ES cell isolation in 1998 (Table 1). Published human ES cell research includes studies on the optimization of the culture environment, characterization of human ES cells, modification of the ES cell genome, and differentiation.

Table 1. Human ES Cell Research Publications

Area of interest	Publications to date	Reference(s)
Derivation	5	[1-5]
Culture Optimization	6	[6-11]
Feeder Layer Alternatives/Replacements	4	*[6-9]*
Media Analysis	1	*[10]*
Freezing	1	*[11]*
Characterization	5	[12-16]
Modification	6	[17-22]
Differentiation into multiple lineages	3	[23-25]
Differentiation into specific lineages	15	[26-40]
Neural	4	*[26-29]*
Cardiac	5	*[30-34]*
Endothelial (Vascular)	1	*[35]*
Hematopoietic (Blood)	2	*[36, 37]*
Pancreatic (Islet-like)	1	*[38]*
Hepatic (liver)	1	*[39]*
Trophoblast	1	*[40]*

Culture Optimization for Human ES Cells.

Improvement of culture conditions to enable large-scale production and reduce safety concerns has been a major research focus. The first two research groups that described the derivation of human ES cell lines examined long-term proliferation, karyotypic stability, developmental potential, and cell surface marker expression by ES cells [1, 2]. Because these first human ES cell lines remain the most extensively characterized, most subsequent research has utilized them. These human ES cell lines were derived on mouse fibroblast feeder layers in the presence of fetal bovine serum. The exposure to these and other sources of animal proteins has raised concern that some yet unidentified pathogen(s) may have been transferred to the ES cells by contact with cells or proteins from other species, and that these pathogens could be transferred to patients if these ES cells were to be used for transplantation therapies. Thus, several research groups have been actively working to reduce or eliminate non-human cells or proteins from human ES cell culture.

Significant progress has been made in eliminating serum, and limited progress has been made in eliminating fibroblasts from human ES cell culture. Serum is a complex, poorly defined mixture of components, and there is significant variation between lots [41]. Individual lots of serum, therefore, must be carefully screened for their ability to sustain undifferentiated ES cell growth. If basic fibroblast growth factor is added to a proprietary serum substitute (Gibco BRL® Knockout™ Serum Replacer), it supports human ES cells and significantly reduces the batch variability associated with serum [12]. However, this medium does not eliminate all serum products from human ES cell culture medium, as it still contains a bovine serum albumin component. With this same medium, human ES cells can be cultured without direct contact with feeder layers if the medium is first conditioned by exposure to mouse embryonic fibroblasts [6]. However, the medium still contains bovine serum products and is exposed to fibroblasts, therefore cross-species contamination with pathogens remains a concern.

Recent reports demonstrate that human ES cells can be maintained on feeder layers of human origin. Feeder layers obtained from human bone marrow [9], fetal muscle or skin [7], adult human fallopian tube epithelial cells [7], or human foreskin [8] support human ES cell proliferation and maintenance of normal karyotype and developmental potential. These results led to the growth of human ES cells in the complete absence of non-human products [7].

New human ES cells derived under these conditions would eliminate concerns about cross-species contamination by pathogens, but such cell lines could not, at present, be supported by federal funding in the United States. Growth on human feeder layers is a significant advance because of reduced safety concerns; nonetheless, the preparation of these feeders remains laborious and introduces a significant source of biological variability. The complete elimination of feeder layers and serum from human ES cell culture medium and their replacement by defined, cloned products remains an important goal for the future and is an active area of research for several groups.

Genetic Modification of Human ES Cells

Although the human genome project is essentially completed, we are ignorant about the function of most human genes. Human ES cells provide a powerful new model for identifying the function of any human gene, and this requires efficient methods for genetic modification of human ES cells. Genetic manipulation of human ES cells is essential to elucidate gene function; direct the differentiation of ES cells to specific lineages; purify desired differentiated cell types from mixed populations of ES cell derivatives; use the differentiated derivatives of ES cells as a vehicle for gene therapy; and modulate the immune response to transplanted ES cell derivatives.

Transfection methods routinely used for mouse ES cells generally fail to transfect human ES cells efficiently, but there have now been several approaches developed for human ES cells. Transient [17] and stable [18] integration of plasmids into human ES cells can be accomplished through specific transfection reagents, the best reagents yielding stable (drug-selectable) transfection rates of about 10^{-5}. Recently more labor-intensive, HIV-based, lentivirus vectors have been shown to transduce human ES cells at an efficiency rate of over 90% [20, 21]. This should allow complex mixtures of genes to be screened for specific phenotypic effects by a process termed "expression cloning" [20].

Homologous recombination allows the defined modifications of specific genes in living cells [42, 43] and has been used extensively with mouse ES cells. However, the differences between mouse and human ES cells delayed the development of homologous recombination in human ES cells. Except for viral approaches, high stable transfection efficiencies in human ES cells have been difficult to achieve, and in particular, electroporation protocols established for mouse ES cells do not work in human ES cells [22]. Also, in contrast to mouse ES cells, human ES cells proliferate inefficiently from single cells, making screening procedures to identify rare homologous recombination events difficult [12]. We have recently developed modified electroporation protocols to overcome these problems and have successfully targeted a ubiquitously expressed gene (HPRT1), an ES cell-specific gene (POU5F1), and a tissue-specific gene (Tyrosine hydroxylase: TH) in human ES cells [22, 44]. The overall targeting frequencies for the three genes suggest that homologous recombination is a broadly applicable technique in human ES cells.

Homologous recombination in human ES cells will be important for studying gene function in vitro and for lineage selection. For therapeutic applications in transplantation medicine, controlled modification of specific genes should be useful for purifying specific ES cell-derived, differentiated cell types from a mixed population [45]; altering the antigenicity of ES cell derivatives; and giving cells new properties (such as viral resistance) to combat specific diseases. Homologous recombination in human ES cells might also be used for approaches combining therapeutic cloning with gene therapy [46]. In vitro studies using homologous recombination in human ES cells will be particularly useful for learning more about the pathogenesis of diseases where mouse models have proven inadequate. For example, HPRT-deficient mice fail to demonstrate an abnormal phenotype, yet defects in this gene cause Lesch-Nyhan disease in children [47]. In vitro neural differentiation of HPRT-deficient human ES cells or transplantation of ES cell-derived neural tissue to an animal model could help to understand the pathogenesis of Lesch-Nyhan syndrome.

Human ES Cells as a Model of Early Human Development

The excitement surrounding the prospective role of human embryonic stem (ES) cells in transplantation therapy has often overshadowed a potentially more important role as a basic research tool for understanding the development and function of human tissues. The use of human ES cells is particularly valuable to derive tissue for study that is difficult to obtain otherwise, and for which animal models are inadequate.

Human ES cells offer a new and unique window into the early events of human development, a period critical for understanding infertility, birth defects, and miscarriage. Because manipulation of the early post-implantation human embryo could jeopardize the health of the resulting child, it has never been possible to examine this important period of human development experimentally. Nearly all of what is known about early human development, especially in the early post-implantation period, is based on very rare histological sections of human embryos, or on an imperfect analogy to experimental studies in the mouse. The mouse has been the mainstay of mammalian experimental embryology because of its historical use, well-defined genetics, and favorable reproductive characteristics. However, early mouse development and early human development differ significantly. For example, human and mouse embryos differ in the expression of embryonic antigens; timing of embryonic genome expression; formation, structure, and function of the fetal membranes and placenta; and formation of an embryonic disc instead of an egg cylinder. Thus, if one is interested in the development of a human tissue known to differ significantly from the corresponding mouse tissue, such as the yolk sac or the placenta, studying a human model is desirable.

The first differentiation event in mammalian embryos is the formation of the trophectoderm, the outer epithelial layer of the blastocyst. The trophectoderm is crucial for implantation of the embryo and gives rise to specialized populations of trophoblast cells in the placenta [48, 49]. Mouse and human placentas differ in structure and function, and these differences are clinically significant. For example, the placental hormone chorionic gonadotropin, which has an essential role in establishing and maintaining human pregnancy, is not even produced by the mouse placenta. When formed into chimeras with intact preimplantation embryos, mouse ES cells rarely contribute to the trophoblast, and the manipulation of external culture conditions has, to date, failed to direct mouse ES cells to form trophoblast [50].

Spontaneously differentiated rhesus monkey [51] or human ES cells [1] do secrete modest amounts of chorionic gonadotropin, indicating the differentiation of trophoblast cells [52]. Recently it was discovered that a single growth factor (BMP4) would induce human ES cells to differentiate to a pure population of early trophoblast [40]. These early human trophoblast cells have never before been available for detailed study, and already this new experimental model has provided information about the specific genes that control the early development of the human placenta [40]. The derivation of other early lineages from human ES cells *in vitro* to provide a more complete understanding of early human development is an active area of research.

Cardiovascular Differentiation

Cardiovascular disease is the leading cause of death in the United States, taking the lives of more people each year than the next five leading causes of death combined [53]. Cardiovascular disease and its related disorders affect more than 68 million Americans, at a cost of more than 350 billion dollars annually. Heart disease alone accounts for 229 billion dollars in health care costs each year. Adult heart tissue cannot be expanded in culture, and thus, there are no human heart cell lines available for research. The limited amount of physiological research done directly on human heart cells has generally relied on biopsy samples, which are small, erratically available, and usually obtained from diseased hearts. In contrast, human ES cells are already providing a reliable *in vitro* supply of human heart cells for experimental study [30-34].

Animal models, such as the mouse, have historically been used for the study of the heart. However, there are clinically significant physiological differences between animal and human cardiomyocytes that limit the usefulness of these models. For example, the mechanisms regulating the QT polarization interval—the time required for repolarization of the heart muscle between beats—differ significantly between species. A prolonged QT polarization interval in humans is related to ventricular arrhythmias and cardiac arrest and has been a significant side effect of a wide range of drugs in early human clinical trials. Drugs exhibiting this serious side effect must be withdrawn from clinical trials, and such drugs have been responsible for patients' deaths. Because the mechanisms that regulate repolarization of the heart muscle cells differ appreciably between human and mouse models, screening drugs on mouse hearts does not reliably detect this side effect. Yet, because they do not divide in culture, human heart cells have not been previously available for screening.

Human ES cells differentiate spontaneously to heart muscle cells, and several research groups have reported the characterization of these cells [23, 31, 54]. Human ES cells allowed to differentiate in unattached clumps (termed "embryoid bodies") form synchronized contracting areas that express appropriate cardiac markers [23, 31, 32]. Co-culture of human ES cells with visceral, endoderm-like cells also causes differentiation to cardiomyocytes [34], and 5-aza-2'-deoxycytidine or density gradient separation allows some enrichment of cardiomyocyte populations [32]. Human ES cell-derived cardiomyocytes display many of the functional properties of native cardiomyocytes, including the generation of synchronized action potentials and response to cardioactive drugs [30-33]. Heart cells exhibiting action potentials characteristic of nodal, atrial, and ventricular cardiomyocytes are all present, and the human-specific mechanisms regulating QT interval are functional [30]. Thus, human ES cell-derived heart cells are already useful for drug screening, and their use should make the drug development process quicker, cheaper, and safer.

There is also a great interest in using human ES cell-derived heart cells for transplantation, but this will likely be challenging. Studies in animal models demonstrate that cell transplantation is effective in increasing the myocyte population in damaged or diseased cardiac tissue [55]. However, when heart cells die in a heart attack, it is not because the heart cells themselves are defective, but because the blood supply is cut off. Thus, to be successful, transplanted heart cells would have to integrate functionally with the surrounding heart cells, obtain a new blood supply, and avoid immune rejection. Each of these problems has potential solutions, but will require significant time and effort to solve. Precursors of vascular tissue can also be derived from human ES cells [35], and such cells may be useful in supporting co-transplanted heart cells.

Neural Differentiation

Because of the country's aging population, neural degenerative disorders such as Parkinson's disease and Alzheimer's disease are becoming increasingly prevalent in the United States. Historically, one of the difficulties in studying the pathogenesis of neural disease has been the very limited access to the specific neural cells involved in these diseases. Neural precursor (or stem) cells cultured from fetal and adult brains have been extensively studied, but appear to have limited developmental potential. For example, the sustainable differentiation of neural stem cells to dopaminergic neurons, the cell defective in Parkinson's disease, has not yet been achieved. Mouse ES cells, for example, differentiate efficiently to dopaminergic neurons, and several groups are beginning to apply approaches used with mouse ES cells to human ES cells. Human ES cells should offer an improved supply of neural tissue, for both *in vitro* experimental studies and transplantation therapies.

Human ES cell-derived embryoid bodies produce both neural precursor cells and cells expressing markers of mature neurons and glia [26-29]. The percentage of neural precursors can be enriched by alteration of culture conditions [26-28] or by purification using cell surface markers [28]. Human ES cell-derived neural cells are able to synthesize and respond to neurotransmitters, form synapses and voltage-dependent ion channels capable of generating action potentials, and generate electrical activity [28]. Some human ES cell-derived neurons express tyrosine hydroxylase, the rate-limiting enzyme involved in dopamine synthesis and a marker of dopaminergic neurons [26, 27].

Human ES cell-derived neural precursors transplanted into the mouse brain differentiate into all three types of central nervous system cells (neurons, glia cells, and oliogodendrocytes) [26, 27]. These differentiated cells migrate, following host developmental cues, into various areas of the brain (including cortex, hippocampus, striatum olfactory bulb, septum, thalamus, hypothalamus, and midbrain) [26, 27]. One of the concerns about using human ES cell-derived neural cells in transplantation therapy is the fear that undifferentiated ES cells may be transplanted with the differentiated cells and form teratomas in the host. To date, transplantation of isolated, human ES cell-derived neural precursor cells into mice has not produced teratomas [26, 27], suggesting that appropriate selection procedures can eliminate undifferentiated ES cell contamination. However, longer-term testing is still needed to address the teratoma formation issue more carefully.

Hematopoietic Differentiation

Human ES cells are already providing a sustainable source of hematopoietic cells for *in vitro* studies [36, 37]. Hematopoietic stem cells are by far the most studied adult stem cells, and bone marrow transplants are the most common and effective form of stem cell-based therapy. However, despite several decades of research by hundreds of laboratories, hematopoietic stem cells have not yet been successfully expanded in clinically useful amounts, and these cells must instead be transferred directly from the donor. When cultured *in vitro*, hematopoietic stem cells do not self-renew, but instead differentiate to specific blood cells, and thus quickly disappear. This makes the *in vitro* study of human hematopoiesis difficult, as researchers must continually return to patients to obtain hematopoietic stem cells from bone marrow, peripheral blood, or placental cord blood. Human ES cells can differentiate into hematopoietic precursor cells through co-culture with murine bone marrow or yolk sac cells [36]. Enrichment of ES cell-derived hematopoietic precursors is accomplished by treatment with cytokines or BMP-4 [37]. Cell sorting using hematopoietic-specific cell surface markers yields myeloid, erythroid, and megakaryocyte precursors [36].

There are three major areas where human ES cell hematopoiesis should impact human medicine. First, because human ES cells can be expanded without limit, human hematopoiesis can be studied without the need to continually return to patients for tissue donations. The knowledge of these *in vitro* studies is likely to improve therapies based on adult hematopoietic stem cells. Second, human ES cell-derived blood cells could be used either in bone marrow transplants, or as a source of blood products such as red blood cells and platelets. And third, ES cell-derived hematopoietic stem cells could aid in ES cell-based transplantation therapies for other (non-hematopoietic) tissues. Transplantation of ES cell-derived hematopoietic stem cells could be used to reduce or eliminate immune rejection by creating hematopoietic chimerism in patients receiving co-transplantation of other human ES cell-derived tissues [45, 56-58].

Pancreatic Differentiation

Type 1 diabetes offers one of the most promising applications of human ES cell-based transplantation therapy. The destruction of pancreatic islet β-cells results in type 1 diabetes. β-cells produce insulin, and as their numbers dwindle, the ability to appropriately control blood glucose levels is lost. Even with current insulin therapies, type 1 diabetes reduces a patient's life expectancy by 10 to 15 years, and these patients often develop serious complications such as blindness and kidney failure [59]. Recently, the transplantation of β-cells from cadavers has proven to be an effective treatment for some forms of uncontrollable diabetes, but the source of tissue for transplantation is severely limiting and will never come close to meeting the demands of over one million people with type 1 diabetes in the United States. Spontaneous *in vitro* differentiation of human ES cells reveals a percentage of cells that produce insulin and express other β-cells specific markers, offering hope of a scalable source of β-cells for transplantation [38]

The challenges for using human ES cell-derived β-cells for transplantation are significant and parallel those that face the entire field of ES cell-based transplantation therapies. First, pancreatic development is incompletely understood, and it is not yet possible to direct ES cells to β-cells efficiently. However, given the pace of advances in developmental biology over the last decade, it is likely that in the next five to ten years, it will be possible to routinely generate clinically useful quantities of β-cells from human ES cells. Second, integration into the body in a form that restores function of the damaged tissue is essential. This is easier for β-cells than for most cell types, as the function that must be restored is secretion of insulin into the blood stream in response to high glucose, and this function does not require a complex physical connection between the transplanted and host tissues. Indeed, the clinical trials using cadaver-derived β-cells have transplanted the cells into the liver, and the cells function in that site. Third, transplanted β-cells must not be rejected by the immune system. Although the transplantation of β-cells has been clinically successful, the severe immunosuppressive therapy required may make the procedure inappropriate for the average diabetic patient. Importantly, β-cells derived from adult stem cells from the patient, or even from ES cells derived through "therapeutic cloning" using a nucleus from the patient, would not solve the immune rejection problem for diabetes. Type 1 diabetes is an autoimmune process, and unless that immune response is altered the very process that made the patient diabetic in the first place would destroy transplanted β-cells genetically identical to the patient's. Finally, neoplastic transformation of the transplanted cells is a serious concern for any cell-based therapy in which the cells are first cultured extensively. All actively dividing cells accumulate mutations over time, and the potential exists that enough mutations could accumulate to make some cells tumor cells.

None of the challenges facing ES cell-based transplantation therapies are insurmountable, and indeed, type 1 diabetes is an excellent candidate for treatment using this approach. However, the challenges do underscore both the importance of careful preclinical testing, particularly in non-human primates, and the amount of work still to be done before people's lives will be improved by these therapies.

Conclusions

Since their initial derivation, there has been significant progress in culture optimization, characterization, genetic modification, and differentiation of human ES cells. However, ethical and political controversy continues to impede progress in human ES cell research. The decision by President George W. Bush, restricting federal funding to human ES cell lines derived before August 9, 2001, created a distribution bottleneck that is just now beginning to be resolved. Although these initial cell lines may support much of the basic research now being conducted, the very first cell lines were originally derived for research purposes, with the expectation that future cell lines would more appropriately address legitimate safety concerns for therapeutic applications. In spite of the slow start, the diversity of investigators already contributing to human ES cell research is, nonetheless, promising and suggests that the initial lag phase for the human ES cell field is already coming to an end and that an exponential growth phase is beginning. During the next year or two, it is likely that the purification of specific, therapeutically useful human ES cell derivatives, such as dopaminergic neurons, will be published, and that defined culture conditions eliminating the need for both feeder layers and non-human proteins will be developed. When these events occur, President Bush's compromise will be particularly damaging to the field, and there will be an even greater need to derive new cell lines.

References

1.   Thomson JA, Itskovitz-Eldor J, Shapiro SS, Waknitz MA, Swiergiel JJ, Marshall VS, Jones JM. Embryonic stem cell lines derived from human blastocysts. Science 1998; 282: 1145-1147.

2.   *Reubinoff BE, Pera MF, Fong CY, Trounson A, Bongso A. Embryonic stem cell lines from human blastocysts: somatic differentiation in vitro. Nat Biotechnol 2000; 18: 399-404.*

3.   *Lanzendorf SE, Boyd CA, Wright DL, Muasher S, Oehninger S. Use of human gametes obtained from anonymous donors for the production of human embryonic stem cell lines. Fertility and Sterility 2001; 76: 132-137.*

4.   He Z, Huang S, Li Y, Zhang Q. Human embryonic stem cell lines preliminarily established in China. Zhonghua Yi Xue Za Zhi 2002; 82: 1314-1318.

5.   Amit M, Itskovitz-Eldor J. Derivation and spontaneous differentiation of human embryonic stem cells. J Anat 2002; 200: 225-232.

6.   Xu C, Inokuma MS, Denham J, Golds K, Kundu P, Gold JD, Carpenter MK. Feeder-free growth of undifferentiated human embryonic stem cells. Nat Biotechnol 2001; 19: 971-974.

7.   Richards M, Fong CY, Chan WK, Wong PC, Bongso A. Human feeders support prolonged undifferentiated growth of human inner cell masses and embryonic stem cells.[comment]. Nature Biotechnology 2002; 20: 933-936.

8.   Amit M, Margulets V, Segev H, Shariki K, Laevsky I, Coleman R, Itskovitz-Eldor J. Human Feeder Layers for Human Embryonic Stem Cells. Biol Reprod 2003; 68: 2150-2156.

9.   Cheng L, Hammond H, Ye Z, Zhan X, Dravid G. Human adult marrow cells support prolonged expansion of human embryonic stem cells in culture. Stem Cells 2003; 21: 131-142.

10.   Lim JW, Bodnar A. Proteome analysis of conditioned medium from mouse embryonic fibroblast feeder layers which support the growth of human embryonic stem cells. Proteomics 2002; 2: 1187-1203.

11.   Reubinoff BE, Pera MF, Vajta G, Trounson AO. Effective cryopreservation of human embryonic stem cells by the open pulled straw vitrification method. Hum Reprod 2001; 16: 2187-2194.

12.   Amit M, Carpenter MK, Inokuma MS, Chiu C, Harris CP, Waknitz MA, Itskovitz-Eldor J, Thomson JA. Clonally derived human embryonic stem cell lines maintain pluripotency and proliferative potential for prolonged periods of in vitro culture. Developmental Biology 2000; 227: 271-278.

13.   Henderson JK, Draper JS, Baillie HS, Fishel S, Thomson JA, Moore H, Andrews PW. Preimplantation human embryos and embryonic stem cells show comparable expression of stage-specific embryonic antigens. Stem Cells 2002; 20: 329-337.

14.   Drukker M, Katz G, Urbach A, Schuldiner M, Markel G, Itskovitz-Eldor J, Reubinoff B, Mandelboim O, Benvenisty N. Characterization of the expression of MHC proteins in human embryonic stem cells. Proc Natl Acad Sci U S A 2002; 99: 9864-9869.

15.   Sathananthan H, Pera M, Trounson A. The fine structure of human embryonic stem cells. Reprod Biomed Online 2002; 4: 56-61.

16.   Moore FL, Jaruzelska J, Fox MS, Urano J, Firpo MT, Turek PJ, Dorfman DM, Pera RA. Human Pumilio-2 is expressed in embryonic stem cells and germ cells and interacts with DAZ (Deleted in AZoospermia) and DAZ-like proteins. Proceedings of the National Academy of Sciences of the United States of America 2003; 100: 538-543.

17.   Lebkowski JS, Gold J, Xu C, Funk W, Chiu CP, Carpenter MK. Human embryonic stem cells: culture, differentiation, and genetic modification for regenerative medicine applications. Cancer J 2001; 7: S83-93.

18.   Eiges R, Schuldiner M, Drukker M, Yanuka O, Itskovitz-Eldor J, Benvenisty N. Establishment of human embryonic stem cell-transfected clones carrying a marker for undifferentiated cells. Current Biology 2001; 11: 514-518.

19.   Pfeifer A, Ikawa M, Dayn Y, Verma I. Transgenesis by lentiviral vectors: Lack of gene silencing in mammalian embryonic stem cells and preimplantation embryos. PNAS 2002; 99: 2140-2145.

20.   Ma Y, Ramezani A, Lewis R, Hawley R, Thomson JA. High-level sustained transgene expression in human embryonic stem cells using lentiviral vectors. Stem Cells 2003; 21: 111-117.

21.   Gropp M, Itsykson P, Singer O, Ben-Hur T, Reinhartz E, Galun E, Reubinoff BE. Stable genetic modification of human embryonic stem cells by lentiviral vectors. Mol Ther 2003; 7: 281-287.

22.   Zwaka TP, Thomson JA. Homologous recombination in human embryonic stem cells. Nat Biotechnol 2003; 21: 319-321.

23.   Itskovitz-Eldor J, Schuldiner M, Karsenti D, Eden A, Yanuka O, Amit M, Soreq H, Benvenisty N. Differentiation of human embryonic stem cells into embryoid bodies compromising the three embryonic germ layers. Mol Med 2000; 6: 88-95.

24.   Schuldiner M, Yanuka O, Itskovitz-Eldor J, Melton DA, Benvenisty N. Effects of eight growth factors on the differentiation of cells derived from human embryonic stem cells. Proc Natl Acad Sci U S A 2000; 97: 11307-11312.

25.   Goldstein RS, Drukker M, Reubinoff BE, Benvenisty N. Integration and differentiation of human embryonic stem cells transplanted to the chick embryo. Dev Dyn 2002; 225: 80-86.

26.   Zhang SC, Wernig M, Duncan ID, Brustle O, Thomson JA. In vitro differentiation of transplantable neural precursors from human embryonic stem cells. Nat Biotechnol 2001; 19: 1129-1133.

27.   Reubinoff BE, Itsykson P, Turetsky T, Pera MF, Reinhartz E, Itzik A, Ben-Hur T. Neural progenitors from human embryonic stem cells. Nat Biotechnol 2001; 19: 1134-1140.

28.   Carpenter MK, Inokuma MS, Denham J, Mujtaba T, Chiu CP, Rao MS. Enrichment of neurons and neural precursors from human embryonic stem cells. Exp Neurol 2001; 172: 383-397.

29.   Schuldiner M, Eiges R, Eden A, Yanuka O, Itskovitz-Eldor J, Goldstein RS, Benvenisty N. Induced neuronal differentiation of human embryonic stem cells. Brain Research 2001; 913: 201-205.

30.   He JQ, Ma Y, Lee Y, Thomson JA, Kamp TJ. Human Embryonic Stem Cells Develop Into Multiple Types of Cardiac Myocytes. Action Potential Characterization. Circ Res 2003; 5: 5.

31.   Kehat I, Kenyagin-Karsenti D, Snir M, Segev H, Amit M, Gepstein A, Livne E, Binah O, Itskovitz-Eldor J, Gepstein L. Human embryonic stem cells can differentiate into myocytes with structural and functional properties of cardiomyocytes. J Clin Invest 2001; 108: 407-414.

32.   Xu C, Police S, Rao N, Carpenter MK. Characterization and enrichment of cardiomyocytes derived from human embryonic stem cells. Circ Res 2002; 91: 501-508.

33.   Kehat I, Gepstein A, Spira A, Itskovitz-Eldor J, Gepstein L. High-resolution electrophysiological assessment of human embryonic stem cell-derived cardiomyocytes: a novel in vitro model for the study of conduction. Circ Res 2002; 91: 659-661.

34.   Mummery C, Ward-Van Oostwaard D, Doevendans P, Spijker R, Van Den Brink S, Hassink R, Van Der Heyden M, Opthof T, Pera M, De La Riviere AB, Passier R, Tertoolen L. Differentiation of human embryonic stem cells to cardiomyocytes: role of coculture with visceral endoderm-like cells. Circulation 2003; 107: 2733-2740.

35.   Levenberg S, Golub JS, Amit M, Itskovitz-Eldor J, Langer R. Endothelial cells derived from human embryonic stem cells. Proc Natl Acad Sci U S A 2002; 99: 4391-4396.

36.   Kaufman DS, Hanson ET, Lewis RL, Auerbach R, Thomson JA. Hematopoietic colony-forming cells derived from human embryonic stem cells. Proc Natl Acad Sci U S A 2001; 98: 10716-10721.

37.   Chadwick K, Wang L, Li L, Menendez P, Murdoch B, Rouleau A, Bhatia M. Cytokines and BMP-4 promote hematopoietic differentiation of human embryonic stem cells. Blood 2003; 17: 17.

38.   Assady S, Maor G, Amit M, Itskovitz-Eldor J, Skorecki KL, Tzukerman M. Insulin production by human embryonic stem cells. Diabetes 2001; 50: 1691-1697.

39.   Rambhatla L, Chiu CP, Kundu P, Peng Y, Carpenter MK. Generation of hepatocyte-like cells from human embryonic stem cells. Cell Transplant 2003; 12: 1-11.

40.   Xu RH, Chen X, Li DS, Li R, Addicks GC, Glennon C, Zwaka TP, Thomson JA. BMP4 initiates human embryonic stem cell differentiation to trophoblast. Nat Biotechnol 2002; 20: 1261-1264.

41.   McKiernan SH, Bavister BD. Different lots of bovine serum albumin inhibit or stimulate in vitro development of hamster embryos. In Vitro Cell Dev Biol 1992; 28A: 154-156.

42.   Smithies O, Gregg RG, Boggs SS, Koralewski MA, Kucherlapati RS. Insertion of DNA sequences into the human chromosomal beta-globin locus by homologous recombination. Nature 1985; 317: 230-234.

43.   Thomas KR, Capecchi MR. Site-directed mutagenesis by gene targeting in mouse embryo-derived stem cells. Cell 1987; 51: 503-512.

44.   Zwaka TP, Thomson JA. Homologous recombination in human embryonic stem cells. In: Lanza R (ed.) The handbook of embryonic stem cells. Vol 2: Embryonic stem cells. San Deigo: Elsevier science; 2003: (in press).

45.   Odorico JS, Kaufman DS, Thomson JA. Multilineage differentiation from human embryonic stem cell lines. Stem Cells 2001; 19: 193-204.

46.   Rideout WM, 3rd, Hochedlinger K, Kyba M, Daley GQ, Jaenisch R. Correction of a genetic defect by nuclear transplantation and combined cell and gene therapy. Cell 2002; 109: 17-27.

47.   Finger S, Heavens RP, Sirinathsinghji DJ, Kuehn MR, Dunnett SB. Behavioral and neurochemical evaluation of a transgenic mouse model of Lesch-Nyhan syndrome. Journal of Neurological Science 1988; 86: 203-213.

48.   Fisher SJ. The placenta dilemma. Semin Reprod Med 2000; 18: 321-326.

49.   Cross JC. Genetic insights into trophoblast differentiation and placental morphogenesis. Semin Cell Dev Biol 2000; 11: 105-113.

50.   Beddington RS, Robertson EJ. An assessment of the developmental potential of embryonic stem cells in the midgestation mouse embryo. Development 1989; 105: 733-737.

51.   Thomson JA, Kalishman J, Golos TG, Durning M, Harris CP, Becker RA, Hearn JP. Isolation of a primate embryonic stem cell line. Proceedings of the National Academy of Sciences, U.S.A. 1995; 92: 7844-7848.

52.   Muyan M, Boime I. Secretion of chorionic gonadotropin from human trophoblasts. Placenta 1997; 18: 237-241.

53.   American Heart Association. Heart disease and stroke statistics — 2003 update. Dallas, Tex: American Heart Association; 2002.

54.   Mummery C, Ward D, van den Brink CE, Bird SD, Doevendans PA, Opthof T, Brutel de la Riviere A, Tertoolen L, van der Heyden M, Pera M. Cardiomyocyte differentiation of mouse and human embryonic stem cells. J Anat 2002; 200: 233-242.

55.   Soonpaa MH, Daud AI, Koh GY, Klug MG, Kim KK, Wang H, Field LJ. Potential approaches for myocardial regeneration. Ann N Y Acad Sci 1995; 752: 446-454.

56.   Gandy KL, Weissman IL. Tolerance of allogeneic heart grafts in mice simultaneously reconstituted with purified allogeneic hematopoietic stem cells. Transplantation 1998; 65: 295-304.

57.   Spitzer TR, Delmonico F, Tolkoff-Rubin N, McAfee S, Sackstein R, Saidman S, Colby C, Sykes M, Sachs DH, Cosimi AB. Combined histocompatability leukocyte antigen-matched donor bone marrow and renal transplantation for multiple myeloma with end-stage renal disease: The induction of allograft tolerance through mixed lymphohematopeoietic chimerism. Transplantation 1999; 68: 480-484.

58.   Wekerle T, Sykes M. Mixed chimerism as an approach for the induction of transplantation tolerance. Transplantation 1999; 68: 459-467.

59.   The effect of intensive treatment of diabetes on the development and progression of long-term complications in insulin-dependent diabetes mellitus. The Diabetes Control and Complications Trial Research Group. N Engl J Med 1993; 329: 977-986.

Appendix J.

Multipotent Adult Progenitor Cells: An Update

CATHERINE M. VERFAILLIE, M.D.
*Division of Hematology, Department of Medicine,
and Stem Cell Institute,
University of Minnesota*

INTRODUCTION:

In this paper, we want to provide updated information regarding a rare cell population, we have named, multipotent adult progenitor cells or MAPC. In 2001-2002, we published a series of papers demonstrating that while attempting to select and culture mesenchymal stem cells (MSC) from human and subsequently mouse and rat bone marrow (BM), we accidentally identified a rare population of cells that has characteristics unlike most adult somatic stem cells in that they appear to proliferate without senescence, and have pluripotent differentiation ability *in vitro* and *in vivo* [1,2].

Phenotype of Bone Marrow MAPC: MAPC can be cultured from human, mouse and rat bone marrow (BM). Unlike MSC, MAPC do not express major histocompatibiliy (MHC)- class I antigens, do not express, or express only low levels of, the CD44 antigen, and are CD105 (also endoglin, or SH2) negative [1,2]. Unlike hematopoietic stem cells (HSC), MAPC do not express CD45, CD34, and cKit [1,2], but like HSC, MAPC express Thy1, AC133 (human MAPC) and Sca1 (mouse) albeit at low levels [1,2] .In the mouse, MAPC express low levels of stage specific embryonic antigen (SSEA)-1, and express low levels of the transcription factors Oct4 and Rex1, known to be important for maintaining embryonic stem (ES) cells undifferentiated [3] and to be down-regulated when ES cells undergo somatic cell commitment and differentiation [2].

MAPC can also be isolated from other tissues, and other species:

We also showed that MAPC can be cultured from mouse brain and mouse muscle [4]. Of note, the differentiation potential and expressed gene profile of MAPC derived from the different tissues

295

appears to be highly similar. These studies used whole brain and muscle tissue as the initiating cell population, therefore containing more than neural cells and muscle cells, respectively. The implications of this will be discussed below. Studies are ongoing to determine if cultivation of MAPC from other organs is possible, and whether culture of MAPC, like ES cells, is mouse-strain dependent. Initial studies suggest that a population of MAPC-like cells can also be cultured from bone marrow from cynomologous monkeys (unpublished observations)(studies done by our collaborator Felipe Prosper, University of Navarra, Pamplona, Spain) and from bone marrow of dogs (unpublished observations)(studies done at the University of Minnesota).

Non-senescent nature of MAPC:

Unlike most adult somatic stem cells, MAPC proliferate without obvious signs of senescence, and have active telomerase. In humans, the length of MAPC telomeres is 3-5kB longer than in neutrophils and lymphocytes, and telomere length is not different when MAPC are derived from young or old donors [1]. This suggests that MAPC are derived from a population of cells that either has active telomerase in vivo, or that is highly quiescent in vivo, and therefore has not yet incurred telomere shortening in vivo. In human MAPC cultures we have not yet seen cytogenetic abnormalities. As human MAPC are however undergoing symmetrical cell divisions, it remains possible that despite lack of gross cytogentic changes, minor mutations accumulate over time. We are therefore planning to use comparative genomic hybridization to address the question at what time genetic abnormalities occur, if they do. Initial results from gene array analysis suggest that MAPC, like ES cells, have a large number of DNA repair genes expressed (unpublished observations), which may protect them from more frequent genetic abnormalities in view of the fact that they undergo multiple sequential symmetrical cell divisions.

However, several subpopulations of mouse MAPC, and to a lesser extent rat MAPC, have become aneuploid, even though additional subpopulations thawed subsequently were cytogenetically normal. Aneuploidy is seen more frequently once mouse (and rat) MAPC have been expanded for >60-70 population doublings and following repeated cryopreservtions and thawing episodes. This characteristic of mouse MAPC is not dissimilar from other mouse cell populations, including mouse ES cells.

Stringent culture conditions required for maintenance of the undifferentiated state of MAPC:

Culture of MAPC is, however, technically demanding. Major factors that play a role in successful maintenance of MAPC include cell density, CO_2 concentration and pH of the medium, lot of fetal calf serum that is used, and even the type of culture plastic that is used. Control of cell density appears to be species specific: mouse, rat and perhaps cynomologous monkey MAPC need to be maintained at densities between 500 and 1,000 cells / cm2, whereas human and perhaps dog MAPC need to be maintained between 1,500 and 3,000 cells/ cm^2. The reason why MAPC tend to differentiate to the default MSC lineage when maintained at higher densities is not known. However, for MAPC to have clinical relevance, this will need to be overcome. Gene array and proteomics studies are ongoing to identify the contact and / or soluble factors that may be responsible for causing differentiation when MAPC are maintained at higher densities. These very demanding technical skills can however be "exported" from the University of Minnesota as, after training at the University of Minnesota, investigators at the University of Tokai, Japan (manuscript submitted) and investigators at the University of Gent, Belgium have successfully isolated MAPC from human bone marrow, and investigators at the University of Navarra, Spain, have successfully isolated MAPC from rat bone marrow.

IN VITRO DIFFERENTIATION POTENTIAL OF MAPC:

We published last year that human, mouse and rat MAPC can be successfully differentiated into typical mesenchymal lineage cells, including osteoblasts, chondroblasts, adipocytes and skeletal myoblasts [1]. In addition, human, mouse and rat MAPC can be induced to differentiate into cells with morphological, phenotypic and functional characteristics of endothelial cells [5], and morphological, phenotypic and functional characteristics of hepatocytes [6].

Neuroectodermal differentiation:

Since then, we have also been able to induce differentiation of MAPC from mouse bone marrow into cells with morphological, phenotypic and functional characteristics of neuroectodermal cells[7]. Differentiation of MAPC to cells with neuroectodermal characteristics occurred by initial culture in the presence of basic fibroblast growth factor (bFGF) as the sole cytokine, followed by culture with FGF-8b and sonic hedgehog (SHH), and then brain derived neurotrophic factor (BDNF) [8,9]. Differentiation using these sequential cytokine stimuli was associated with activation of transcription factors known to be important in neural commitment *in vivo* and differentiation from NSC and mES cells *in vitro*. Cells staining positive for astrocyte, oligodendrocyte and neuronal markers were detected. Neuron-like cells became polarized, and as has been described in most studies in which ES cells or NSC were differentiated *in vitro* to a mid-brain neuroectodermal fate using FGF8 and SHH, approximately 25% of cells stained positive for dopaminergic markers, 25% for serotonergic markers, and 50% for GABA-ergic markers. Subsequent addition of astrocytes induced further maturation and prolonged survival of the MAPC-derived neuron-like cells, which now also acquired electrophysiological characteristics consistent with neurons, namely voltage gated sodium channels and synaptic potentials [10,11].

Muscle differentiation:

In addition, we now have convincing evidence that MAPC can differentiate into cells with phenotypic as well as functional characteristics of smooth muscle cells (manuscript in preparation). Interestingly, the lineage that continues to be elusive is cardiac myoblasts, despite the fact that mouse MAPC injected in the blastocyst contribute to the cardiac muscle [2]. Although a number of *in vitro* differentiation conditions induce expression of Nkx2.5, GATA4, and myosin heavy chain mRNA and proteins [12-14], we have been unable to induce differentiation of MAPC to cells with the typical functional characteristic of cardiac myoblasts, i.e. spontaneous rhythmic contractions or beating, a differentiation path that is almost a default differentiation pathway for mouse ES cells. The reason for the lack of functional cardiac myoblast properties is currently unknown.

Another important cell lineage that has not yet been generated is insulin-producing cells, even though initial studies suggest that differentiation to cells expressing at least early pancreatic and endocrine pancreas transcription factors can be obtained.

In vitro differentiation of MAPC as model system for gene discovery:

A last comment regarding *in vitro* differentiation of MAPC is that, in contrast to differentiation of ES cells *in vitro*, the final differentiated cell product derived from MAPC is commonly >70-80% pure. This should allow using these *in vitro* differentiation models for gene and drug discovery. For instance, in a recently published study [15] we compared the expressed gene profile in human MAPC induced to differentiate to osteoblasts and chondroblasts, two closely related cell lineages. We could demonstrate that although a large number of genes are co-regulated when MAPC differentiate to these two lineages, specificity in differentiation can readily be detected. For instance a number of known and yet to be fully characterized transcription factor mRNAs were differentially expressed during the initial phases of differentiation. Studies are ongoing to further define the role of these genes in lineage specific differentiation. These studies exemplify however the power of this model system to study lineage specific differentiation *in vitro*.

DEGREE OF PLURIPOTENCY OF MAPC:

We have shown that transfer of 10-12 mouse MAPC into mouse blastocysts results in the generation of chimeric mice. When 10-12 MAPC, expanded for 50-55 population doublings, were injected approximately 80% of offspring were chimeric, with the degree of chimerisms varying between 1–40% [4]. Cells found in different organs acquire phenotypic characteristics of the tissue. For instance MAPC derived cells detected in the brain of chimeric animals differentiate appropriately into region specific neurons, as well as astrocytes and oligodendrocytes [16]. More recent studies using MAPC from later population doublings have shown that the frequency of chimerism decreases when MAPC are maintained for longer time in culture, even though animals with chimerism of more than 70% could be obtained (unpublished observations). These studies indicate that like ES cells, MAPC can give rise to most if not all somatic cell types of the mouse. Whether MAPC can do this without help of other cells in the inner cell mass, i.e. can generate a mouse by tetraploid complementation [17], is not yet known. Also not yet known is whether MAPC contribute to the germ line when injected in the blastocyst.

POST-NATAL CONTRIBUTION TO TISSUES:

Neither human nor mouse MAPC injected into the muscles of severe combined immunodeficient (SCID) mice have led to the development of teratomas (unpublished observations). Likewise, we have not yet detected donor-derived tumor formation following IV injection of human or mouse MAPC in NOD-SCID animals. However, when mouse undifferentiated MAPC are administered IV to NOD-SCID mice, engraftment in the hematopoietic system as well as epithelia of gut, liver and lung is seen [2]. Preliminary studies using human MAPC suggest that a similar pattern of engraftment may occur, even though the level of contribution to blood, liver, gut and lung is lower (unpublished observations). Noteworthy is the fact that neither mouse nor human MAPC appear to contribute to other tissues when injected IV, except to endothelium (see below). Although PCR analysis for human DNA in human – mouse transplants or for b-galactosidase in mouse-mouse transplants yielded positive signals in many tissues, we believe that this is mainly due to contaminating blood cells. When tissues were carefully examined for tissue specific differentiated MAPC progeny, we could not detect MAPC-progeny in brain, skeletal muscle, cardiac muscle, skin or kidneys. Lack of engraftment in brain, skeletal and cardiac muscle may be due to the fact that transplants were done in non-injured animals, where the blood brain barrier is intact, and where little or no cell turnover is expected in muscle. More difficult to explain is the absence of MAPC-derived progeny in skin, possibly the organ with the greatest cell turnover. Studies are ongoing to trace the homing behavior of MAPC following infusion in non-injured animals and injured animals, which may shed light on these observations.

In vivo differentiation into skeletal muscle:

Muguruma et al have also shown that undifferentiated human MAPC injected in the muscle of non-obese diabetic (NOD)-SCID mice differentiate into cells that stain positive for muscle transcription factors and muscle cytoskeletal proteins (manuscript submitted). Similar results were seen in Minnesota. We also found that pre-treatment of human MAPC with 5-azacytidine, required to induce muscle differentiation in vitro, enhanced the degree of engraftment of human cells in mouse muscle, suggesting that pre-differentiation of MAPC may under certain circumstances enhance the level of engraftment (unpublished observations).

Contribution to endothelium in vivo:

When endothelial cells generated from human MAPC by incubation in vitro with vascular endothelial growth factor (VEGF) [5] were infused in animals in which a tumor had been implanted underneath the skin, we detected enhanced tumor growth and found that up to 30% of the tumor vasculature was derived from the human endothelial cells. Likewise, wounds in the ears of these animals as a result of ear tagging contained human endothelial cells. One of the animals developed a host-tumor, an occurrence seen frequently in aging NOD-SCID mice. We detected contribution of MAPC-derived endothelium to tumor vessels [2]. Likewise, one of the NOD-SCID mice that received human MAPC developed a host thymic lymphoma. Human MAPC, like mouse MAPC, appeared to differentiate into endothelial cells that contribute to tumor angiogenesis.

Engraftment of MAPC in stroke model:

In yet another *in vivo* study [18] we evaluated the effect of human MAPC in a rat stroke model. Cortical brain ischemia was produced in male rats by permanently ligating the right middle cerebral artery distal to the striatal branch. Animals were placed on cyclosporine-A and 2 weeks later, 2×10^5 human MAPC were injected around the infarct zone. As controls, animals received normal saline or MAPC conditioned medium. Limb placement test and tactile stimulation test were blindly assessed 1 week before brain ischemia, 1 day before transplantation, and at 2 and 6 weeks after grafting. The limb placement test included eight subtests described by Johansson and coworkers [19]. In a tactile stimulation test [20], a small piece of adhesive tape was rapidly applied to the radial aspect of each forepaw. The rats were then returned to their home cages, and the order of the tape removal (i.e., left versus right) was recorded. Three to five trials were conducted on each test day. Each trial was terminated when the tapes were removed from both forepaws or after 3 min. Animals were subsequently sacrificed to determine the fate of the human cells injected in the brain. After 2 and 6 weeks, animals that received human MAPC scored statistically significantly better in the limb placement test as well as tactile stimulation test compared with animals that received only cyclosporine-A (CSA), or were injected with normal saline or MAPC conditioned medium. The level of recuperation of motor and sensory function was 80% of animals without stroke. When the brain was examined for the presence and differentiation of human MAPC to neuroectodermal cells, we found that human MAPC were present, but remained rather immature. Therefore, we cannot attribute the motor and sensory improvement to region specific differentiation to neuronal cells and integration of neurons derived from MAPC in the host brain. Rather the improvement must be caused by trophic effects emanated by the human MAPC to either improve vascularization of the ischemic area, to support survival of the remaining endogenous neurons, or to recruit neuronal progenitors from the host brain. These possibilities are currently being evaluated.

POSSIBLE MECHANISMS UNDERLYING THE PHENOMENON OF MULTIPOTENT ADULT PROGENITOR CELLS:

Currently we do not fully understand the mechanism(s) underlying the culture selection of MAPC. We have definitive data to demonstrate that the pluripotency of MAPC is not due to co-culture of several stem cells.

Pluripotency cannot be attributed to multiple stem cells:

First, using retroviral marking studies we have definitive proof that a single cell can differentiate *in vitro* to cells of mesoderm, both mesenchymal and non-mesenchymal, neuroectoderm and hepatocyte-like cells, and this for human [1,6], mouse and rat MAPC [2,6]. Second, we have shown that a single mouse MAPC is sufficient for generation of chimeric animals [2]. Indeed, we published that 1/3 animals born from blastocysts in which a single MAPC was injected were chimeric with chimerism degrees varying between 1 and 45%. This rules therefore out that the pluripotent nature of these cells is due to co-existence in culture of multiple somatic stem cells.

Cell fusion is not likely explanation:

A second possibility for the greater degree of differentiation potential would be that cells undergo fusion and acquire via this mechanism greater pluripotency. Fusion has been shown to be responsible for apparent ES characteristics of marrow and neural stem cells [21,22] that had been cocultured with ES cells *in vitro*, and more recently for the apparent lineage switch of bone marrow cells to hepatocytes when hematopoietic cells were infused in animals with hereditary tyrosinemia due to lack of the fumarylacetoacetate hydroxylase (FAH) gene [23][Wang et al, Nature 2003]. In the former two studies, the majority of genes expressed in the marrow or neural cell that fused with ES cells were silenced, and the majority of the genes expressed in ES cell were persistently expressed. Likewise for the bone marrow-hepatocyte fusion, the majority of genes expressed normally in hematopoioetic cells (except the FAH gene) were silenced whereas genes expressed in hepatocytes predominated. Finally, the cells generated were in general tetraploid or aneuploid.

We do not believe that this phenomenon underlies the observation that MAPC are pluripotent. Cultivation and differentiation *in vitro* (in general, except the final differentiation step for neuroectoderm) does not require that MAPC are co-cultured with other cells, making the likelihood that MAPC are the result of fusion very low. Smith et al suggested in a recent commentary that MAPC could be caused by fusion of multiple cell types early on during culture leading to reprogramming of the genetic information and pluripotency [REF]. However, we have no evidence that MAPC are tetraploid or aneuploid early during culture, making this possibility less likely. Nevertheless, studies are ongoing to rule this out. The in vivo studies were not set up to fully be capable of ruling out this possibility.

However, a number of findings suggest that fusion may not likely be the cause for the engraftment seen postnatally, nor the chimerism in the blastocyst injection experiment. The frequency of the fusion event described for the ES-BM, ES-NSC, and HSC-hepatocyte fusion was in general very low, i.e. 1/100,000 cells. Expansion of such fused cells could only be detected when drug selection was applied in the in vitro systems, and withdrawal of NTBC (2-(2-nitro-4-trifluoro-methylbenzoyl)-1,3-cyclohexanedione) in the FAH mouse model was used to select for cells expressing the FAH gene. The percent engraftment seen in our post-natal transplant models was in the range of 1% - 9%. The frequency of chimerism seen in blastocyst injection studies ranged between 33% and 80% when 1 and 1 and 10-12 MAPC were injected, respectively. These frequencies are significantly higher than what has been described for fusion events with ES cells *in vitro,* and in the HSC-hepatocyte fusion studies *in vivo.*

Furthermore, in contrast to what was described in the papers indicating that fusion may be responsible for apparent plasticity, all *in vivo* studies done with MAPC were done without selectable pressure, mainly in non-injured animals. Therefore, it is less likely that the pluripotent behavior of MAPC in vivo is due to fusion between the MAPC and the tissues where they engraft / contribute to. However, specific studies are currently being designed to formally rule this out.

Primitive ES-like cells that persist vs. de-differentiation:

Currently, we do not have proof that MAPC exist as such *in vivo*. Until we have positive selectable markers for MAPC, this question will be difficult to answer. If the cell exists *in vivo*, one might hypothesize that it is derived for instance from primordial germ cells that migrated aberrantly to tissues outside the gonads during development. It is, however, also possible that removal of certain (stem) cells from their *in vivo* environment results in "reprogramming" of the cell to acquire greater pluripotency. The studies on human MAPC suggest that such a cell that might undergo a degree of reprogramming is likely a protected (stem) cell *in vivo*, as telomere length of MAPC from younger and older donors is similar, and significantly longer than what is found in hematopoietic cells from the same donor. The fact that MAPC can be isolated from multiple tissues might argue that stem cells from each tissue might be able to be reprogrammed. However, as was indicated above, the studies in which different organs were used as the initiating cell population for generation of MAPC did not purify tissue specific cells or stem cells. Therefore, an alternative explanation is that the same cells isolated from bone marrow that can give rise to MAPC in culture might circulate, and be collected from other organs. However, we have until now been unsuccessful in isolating MAPC from blood or from umbilical cord blood, arguing against this phenomenon. Finally, cells selected from the different organs could be the same cells resident in multiple organs, such as MSC that are present in different locations, or cells associated with tissues present in all organs such as for instance blood vessels. Studies are ongoing to determine which of these many possibilities is correct.

CONCLUSION:

We believe that MAPC would have clinical relevance whether they exist *in vivo*, or are created *in vitro*. However, understanding the nature of the cell will have impact on how one would approach their clinical use. If they exist *in vivo*, it will be important to learn where they are located, and to determine whether their migration, expansion and differentiation in a tissue specific manner can be induced and controlled *in vivo*. If they are a culture creation, understanding the mechanism underlying the reprogramming event will be important as that might allow this phenomenon to happen on a more routine and controlled basis.

Either way, a long road lies ahead before MAPC might be applicable in clinical trials. Hurdles to be overcome include development of robust culture systems that will allow automatization. Like of other stem cells, including ES cells, we will need to determine in preclinical models whether undifferentiated vs. lineage committed vs. terminally differentiated cells should be used to treat a variety of disorders. If lineage committed or terminally differentiated cells will be needed, robust clinical scale differentiation cultures will need to be developed. Furthermore, studies will need to be performed to demonstrate whether potentially contaminating undifferentiated MAPC will interfere with engraftment, and / or differentiate inappropriately in vivo. Likewise, studies aimed at determining what level of HLA-mismatch will be tolerated in transplantations, whether tolerization via hematopoietic engraftment from MAPC will be required. As is also the case for other extensively cultured cells, we will need to further determine if prolonged expansion leads to genetic abnormalities in cells that might lead to malignancies when transplanted in vivo.

As a final remark, MAPC appear to have pluripotent potential both *in vitro* and *in vivo*. Furthermore, they appear to proliferate without obvious senescence when maintained under very stringently controlled culture conditions. Because of these reasons, some have argued that they might be a viable alternative to ES cells. However, at this stage of the research, I feel that such a conclusion is premature. Whether MAPC have equal longevity as ES cells, and have the ability to create all >200 cell types in the body is still not known. Moreover, there appear to be certain cell types that are more readily generated from ES cells compared with MAPC, such as for instance cardiac myoblasts, whereas it appears for instance more easy to generate hepatocyte like cells from MAPC than ES cells. Therefore, I continue to strongly believe that strict comparative studies between the two cell populations are needed to determine the true potential of the cells, and that the scientific insights gained from these studies should be used to determine which of the cells will be suitable for use in the clinical setting.

REFERENCES

1. Reyes M, Lund T, Lenvik T, Aguiar D, Koodie L, Verfaillie CM. Purification and ex vivo expansion of postnatal human marrow mesodermal progenitor cells. Blood. 2001;98:2615-2625

2. Jiang Y, Jahagirdar B, Reyes M, Reinhardt RL, Schwartz RE, Chang H-C, Lenvik T, Lund T, Blackstad M, Du J, Aldrich S, Lisberg A, Kaushal S, Largaespada DL, Verfaillie CM. Pluripotent nature of adult marrow derived mesenchymal stem cells. Nature. 2002;418:41-49

3. Niwa H, Miyazaki J, Smith AG. Quantitative expression of Oct-3/4 defines differentiation, dedifferentiation or self-renewal of ES cells. Nat Genet. 2000;24:372-376

4. Jiang Y, Vaessen B, Lenvik T, Blackstad M, Reyes M, Verfaillie CM. Multipotent progenitor cells can be isolated from post-natal murine bone marrow, muscle and brain. Exp Hematol. 2002;30(8):896-904

5. Reyes M, Dudek A, Jahagirdar B, Koodie K, Marker PH, Verfaillie CM. Origin of endothelial progenitors in human post-natal bone marrow. J Clin Invest. 2002;109:337-346

6. Schwartz RE, Reyes M, Koodie L, Jiang Y, Blackstad M, Lund T, Lenvik T, Johnson S, Hu W-S, Verfaillie CM. Multipotent adult progenitor cells from bone marrow differentiate into functional hepatocyte-like cells. J Clin Invest. 2002;109(10):1291-1302

7. Jiang Y, Henderson D, Blackstadt M, Chen A, Miller FF, Verfaillie CM. Neuroectodermal differentiation from mouse multipotent adult progenitor cells. Proc. Natl. Acad. Sci U S A. 2003;In Press

8. Ling Z, Potter E, Lipton J, Carvey P. Differentiation of mesencephalic progenitor cells into dopaminergic neurons by cytokines. Exp Neurol. 1998;149:411-423

9. Lee SH, Lumelsky N, Studer L, Auerbach JM, McKay RD. Efficient generation of midbrain and hindbrain neurons from mouse embryonic stem cells. Nat Biotechnol. 2000;18:675-679

10. Wagner J, Akerud P, Castro DS, Holm PC, Canals JM, Snyder EY, Perlmann T, Arenas E. Induction of a midbrain dopaminergic phenotype in Nurr1-overexpressing neural stem cells by type 1 astrocytes. Nat Biotech. 1999;17:653-659

11. Song H, Stevens CF, Gage FH. Astroglia induce neurogenesis from adult neural stem cells. Nature. 2002;417:39-44

12. Tanaka M, Chen Z, Bartunkova S, Yamasaki N, Izumo S. The cardiac homeobox gene Csx/Nkx2.5 lies genetically upstream of multiple genes essential for heart development. Development. 1999;126:1269-1280

13. Laverriere AC, MacNeill C, Mueller C, Poelmann RE, Burch JB, Evans T. GATA-4/5/6, a subfamily of three transcription factors transcribed in developing heart and gut. J Biol Chem. 1994;269:23177-23184

14. Klug MG, Soonpaa MH, Koh GY, Field LJ. Genetically selected cardiomyocytes from differentiating embronic stem cells form stable intracardiac grafts. J Clin Invest. 1996;98:216-224

15. Qi H, Aguiar DJ, Williams SM, La Pean A, Pan W, Verfaillie CM. Identification of genes responsible for osteoblast differentiation from human mesodermal progenitor cells. Proc. Natl Acad. Sci U S A. 2003 Mar18;100(6):3305-3310

16. Keene CD, Ortiz-Gonzalez XR, Jiang Y, Largaespada DA, Verfaillie CM, Low WC. Neural differentiation and incorporation of bone marrow-derived multipotent adult progenitor cells after single cell transplantation into blastocyst stage mouse embryos. Cell Transplant. 2003;12(3):201-213

17. Wang ZQ, Kiefer F, Urbanek P, Wagner EF. Generation of completely embryonic stem cell-derived mutant mice using tetraploid blastocyst injection. Mech Dev. 1997;62:137-145

18. Zhao L-R, Duan W-M, Reyes M, Keene CD, Verfaillie CM, Low WC. Human bone marrow stem cells exhibit neural phenotypes and ameliorate neurological deficits after grafting into the ischemic brain of rats. Exp Neurol. 2002;174:11-20

19. Ohlsson AL, Johansson BB. Environment influences functional outcome of cerebral infarction in rats. Stroke. 1995;4:644-649

20. Netto CA, Hodges JD, Sinden JD, LePeillet E, Kershaw T, Schallert T, Whishaw IQ. Bilateral cutaneous stimulation of the somatosensory system in hemidecorticate rats. Behav. Neurosci. 1984;98:518-540

21. Ying QY, Nichols J, Evans EP, Smith AG. Changing potency by spontaneous fusion. Nature. 2002;416:545-548

22. Terada N, Hamazaki T, Oka M, Hoki M, Mastalerz DM, Nakano Y, Meyer EM, More Ll, Petersen BE, Scott EW. Bone marrow cells adopt the phenotype of other cells by spontaneous cell fusion. Nature. 2002;416:542-545

23. Lagasse E, Connors H, Al-Dhalimy M, Reitsma M, Dohse M, Osborne L, Wang X, Finegold M, Weissman IL, Grompe M. Purified hematopoietic stem cells can differentiate into hepatocytes in vivo. Nat Med. 2000;6:1229-1234

Appendix K.

Adult Stem Cells

DAVID A. PRENTICE, PH.D.
Professor of Life Sciences at Indiana State University, Terre Haute, Indiana

Within just a few years, the possibility that the human body contains cells that can repair and regenerate damaged and diseased tissue has gone from an unlikely proposition to a virtual certainty. Adult stem cells have been isolated from numerous adult tissues, umbilical cord, and other non-embryonic sources, and have demonstrated a surprising ability for transformation into other tissue and cell types and for repair of damaged tissues. This paper will examine the published literature regarding the identity of adult stem cells and possible mechanisms for their observed differentiation into tissue types other than their tissue of origin. Reported data from both human and animal studies will be presented on the various tissue sources of adult stem cells and the differentiation and repair abilities for each source, especially with regards to current and potential therapeutic treatments.

Adult stem cells have received intense scrutiny over the past few years due to surprising discoveries regarding heretofore unknown abilities to form multiple cell and tissue types, as well as the discovery of such cells in an increasing number of tissues. The term "adult stem cell" is somewhat of a misnomer, because the cells are present even in infants and similar cells exist in umbilical cord and placenta. More accurate terms have been proposed, such as tissue stem cells, somatic stem cells, or post-natal stem cells. However, because of common usage this review will continue to use the term adult stem cell.

This paper will review the literature related to adult stem cells, including current and potential clinical applications (with apologies to the many who are not cited, due to the exponential increase in papers regarding adult stem cells and the limitations of this review.) The focus will be on human adult stem cells, but will also include results from animal studies which bear on the potential of adult stem cells to be used therapeutically for patients.

This paper will not attempt to review the literature related to hematopoietic stem cells, *i.e.*, the bone marrow stem cell that is the immediate precursor for blood cells, and the formation of typical blood cells. Nor will this paper review the substantial literature regarding clinical use of bone marrow or bone marrow stem cell transplants for hematopoietic conditions such as various cancers and anemias, nor the striking clinical results seen for conditions such as scleromyxedema, multiple sclerosis, systemic lupus, arthritis, Crohn's disease, etc.[1] In these instances, the stem cells are used primarily to replace the hematopoietic system of the patient, after ablation of the patient's own bone marrow hematopoietic system. Finally, multipotent adult progenitor cells (MAPC's), a bone marrow stem cell that has shown significant abilities at proliferation in culture and differentiation into other body tissues,[2] have been reviewed by Dr. Catherine Verfaillie in a separate paper for the President's Council on Bioethics, and the reader is directed to that review for more information.

Key questions regarding adult stem cells are: (1) their identity, (2) their tissue source of origin, (3) their ability to form other cell or tissue types, and (4) the mechanisms behind such changes in differentiation and effects on tissues and organs. Historically only a few stem cells were recognized in humans, such as the hematopoietic stem cell which produces all of the blood cell types, the gastrointestinal stem cell associated with regeneration of the gastrointestinal lining, the stem cell responsible for the epidermal layer of skin, and germ cell precursors (in the adult human, the spermatogonial stem cell.) These stem cells were considered to have very limited repertoires, related to replenishment of cells within their tissue of origin. These limitations were considered to be a normal part of the developmental paradigm in which cells become more and more restricted in their lineage capabilities, leading to defined and specific differentiated cells in body tissues. Thus, discovery of stem cells in other tissues, or with the ability to cross typical lineage boundaries, is both exciting and confusing because such evidence challenges the canonical developmental paradigm.

STEM CELL MARKERS

Identification of cells typically relies on use of cell surface markers—cellular differentiation (CD) antigens—that denote the expression of particular proteins associated with genomic activity related to a particular differentiation state of the cell. Identification also has relied on morphological and molecular indications of function, such as expression of specific enzymes. Since stem cells by definition have not yet taken on a specific differentiated function, their identification has relied primarily on use of cell surface markers, and only secondarily on production of differentiated products in various tissues. One stated goal has been to isolate a single putative adult stem cell, characterized fully by specific markers and molecular characteristics, and then to follow the differentiation of this single cell (and/or its progeny) to show that it indeed has multipotent or pluripotent capabilities (clonogenic ability). For bone marrow stem cells, selection of putative adult stem cells has usually excluded typical markers for hematopoietic lineages (lin⁻), CD45, CD38, with inclusion or exclusion of the hematopoietic marker CD34 and inclusion of the marker c-kit (CD117). Other proposed markers for adult stem cells are AC133-2 (CD133), which is found on many stem cell populations,[3] and $C1qR_p$, the receptor for complement molecule C1q,[4] found on a subset of CD34$^{+/-}$ human stem cells from bone marrow and umbilical cord blood. When transplanted into immunodeficient mice, $C1qR_p$-positive human stem cells formed not only hematopoietic cells but also human hepatocytes. Other methods of isolation and identification include the ability of putative stem cells to exclude fluorescent dyes (rhodamine 123, Hoechst 33342), allowing isolation by fluorescence-activated cell sorter (FACS) of a "side population" of cells within a tissue that have stem cell characteristics. Expression of the *Bcrp1* gene (ABCG2 gene in humans) is apparently responsible for this dye exclusion, and could provide a common molecular expression marker for stem cells[5]. A study of expressed genes from a single cell-derived colony of human mesenchymal stem cells identified transcripts from numerous cell lineages,[6] and a similar attempt at profiling the gene expression of human neural stem cell in culture with leukemia inhibitory factor (LIF) has been done,[7] perhaps providing an expressed molecular milieu which could identify candidate stem cells. Attempts to determine the complete molecular signature of gene expression common to human and mouse stem cells have shown over 200 common genes between hematopoietic and neural stem cells, with some considerable overlap with mouse embryonic stem cells as well.[8] The function of many of these genes is as yet unknown, but may provide distinctive markers for identification of adult stem cells in different tissues.

However, dependence on particular markers for prospective identification and isolation of adult stem cells seems unreliable. In particular, the use of specific hematopoietic markers such as the presence or absence of CD34, has yielded mixed results in terms of the identification of putative stem cells. There is evidence that the expression of CD34 and CD133 can actually change over time, and its expression may be part of a cycling phenomenon among human hematopoietic and mesenchymal stem cells in the bone marrow and peripheral blood, and perhaps in other tissues,[9] i.e., an isolated CD34+ cell may become CD34-, and then reacquire CD34 expression. Likewise, a systematic analysis of the cell surface markers and differentiation potential of supposedly distinct isolated populations of human bone marrow stem cells revealed no differences in practice between the cell populations.[10] Moreover, an analysis of genetic and ultrastructural characteristics of human mesenchymal stem cells undergoing differentiation and dedifferentiation has revealed reversibility in the characteristics studied.[11] Thus, any attempt to isolate a single type of adult stem cell for study may not actually capture the intended cell, or may, by using a particular set of isolation or growth conditions, alter its gene expression. This idea has been elaborated by Thiese and Krause,[12] who note that this "uncertainty principle" means any attempt to isolate and characterize a cell necessarily alters its environment, and thereby potentially its gene expression, identity, and potential ability to differentiate along various lineages. Likewise, the stochastic nature of cell differentiation in such dynamic and interacting systems means that attempts to delineate differentiation pathways must include descriptions of each parameter associated with the conditions used, and still may lead only to a probabilistic outcome for differentiation of a stem cell into a particular tissue. Blau et al.[13] have raised the question of whether there may be a "universal" adult stem cell, residing in multiple tissues and activated dependent on cellular signals, e.g., tissue injury. When recruited to a tissue, the stem cell would take its cues from the local tissue milieu in which it finds itself (including the soluble growth factors, extracellular matrix, and cell-cell contacts.) Examples of such environmental influences on fate choice have been noted previously.[14] Thus, it may not be surprising to see examples of cells isolated using the same marker set showing disparate differentiative potentials,[15,16,17,18] based on the context of the isolation or experimental conditions, or to see cells with different marker sets showing similar differentiation. In the final analysis, description of a "stem cell", its actual tissue of origin, and even its differentiation ability, may be a moving target describable only within the context of the particular experimental paradigm used, and may require asking the correct questions in context of the cell's identity and abilities not clonally but rather within a population of cells, and within a certain environment.[12,19]

Given the uncertainties involved in isolating and identifying particular adult stem cells, Moore and Quesenberry[20] suggest that we consider an adult stem cell's functional ability to be, at a minimum, taking on the morphology and cell markers of a differentiated tissue, supplemented by any further functional activity and interaction within a tissue. Certainly a physiological response by improvement of function in a damaged organ system is an indication of a functional response.[19,20] As will be discussed later, the function and therapeutic benefit may not necessarily require direct differentiation and integration of an adult stem cell into a desired tissue, but could be accomplished by stimulation of endogenous cells within the tissue.

DIFFERENTIATION MECHANISMS

Several possible mechanisms have been proposed for differentiation of adult stem cells into other tissues. One mechanism that has received attention lately is the possibility of cell fusion, whereby the stem cell fuses with a tissue cell and takes on that tissue's characteristics. In vitro experiments using fusion of somatic cells with embryonic stem cells and embryonic germ cells[21] have demonstrated that the cell hybrid can take on characteristics of the more primitively developed cell. However, given that such characteristics of spontaneous cell fusion hybrids in vitro have been known for quite some time,[22] and that a cell fusion hybrid does not explain in vitro differentiation of adult stem cells unexposed to tissues, the experiments could not verify this as a possible mechanism for adult stem cell differentiation. More recently, in vivo experiments have shown that for liver,[23] formation of a cell fusion hybrid is a viable explanation for some of the differentiation as well as repair of liver damage seen in these experiments. In an in vitro experiment where human mesenchymal stem cells were co-cultured with heat-shocked small airway epithelial cells, a mixed answer was obtained—some of the stem cells differentiated directly into epithelial cells, while others formed cell fusion hybrids to repair the damage.[24] The ability to form cell hybrids in some tissues may be a useful mechanism for repair of certain types of tissue damage or for delivery of therapeutic genes to a tissue.[25] The reprogramming of cellular gene expression via hybrids is not unlike a novel method reported recently for transdifferentiation of somatic cells. In this method, fibroblasts were soaked in the cytoplasm and nucleoplasm of a lysed, differentiated T lymphocyte cell, taking up factors from the exposed "soup" of the cellular contents of the differentiated cell, and began expressing functional characteristics of a T cell.[26]

In contrast to the results discussed above, other experiments have shown no evidence that cell fusion plays a role in differentiation of adult stem cells into other tissue types. For example, using human subjects it was shown that human bone marrow cells differentiated into buccal epithelial cells *in vivo* without cell fusion,[27] and human cord blood stem cells formed hepatocytes in mouse liver without evidence of cell fusion.[28] In these cases it appears that the adult stem cells underwent changes in gene expression and directly differentiated into the host tissue cell type, integrating into the tissue. It is likely that the mechanism of adult stem cell differentiation may vary depending on the target tissue, or possibly on the state of the adult stem cell used, especially given that normal functioning liver typically shows cell fusion hybrids, with cell fusion functioning as a mechanism for most of the differentiation and repair in tissues such as liver, and direct differentiation (transdifferentiation) into other cell types functioning in other tissues. Much remains to be determined regarding the mechanisms associated with adult stem cell differentiation.

Keeping in mind the uncertainties noted above for identification of a particular adult stem cell and its initial tissue of origin, the majority of this review will focus on some of the evidence for adult stem cell differentiation into other tissues. The cells will be categorized based on general tissue of isolation, with the primary emphasis on human adult stem cells, supplemented with information from animal studies.

BONE MARROW STEM CELLS

Bone marrow contains at least two, and likely more,[2,29] discernable stem cell populations. Besides the hematopoietic stem cell which produces blood cell progeny, a cell type termed mesenchymal or stromal also exists in marrow. This cell provides support for hematopoietic and other cells within the marrow, and has also been a focus for possible tissue repair.[30] Isolation is typically based on some cell surface markers, but also primarily on the ability of these cells to form adherent cell layers in culture. Human mesenchymal stem cells have been shown to differentiate *in vitro* into various cell lineages including neuronal cells,[31,32] as well as cartilage, bone, and fat lineages.[33] *In vivo*, human adult mesenchymal stem cells transferred *in utero* into fetal sheep can integrate into multiple tissues, persisting for over a year. The cells differentiated into cardiac and skeletal muscle, bone marrow stromal cells, fat cells, thymic epithelial cells, and cartilage cells. Analysis of a highly purified preparation of human mesenchymal stem cells[34] indicated that they could proliferate extensively in culture, constitutively expressing the telomerase enzyme, and even after extensive culture retained the ability to differentiate *in vitro* into bone, fat, and cartilage cells. Isolated colonies of the cells formed bone when injected into immunodeficient mice. Expanding on their previous *in vitro* work with rat and human mesenchymal/stromal stem cells, Woodbury et al.[35] performed molecular analyses of rat stromal stem cells and found that the cells express genes associated with all three primary germ layers—mesodermal, ectodermal, and endodermal—as well as a gene associated for germinal cells. The gene expression pattern was also seen in a clonal population of cells, indicating that it was not due to an initial mixed population of cells, but was the typical gene expression pattern of the stromal cells. The results suggested that the stromal stem cells were already multidifferentiated and that switching to a neuronal differentiation pattern involved quantitative regulation of existing gene expression patterns. Koc et al.[36] have used infusion of allogeneic donor mesenchymal stem cells in an attempt to correct some of the skeletal and neurological defects associated with Hurler syndrome (mucopolysaccharidosis type-IH) and metachromatic leukodystrophy (MLD). A total of 11 patients received donor mesenchymal stem cells, expanded from bone marrow aspirate. Four patients showed significant improvements in nerve conduction velocities, and all patients showed maintenance or slight improvement in bone mineral density.

Bone marrow-derived cells in general have shown ability to form many tissues in the body. For example, bone marrow-derived stem cells *in vivo* appear able to form neuronal tissues,[18,37] and a single adult bone marrow stem cell can contribute to tissues as diverse as marrow, liver, skin, and digestive tract.[16] One group has now developed a method for large-scale generation of neuronal precursors from whole adult rat bone marrow.[38] In this procedure, treatment of unfractionated bone marrow in culture with epidermal growth factor and basic fibroblast growth factor gave rise to neurospheres with cells expressing neuronal markers.

In vivo studies using fluorescence and genetic tracking of adult stem cells in animals, and tracking of the Y chromosome in humans, has shown that bone marrow stem cells can contribute to numerous adult tissues. Follow-up of patients receiving adult bone marrow stem cell transplants has allowed tracking of adult stem cells within humans, primarily by identification of Y chromosome-bearing cells in female patients who had received bone marrow stem cells from male donors. Biopsy or postmortem samples show that some of the transplanted bone marrow stem cells could form liver, skin, and digestive tract cells,[39] as well as participate in the generation of new neurons within the human brain.[40] Bone marrow stem cells have also been shown to contribute to Purkinje cells in the brains of adult mice[41] and humans[42]. Generation of this particular type of neural cell is significant in that new Purkinje cells do not normally appear to be generated after birth.

Regeneration or replacement of dead or damaged cells is the primary goal of regenerative medicine and one of the prime motivations for study of stem cells. It is thus of significant interest that bone marrow stem cells have shown the ability to produce therapeutic benefit in animal models of stroke. In mice, fluorescence-tracked bone marrow derived stem cells expressed neuronal antigens and also incorporated as endothelial cells, possibly producing therapeutic benefit by allowing increased blood flow to damaged areas of the brain.[43] In rats, intravenous (IV) administration of rat[44] or human[45] bone marrow stromal cells resulted in significant behavioral recovery after stroke. Interestingly, only a small percentage of the stromal stem cells appeared to incorporate into the damaged brain as neuronal cells (1-5% in the case of the human marrow stromal cells), but the levels of neurotrophin growth factors within the brains increased and were possibly the signal for repair of damaged brain tissue, perhaps by stimulation of endogenous neuronal precursors. It is also of interest that the marrow stromal cells were injected IV and not intracerebrally, indicating that the stem cells somehow "homed" to the site of tissue damage. Most studies showing adult stem cell differentiation into other tissues show an increased incorporation of cells, or even an absolute requirement for differentiation, relying on tissue damage to initiate the differentiation. This may indicate that without a "need" for replacement and repair, there is little or no activation of adult stem cells. The recruitment and homing of adult stem cells to damaged tissues are fascinating but relatively unexplained phenomena. One report[46] indicates that recruitment of quiescent stem cells from bone marrow to the circulation requires release of soluble c-kit ligand (stem cell factor), but the range of factors necessary for recruitment and homing to organs other than bone marrow is unknown at this time and warrants increased investigation.

Bone marrow stem cells have also shown the ability to participate in repair of damaged retinal tissues. When bone marrow stem cells were injected into the eyes of mice, they associated with retinal astrocytes and extensively incorporated into the vascular (blood vessel) network of the eye.[47] The cells could also rescue and maintain normal vasculature in the eyes of mice with a degenerative vascular disease. In another animal study, bone marrow derived stem cells were observed to integrate into injured retina and differentiated into retinal neuronal cells.[48] Stromal stem cells have also shown capability in mice to repair spinal cord which was demyelinated.[49] One of the problems related to spinal cord injury is loss of the protective myelin sheath from spinal cord after injury. A mixed bone marrow stem cell fraction was injected into the area of damage in the spinal cord, and remyelination of the area was seen. In another mouse study, marrow stromal cells injected into injured spinal cord formed guiding strands within the cord;[50] interestingly, the effect was more pronounced when the stromal cells were injected 1 week after injury rather than immediately after injury.

Because bone marrow stem cells are of mesodermal lineage, it is not surprising that they show capabilities at forming other tissues of mesodermal origin. Human marrow stromal cells, which have been shown to form cartilage cells, have been used in an *in vitro* system to define many of the molecular events associated with formation of cartilage tissue.[51] Bone marrow derived stem cells have also been shown capable of regenerating damaged muscle tissue.[52] In an elegant study following genetically marked bone marrow stem cells in mice, LaBarge and Blau were able to document multiple steps in the progression of the stem cells to form muscle fibers and repair muscle damage.[53] The ability of human bone marrow derived stem cells to form muscle cells and persist in the muscle was recently documented. In this case, a patient had received a bone marrow transplant at age 1, and developed Duchenne muscular dystrophy at age 12. Biopsies at age 14 showed donor nuclei integrated within 0.5-0.9% of the muscle fibers of the patient, indicating the ability of donated marrow cells to persist in tissue over long periods of time.[54]

Bone marrow stem cells have also shown capability at forming kidney cells. Studies following genetically marked bone marrow stem cells in rats[55] and mice[56] showed that the stem cells could form mesangial cells to repopulate the glomerulus of the kidney. In the mouse study, formation of cell fusion products was ruled out as a mechanism for differentiation of the bone marrow stem cells. Other animal studies have shown contribution of bone marrow stem cells to repair of damaged renal tubules in the kidney;[57,58] taken together, animal studies indicate that bone marrow stem cells can participate in restoring damaged kidney tissue.[59]

Liver was one of the earliest tissues recognized as showing potential contribution to differentiated cells by bone marrow stem cells. Bone marrow stem cells have been induced to form hepatocytes in culture[60] and liver-specific gene expression has been induced *in vitro* in human bone marrow stem cells.[61] *In vivo*, bone marrow stem cells were able to incorporate into liver as hepatocytes and rescue mice from a liver enzyme deficiency, restoring normal liver function.[62] Bone marrow stem cells also repopulated liver after irradiation of mice to destroy their bone marrow.[63] Examination of livers of female patients who had received male bone marrow transplants, and male patients who had received female liver transplants, showed that similar repopulation of liver from bone marrow stem cells could take place in humans.[64] Examination of the kinetics of liver repopulation by bone marrow stem cells in a mouse model indicated that the replacement was slow, with only small numbers of cells replaced by the bone marrow stem cells.[65] As noted previously, two recent studies have found that replenishment of liver by bone marrow stem cells occurs primarily via cell fusion hybrid formation, even in repair of liver damage.[23] A side-population of stem cells has been identified in mouse liver, similar to that seen in bone marrow. This hepatic side-population, which contributes to liver regeneration, can be replenished by side-population bone marrow stem cells.[66]

Pancreas and liver arise from adjacent endoderm during embryological development, and show relatedness in some gene expression and interconversion in some instances. Bone marrow derived cells have shown the ability to form pancreatic cells in animal studies. Mouse bone marrow stem cells containing a genetic fluorescent marker that is only expressed if insulin is expressed were transplanted into irradiated female mice.[67] Within 6 weeks of transplant, fluorescent donor cells were observed in pancreatic islets; donor cells identified in bone marrow and peripheral blood did not show fluorescence. *In vitro*, the bone marrow derived cells showed glucose-dependent insulin secretion as well. Bone marrow derived stem cells have also demonstrated the ability to induce regeneration of damaged pancreas in the mouse.[68] Mice with experimentally induced hyperglycemia from pancreatic damage were treated with bone marrow derived stem cells expressing the c-kit marker. Interestingly, only a low percentage of donor cells were identified as integrating into the regenerating pancreas, with most of the regeneration due to induced proliferation and differentiation of endogenous pancreatic cell precursors, suggesting that the bone marrow stem cells provided growth signals for the tissue regeneration.

Heart, as a mesodermally-derived organ, is a likely candidate for regeneration with bone marrow derived stem cells. Numerous references now document the ability of these adult stem cells to contribute to regeneration of cardiac tissue and improve performance of damaged hearts. In animal studies, for example, rat[69], mouse[70,71,72] and human[73,74] stem cells have been identified as integrating into cardiac tissue, forming cardiomyocytes and/or cardiac blood vessels, regenerating infarcted heart tissue, and improving cardiac function. In mice, bone marrow derived stem cells injected into old animals seems capable of restoring cardiac function,[75] apparently through increased activity for cardiac blood vessel formation. One fascinating study using xenogeneic (cross-species) transplants suggests that stromal cells may show immune tolerance by the host.[76] Mouse marrow stromal cells were transplanted into fully immunocompetent rats, and contributed formation of cardiomyocytes and cardiac vessels. Even after 13 weeks, the mouse cells were not rejected by the rat hosts. Evidence has accumulated from postmortem studies that bone marrow stem cells can contribute to cardiomyocytes after damage to the human heart as well.[77,78] The evidence has led numerous groups to use bone marrow derived stem cells in treatment of patients with damaged cardiac tissue.[79,80,81,82] Results from these clinical trials indicate that bone marrow derived stem cells, including cells from the patients themselves, can regenerate damaged cardiac tissue and improve cardiac performance in humans. In terms of restoring angiogenesis and improving blood circulation, results in patients are not limited to the heart. Tateishi-Yuyama et al.[83] have shown that bone marrow derived stem cells from the patients themselves can improve blood circulation in gangrenous limbs, in many cases obviating the need for amputation.

Bone marrow derived adult stem cells have also been found to contribute to various other adult tissues. Animal studies indicate evidence that bone marrow stem cells can contribute as progenitors of lung epithelial tissue[84], and mesenchymal stem cells can home to damaged lung tissue, engraft, and take on an epithelial morphology, participating in repair and reduction of inflammation.[85] Bone marrow derived stem cells also have been shown to contribute to regeneration of gastrointestinal epithelia in human patients.[86] A recent study in mice has indicated that bone marrow stem cells can also participate in cutaneous healing, contributing to repair of skin after wounding.[87]

PERIPHERAL BLOOD STEM CELLS

There is abundant evidence that bone marrow stem cells can leave the marrow and enter the circulation, and specific mobilization of bone marrow stem cells is used to harvest stem cells more easily for various bone marrow stem cell treatments.[88] Therefore, it is not surprising that adult stem cells have been isolated from peripheral blood. Mobilized stem cells in peripheral blood have been administered intravenously in a rat model of stroke, ameliorating some of the behavioral deficits associated with the damaged neural tissue[89], leading to a proposal that stem cell mobilization in patients might be used as a treatment for stroke in humans.[90] Mobilized stem cells have also been used in cardiac regeneration in mice[72]. Two recent studies have found that human peripheral blood stem cells exhibiting pluripotent properties can be isolated from unmobilized human blood. One study showed that the isolated cells were adherent, similar to marrow mesenchymal cells, and could be induced to differentiate into cells from all three primary germ layers, including macrophages, T lymphocytes, epithelial cells, neuronal cells, and liver cells.[91] The other study showed induction of the peripheral blood stem cells could produce hematopoietic, neuronal, or cardiac cells in culture.[92] In the latter study, undifferentiated stem cells were negative for both major histocompatability antigens (MHC) I and II, expressed high levels of the Oct-4 gene (usually associated with pluripotent capacity in other stem cells), and could form embryoid body structures in culture.

NEURONAL STEM CELLS

One extremely interesting finding of the past few years has been the discovery of neuronal stem cells, indicating that cell replenishment was possible within the brain (something previously considered impossible.) Neuronal stem cells have been isolated from various regions of the brain including the more-accessible olfactory bulb[93] as well as the spinal cord[94], and can even be recovered from cadavers soon after death.[95] Evidence now exists that neuronal stem cells can produce not only neuronal cells but also other tissues, including blood and muscle.[96,97,98,99,100,101] Animal studies have shown that adult neural stem cells can participate in repair of damage after stroke, either via endogenous neuronal precursors[102] or transplanted neural stem cells.[103] Evidence indicates that endogenous neurons and astrocytes may also secrete growth factors to induce differentiation of endogenous precursors.[104] In addition, two studies now provide suggestive evidence that neural stem cells/neural progenitor cells may show low immunogenicity, being immunoprivileged on transplant,[105] and raising the possibility for use of donor neural stem cells to treat degenerative brain conditions.

Pluchino et al.[106] recently used adult neural stem cells to test potential treatment of multiple sclerosis lesions in the brain. Using a mouse model of chronic multiple sclerosis—experimental immune encephalitis—they injected neural stem cells either intravenously or intracerebrally into affected mice. Donor cells entered damaged, demyelinated regions of the brain and differentiated into neuronal cells. Remyelination of brain lesions and recovery from functional impairment were seen in the mice. Neural stem cells have also been used to investigate potential treatments for Parkinson's disease. Using experimentally-lesioned animals as models for Parkinson's disease, human neural stem cells have been observed to integrate and survive for extended periods of time.[107] Dopaminergic cells (the cells degenerated in Parkinson's disease) can be induced in these systems,[108] and neural stem cells are capable of rescuing and preventing the degeneration of endogenous dopaminergic neurons,[109,110] also producing improved behavioral performance in the animals. In these studies, the data suggest that the transplanted neural stem cells did not participate to a large extent in direct formation of dopaminergic neurons, but rather secreted neuroprotective factors and growth factors that stimulated the endogenous neural cells. In this respect, infusion of transforming growth factor into the brains of Parkinson's mice induced proliferation and differentiation of endogenous neuronal precursors in mouse brain.[111] Following this potential for stimulation of endogenous neuronal cells, Gill et al. recently reported on a Phase I trial in which glial derived neurotrophic factor (GDNF) was infused into the brains of five Parkinson's patients.[112] After one year there was a 61% increase in the activities of daily living score, and an increase in dopamine storage observed in the brain. In a tantalizing clinical application with direct injection of neural stem cells, a Parkinson's patient was implanted with his own neural stem cells, resulting in an 80% reduction in symptoms at one year after treatment.[113] Further clinical trials are underway.

The olfactory ensheathing glial (OEG) cell from olfactory bulb has been used extensively in studies regarding spinal cord injury and axon regrowth. Human OEG cells can be expanded in number in culture and induced to produce all three main neural cell types.[93] Transplant of the cells into animal models of spinal cord injury has shown that the cells can effect remyelination of demyelinated spinal cord axons,[114] and provide functional recovery in paraplegic rats,[115] including in transected spinal cords.[116]   Another study has found that infusion of growth factors such as GDNF can stimulate functional regeneration of sensory axons in adult rat spinal cord.[117] Interestingly, one group has made use of the similarities between enteric glial cells and OEG cells, and shown that transplanted enteric glial cells can also promote regeneration of axons in the spinal cord of adult rats.[118]   Clinical trials are underway to test the abilities of OEG cells in spinal cord injury patients.   Finally, a significant impediment to recovery from spinal cord injury is the formation of a glial/astrocyte scar at the site of injury, which can prevent growth of axons no matter what the source of the cells.   Menet et al. have shown, using a mutant mouse model, that much of the scar can be prevented by inhibition of glial fibrillary acidic protein and vimentin.[119]   In mutant mice that lacked these genes, there was increased sprouting of axons and functional recovery after spinal cord injury.   Thus, endogenous neural cell growth and reconnection might suffice for repair of damage if inhibitory mechanisms can be removed from neural systems.

hNT CELLS

Embryonal carcinoma (EC) cells can be derived from teratocarcinomas of adult patients, and show multipotent differentiation abilities in culture. From one such isolation, a "tamed" (non-tumorigenic) line of cells with neuronal generating capacity has been developed, termed hNT (NT-2) cells.   Because of their capacity to generate neuronal cells, these cells have been studied for possible application in regeneration of neuronal tissues.   The hNT neurons show the ability to generate dopaminergic neurons,[120] and have shown some benefit of transplantation in animal models of amyotrophic lateral sclerosis (ALS, Lou Gehrig's disease).[121]   Early clinical trials using hNT neurons transplanted into stroke patients have shown initial positive results.[122]

MUSCLE STEM CELLS

Muscle contains satellite cells that normally participate in replacement of myoblasts and myofibers.   There are also indications that muscle additionally may harbor other stem cells, either as hematopoietic migrants from bone marrow and peripheral blood, or as intrinsic stem cells of muscle tissue.   Muscle appears to contain a side population of stem cells, as seen in bone marrow and liver, with the ability to regenerate muscle tissue.[123]   Muscle derived stem cells have been clonally isolated and used to enhance muscle and bone regeneration in animals.[124] An isolated population of muscle-derived stem cells has also been shown to participate in muscle regeneration in a mouse model of muscular dystrophy.[125] Stimulation of muscle regeneration from muscle-derived stem cells, as observed in other tissues, is greatly increased after injury of the tissue.[126,127] An interesting use of muscle-derived stem cells has been the regeneration and strengthening of bladder in a rat model of incontinence.[128] Because of the similar nature of muscle cells between skeletal muscle and heart muscle, muscle-derived stem cells have also been proposed for use in repairing cardiac damage,[129] with evidence that mechanical beating is necessary for full differentiation of skeletal muscle stem cells into cardiomyocytes.[130] At least one group has used skeletal muscle cells for clinical application to repair cardiac damage in a patient, with positive results.[131]

LIVER STEM CELLS

As noted before, there are similarities between liver and pancreas which could facilitate interconversion of cells between the two tissues. This concept has been demonstrated using genetic engineering to add a pancreatic development gene to liver cells, converting liver to pancreas.[132] Rat liver stem cells have been converted *in vitro* into insulin-secreting pancreatic cells.[133] When transplanted into immundeficient mice which are a model for diabetes, the converted liver stem cells were able to reverse hyperglycemia in the mice. One other interesting observation regarding liver stem cells has been the possible formation of myocytes in the heart by liver stem cells. A clonal cell line derived from adult male rat liver and genetically tagged was injected into female rats, and marked, Y-chromosome bearing myocytes were identified in the host hearts after six weeks.[134]

PANCREATIC STEM CELLS

Interconversion between pancreas and liver has also been demonstrated starting with pancreatic stem cells, in which mouse pancreatic cells repopulated the liver and corrected metabolic liver disease.[135] For pancreas, however, the possibility of solutions to the scourge of diabetes has been a driving force in efforts to define a stem cell that could regulate insulin in a normative, glucose-dependent fashion. The success of the Edmonton protocol,[136] where cadaveric pancreatic islets are transplanted into patients, has provided a glimmer of hope, but more readily-available sources of insulin-secreting cells are needed. Fortunately, there seems to be no shortage of potential candidates that can form insulin-secreting cells. The pancreas itself appears to contain stem/progenitor cells that can regenerate islets *in vitro* and *in vivo*. Studies indicate that these pancreatic stem cells can functionally reverse insulin-dependent diabetes in mice.[137] Similar pancreatic stem cells have been isolated from humans and shown to form insulin-secreting cells *in vitro*,[138] the hormone glucagon-like peptide-1 appears to be an important inducing factor of pancreatic stem cell differentiation. Interestingly, the same hormone could induce mouse intestinal epithelial cells to convert into insulin-producing cells *in vitro*, and the cells could reverse insulin-dependent diabetes when implanted into diabetic mice.[139] Besides pancreatic and intestinal stem cells, other adult stem cell types showing the ability to secrete insulin and regenerate damaged pancreas include bone marrow[57,58] and liver.[133] Genetic engineering of rat liver cells to contain the pancreatic gene PDX-1 has also been used to generate insulin-secreting cells *in vitro*; the cells could also restore normal blood glucose levels when injected into mice with experimentally-induced diabetes.[140]

CORNEAL LIMBAL STEM CELLS

Corneal limbal stem cells have become commonly used for replacement of corneas, especially in cases where cadaveric donor corneas are insufficient. Limbal cells can be maintained and cell number expanded in culture,[141] grown on amniotic membranes to form new corneas, and transplanted to patients with good success.[142] A recent report indicates that human corneal stem cells can also display properties of functional neuronal cells in culture.[143] Another report found that limbal epithelial cells or retinal cells transplanted into retina of rats could incorporate and integrate into damaged retina, but did not incorporate into normal retina.[144]

MAMMARY STEM CELLS

Reports have indicated that mammary stem cells also exist. Isolated cells from mouse could be propagated *in vitro* and differentiated into all three mammary epithelial lineages.[145] Clonally-propagated cells were induced in culture to generate complex three-dimensional structures similar to that seen *in vivo*. Transcriptional profiling indicated that the mammary stem cells showed similar gene expression profiles to those of bone marrow stem cells. In that respect, there is a report that human and mouse mammary stem cells exist as a side population, as seen for bone marrow, liver, and muscle stem cells.[146] When propagated in culture, the isolated mammary side population stem cells could form epithelial ductal structures.

SALIVARY GLAND

A recent report indicates that stem cells can be isolated by limiting dilution from regenerating rat salivary gland and propagated *in vitro*.[147] Under differing culture conditions, the cells express genes typical of liver or pancreas, and when injected into rats can integrate into liver tissue.

SKIN

Multipotent adult stem cells have been isolated from the dermis and hair follicle of rodents.[148] The cells play a role in maintenance of epidermal and hair follicle structures, can be propagated *in vitro,* and clonally isolated stem cells can be induced to form neurons, glia, smooth muscle, and adipocytes in culture. Dermal hair follicle stem cells have also shown the ability to reform the hematopoietic system of myeloablated mice.[149]

TENDON

A recent report notes the isolation of established stem cell-like lines from mouse tendon. The cells exhibited a mesenchymal morphology, and expressed genes related to osteogenic, chondrogenic, and adipogenic potential, similar to that seen in bone marrow mesenchymal stem cells.[150]

SYNOVIAL MEMBRANE

Stem cells from human synovial membrane (knee joint) have been isolated which show multipotent abilities for differentiation, including evidence of myogenic potential.[151] These stem cells were used in a mouse model of Duchenne muscular dystrophy to test their ability to repair damaged muscle. Stem cells injected into the bloodstream could engraft and incorporate into muscle, taking on a muscle phenotype, and with evidence of muscle repair.[152]

HEART

Beltrami *et al.* analyzed the hearts of post-mortem patients who succumbed 4-12 days after heart attack, and found evidence of dividing myocytes in the human heart. While it is unclear from the study whether the cells were originally cardiomyocytes or were other stem cells which had homed to damaged heart tissue, such as bone marrow stem cells, the evidence indicated dividing cells within the heart.[153]

CARTILAGE

Human cartilage biopsies placed into culture show apparent dedifferentiation into primitive chondrocytes with mesenchymal stem cell appearance.[154] These chondrocytes have been used for transplants to repair articular cartilage damage, and in treatment of children with osteogenesis imperfecta.[155,156,157]

THYMIC PROGENITORS

Bennett *et al.* have reported the isolation of thymic epithelial progenitor cells.[158] Ectopic grafting (under the kidney capsule) of the cells into mice allowed production of all thymic epithelial cell types, as well as attraction of homing T lymphocytes. In separate experiments, Gill *et al.* also isolated a putative thymic progenitor cell from mice and were able to use these cells to reform miniature thymuses when the cells were transplanted under mouse kidney capsule.[159]

DENTAL PULP STEM CELLS

Stem cells have been isolated from human adult dental pulp that could be clonally propagated and proliferated rapidly.[160] Though there were some similarities with bone marrow mesenchymal stem cells, when injected into immunodeficient mice the adult dental pulp stem cells formed primarily dentin-like structures surrounded by pulpy interstitial tissue. Human baby teeth have also been identified as a source of stem cells, designated SHED cells (Stem cells from Human Exfoliated Deciduous teeth).[161] *In vitro*, SHED cells could generate neuronal cells, adipocytes, and odontoblasts, and after injection into immunodeficient mice, the cells were indicated in formation of bone, dentin, and neural cells.

ADIPOSE (FAT) DERIVED STEM CELLS

One of the more interesting sources identified for human stem cells has been adipose (fat) tissue, in particular liposuctioned fat. While there is some debate as to whether the cells originate in the fat tissue or are perhaps mesenchymal or peripheral blood stem cells passing through the fat tissue, they represent a readily-available source for isolation of potentially useful stem cells. The cells can be maintained for extended periods of time in culture, have a mesenchymal-like morphology, and can be induced *in vitro* to form adipose, cartilage, muscle, and bone tissue.[162] The cells have also shown the capability of differentiation into neuronal cells.[163]

UMBILICAL CORD BLOOD

Use of umbilical cord stem cells has seen increasing interest, as the cells have been recognized as a useful source for hematopoietic transplants similar to bone marrow stem cell transplants, including for treatment of sickle cell anemia.[164] Cord blood shows decreased graft-versus-host reaction compared to bone marrow,[165] perhaps due to high interleukin-10 levels produced by the cells.[166] Another possibility for the decreased rejection seen with cord blood stem cell transplants is decreased expression of the beta-2-microglobulin on human cord blood stem cells.[167] Cord blood can be cryopreserved for over 15 years and retain significant functional potency.[168] Cord blood stem cells also show similarities with bone marrow stem cells in terms of their potential to differentiate into other tissue types. Human cord blood stem cells have shown expression of neural markers *in vitro*,[169] and intravenous administration of cord blood to animal models of stroke has produced functional recovery in the animals.[89,170] Infusion of human cord blood stem cells has also produced therapeutic benefit in rats with spinal cord injury,[171] and in a mouse model of ALS.[172] A recent report noted establishment of a neural stem/progenitor cell line derived from human cord blood that has been maintained in culture over two years without loss of differentiation ability.[173] Several reports also note the production of functional liver cells from human cord blood stem cells.[174] Additional differentiative properties of human umbilical cord blood stem cells are likely to be discovered as more investigation proceeds on this source of stem cells.

UMBILICAL CORD MESENCHYME (WHARTON'S JELLY)

While most of the focus regarding umbilical cord stem cells has focused on the cord blood, there are also reports that the matrix cells from umbilical cord contain potentially useful stem cells. Using pigs, this matrix from umbilical cord, termed Wharton's jelly, has been a source for isolation of mesenchymal stem cells. The cells express typical stem cell markers such as c-kit and high telomerase activity, have been propagated in culture for over 80 population doublings, and can be induced to form neurons *in vitro*.[175] When transplanted into rats, the cells expressed neuronal markers and integrated into the rat brain, additionally without any evidence of rejection.[176]

AMNIOTIC STEM CELLS

Amniotic fluid has also been found to contain stem cells that can take on neuronal properties when injected into brain.[177] These stem cells were recently isolated from human amniotic fluid,[178] and were found to express Oct-4, a gene typically associated with expression in pluripotent stem cells.

MESANGIOBLASTS

Mesangioblasts are a multipotent stem cell that has been isolated from large blood vessels such as dorsal aorta.[179] The cells show long term proliferative capacity in culture as well as the capability of differentiation into most mesodermally derived types of tissue. In a recent report, the cells were injected into the bloodstream of mice that are a model for muscular dystrophy,[180] and participated in repair of the muscle tissue.

Adult stem cells in other tissues very likely exist, but this survey of many of the known adult stem cells and their capacities for differentiation and tissue repair can serve as a beginning point for discussion regarding the progress as well as potential of adult stem cells. Some final thoughts on current and potential utilization of adult stem cells follow.

ADULT STEM CELL MOBILIZATION FOR TISSUE REPAIR

An important point to consider as we look ahead regarding utilization of adult stem cells for tissue repair is that it may be unnecessary first to isolate and culture stem cells before injecting them back into a patient to initiate tissue repair. Rather, it may be easier and preferable to mobilize endogenous stem cells for repair of damaged tissue. Initial results regarding this possibility have already been seen in some animal experiments, in which bone marrow and peripheral blood stem cells were mobilized with injections of growth factors and participated in repair of heart and stroke damage.[72,89,90] The ability to mobilize endogenous stem cells, coupled with natural or perhaps induced targeted homing of the cells to damaged tissue, could greatly facilitate use of adult stem cells in simplified tissue regeneration schemes.[181]

GENE THERAPY APPLICATIONS WITH ADULT STEM CELLS

Adult stem cells can provide an efficient vehicle for gene therapy applications, and engineered adult stem cells may allow increased functionality, proliferative capacity, or stimulatory capability to these cells. The feasibility of genetically engineering adult stem cells has been shown, for example, in the use of bone marrow stem cells containing stably inserted genes. The engineered stem cells when injected into mice could still participate in formation and repair of differentiated tissue, such as in lung.[182] As another example, engineered stem cells containing an autoantigen, to induce immune tolerance of T cells to insulin-secreting cells, were shown to prevent onset of diabetes in a mouse model of diabetes,[183] a strategy that may be useful for various human autoimmune diseases. Introduction of the PDX-1 gene into liver stem cells stimulated differentiation into insulin-producing cells which could normalize glucose levels when transplanted into mice with induced diabetes.[140] Simply engineering cells to increase their proliferative capacity can have a significant effect on their utility for tissue engineering and repair. For example, McKee et al.[184] engineered human smooth muscle cells by introducing human telomerase, which greatly increased their proliferative capacity beyond the normal lifespan of smooth muscle cells in culture, while allowing retention of their normal smooth muscle characteristics. These engineered smooth muscle cells were seeded onto biopolymer scaffolds and allowed to grow into smooth muscle layers, then seeded with human umbilical vein endothelial cells. The resulting engineered arterial vessels could be useful for transplants and bypass surgery. Similarly, human marrow stromal cells that were engineered with telomerase increased their proliferative capacity significantly, but also showed enhanced ability at stimulating bone formation in experimental animals.[185] Genetically-engineered human adult stem cells have already been used in successful treatment of patients with genetic disease. Bone marrow stem cells, from infants with forms of severe combined immunodeficiency syndrome (SCID), were removed from the patients, a functional gene inserted, and the engineered cells reintroduced to the same patients. The stem cells homed to the bone marrow, engrafted, and corrected the defect.[186,187,188]

Adult stem cells could also be used to deliver stimulatory or protective factors to tissues and endogenous stem cells. This would utilize the innate homing ability of adult stem cells, but would not necessarily rely on differentiation of the stem cells to participate in tissue replenishment. For example, Benedetti et al. utilized the homing capacity of neural stem cells in brain by engineering mouse neural stem cells with the gene for interleukin-4. Transfer into brain glioblastomas in mice led to the survival of most of the mice, and imaging analysis documented the progressive disappearance of large tumors.[189] Likewise, engineered mesenchymal stem cells were transplanted into the brains of mice that are a model of Niemann-Pick disease; the enzyme acid sphingomyelinase is lost in the disease, resulting in neurological damage and early death. The mesenchymal stem cells were engineered to overexpress the missing enzyme. When injected into brains of the mouse model, the mice showed a delay in onset of neurological abnormalities and an extension of lifespan, suggesting that the stem cells delivered and secreted the necessary enzyme to the brain tissue.[190] Muscle-derived stem cells that were engineered to express the growth factor bone morphogenetic protein-2 were used to stimulate bone healing in mice with skull bone defects. While the muscle-derived stem cells did show differentiation as bone cells, the results indicated that the critical factor was delivery of the secreted growth factor by the stem cells to the areas of bone damage, allowing much more rapid healing than in control animals.[191] As noted previously, neural stem cells show an ability to rescue degenerating neurons, including the dopaminergic neurons whose loss is associated with Parkinson's disease. The delivery of neuroprotective substances is postulated as the most likely explanation for this phenomenon, rather than substantial differentiation by the injected neural stem cells.[109] In support of this hypothesis, when neural stem cells were specifically engineered to overexpress a neurotrophic factor similar to glial derived neurotrophic factor, degeneration of dopaminergic neurons was prevented.[110]

STIMULATING ENDOGENOUS CELLS

The indications from the previous examples suggest that direct stimulation of endogenous stem cells within a tissue may be the easiest, safest, and most efficient way to stimulate tissue regeneration. Such stimulation need not rely on any added stem cells. This approach would circumvent the need to isolate or grow stem cells in culture, or inject any stem cells into the body, whether the cells were derived from the patient or another source. Moreover, direct stimulation of endogenous tissue stem cells with specific growth factors might even preclude any need to mobilize stem cells to a site of tissue damage. A few experimental results suggest that this approach might be possible. One group has reported that use of glial derived neurotrophic factor and neurotrophin-3 can stimulate regeneration of sensory axons in adult rat spinal cord.[117]

Administration of transforming growth factor to the brains of Parkinson's mice stimulated proliferation and differentiation of endogenous neuronal stem cells and produced therapeutic results in the mice,[111] and infusion of glial derived neurotrophic factor into the brains of Parkinson's patients resulted in increased dopamine production within the brain and therapeutic benefit to the patients.[112] And, Zeisberg et al. have found that bone morphogenetic protein-7 (BMP-7) can counteract deleterious cell changes associated with tissue damage. In this latter study, a mouse model of chronic kidney damage was used. Damage to the tissue causes a transition from epithelial to mesenchymal cell types in the kidney, leading to fibrosis. The transition appears to be initiated by the action of transforming growth factor beta-1 on the tissues, and BMP-7 was shown to counteract this signaling in vitro. Systemic administration of BMP-7 in the mouse model reversed the transition in vivo and led to repair of severely damaged renal tubule epithelial cells.[192] These experiments indicate that direct stimulation of tissues by the correct growth factors could be sufficient to prevent or repair tissue damage. The key to such treatments would be identification of the correct stimuli specific to a tissue or cell type.

In summary, our current knowledge regarding adult stem cells has expanded greatly over what was known just a few short years ago. Results from both animal studies and early human clinical trials indicate that they have significant capabilities for growth, repair, and regeneration of damaged cells and tissues in the body, akin to a built-in repair kit or maintenance crew that only needs activation and stimulation to accomplish repair of damage. The potential of adult stem cells to impact medicine in this respect is enormous.

Adult Stem Cells—Addendum (October 2003)
For the President's Council on Bioethics
David A. Prentice

Since initial submission of the commissioned paper, numerous additional published references have documented the abilities of adult stem cells to stimulate regeneration of damaged tissues. Just a few of the most significant are mentioned here. Mesenchymal stem cells engineered to express the *Akt1* gene, when transplanted into mice, demonstrated the ability to repair and restore performance of infarcted heart, essentially to a normal state.[a] Another clinical trial in addition to those mentioned in the paper has shown significant improvement in patients with heart damage, with reduction in the area of damage and improved heart function after adult stem cell treatment.[b] Three more published articles support the existence of a stem cell in the heart and its participation in cardiac regeneration.[c] Stroke damage in rats was repaired using human neural stem cells[d] and prostate was regenerated *in vivo* in mice using adult stem cells.[e] Another report indicates that human mixed bone marrow stem cells can contribute significant amounts of lung tissue in patients[f] and pluripotent stem cells were discovered in the mouse inner ear[g], which can form all 3 primary germ layers and might lead to potential therapies for hearing loss. Finally, bone marrow stem cells were discovered to have a protective as well as regenerative role in diabetes.[h]

ADDENDUM REFERENCES

[a]Mangi AA et al., "Mesenchymal stem cells modified with Akt prevent remodeling and restore performance of infarcted hearts", *Nature Medicine* 9, 1195-1201, Sept 2003

[b]Britten MB et al., "Infarct remodeling after intracoronary progenitor cell treatment in patients with acute myocardial infarction", *Circulation* 108, 2212-2218, Nov 2003

[c]Urbanek K et al., "Intense myocyte formation from cardiac stem cells in human cardiac hypertrophy", *Proceedings of the National Academy of Sciences USA* 100, 10440-10445, 2 Sept 2003; Beltrami AP et al., "Adult cardiac stem cells are multipotent and support myocardial regeneration", *Cell* 114, 763-776, 19 Sept 2003; Oh H et al., "Cardiac progenitor cells from adult myocardium: homing, differentiation, and fusion after infarction", *Proceedings of the National Academy of Sciences USA* 100, 12313-12318, 14 Oct 2003

[d]Jeong S-W et al., "Human neural stem cell transplantation promotes functional recovery in rats with experimental intracerebral hemorrhage", *Stroke* 34, 2258-2263, Sept 2003

[e]Xin L et al., "In vivo regeneration of murine prostate from dissociated cell populations of postnatal epithelia and urogenital sinus mesenchyme", Proceedings of the National Academy of Sciences USA 100, 11896-11903, 30 Sept 2003

[f]Suratt BT et al., "Human pulmonary chimerism after hematopoietic stem cell transplantation", American Journal of Respiratory and Critical Care Medicine 168, 318-322, 2003

[g]Li H et al., "Pluripotent stem cells from the adult mouse inner ear", Nature Medicine 9, 1293-1299, Oct 2003

[h]Li FX et al., "The development of diabetes in E2f1/E2f2 mutant mice reveals important roles for bone marrow-derived cells in preventing islet cell loss", Proceedings of the National Academy of Sciences USA 100, 12935-12940, 28 Oct 2003

REFERENCES

[1] See, for example: Feasel AM et al., "Complete remission of scleromyxedema following autologous stem cell transplantation", Archives of Dermatology 137, 1071-1072; Aug 2001; Mancardi GL et al.; "Autologous hematopoietic stem cell transplantation suppresses Gd-enhanced MRI activity in MS"; Neurology 57, 62-68; 10 July 2001; Traynor AE et al.; "Hematopoietic stem cell transplantation for severe and refractory lupus"; Arthritis Rheumatology 46, 2917-2923; November 2002; Wulffraat NM et al.; "Prolonged remission without treatment after autologous stem cell transplantation for refractory childhood systemic lupus erythematosus"; Arthritis Rheumatology 44, 728-731; March 2001; Burt RK et al., "High-dose immune suppression and autologous hematopoietic stem cell transplantation in refractory Crohn disease", Blood 101, 2064-2066; 2003.

[2] Jiang Y et al; "Pluripotency of mesenchymal stem cells derived from adult marrow"; Nature 418, 41-49; 4 July 2002.

[3] Yu Y et al., "AC133-2, a novel isoform of human AC133 stem cell antigen", Journal of Biological Chemistry 277, 20711-20716; 7 June 2002.

[4] Danet GH et al., "C1qRp defines a new human stem cell population with hematopoietic and hepatic potential", Proceedings of the National Academy of Sciences USA 99, 10441-10445; 6 Aug 2002.

[5] Zhou S et al., "The ABC transporter Bcrp1/ABCG2 is expressed in a wide variety of stem cells and is a molecular determinant of the side-population phenotype", Nature Medicine 7, 1028-1034; Sept 2001.

[6] Tremain N et al; "MicroSAGE Analysis of 2,353 Expressed Genes in a Single Cell derived Colony of Undifferentiated Human Mesenchymal Stem Cells Reveals mRNAs of Multiple Cell Lineages." Stem Cells 19, 408-418; 2001.

[7] Wright LS et al., "Gene expression in human neural stem cells: effects of leukemia inhibitory factor", Journal of Neurochemistry. 86, 179-195; July 2003.

[8] Ivanova NB et al., "A stem cell molecular signature", Science 298, 601-604; 18 Oct 2002; Ramalho-Santos M et al., " "Stemness": Transcriptional profiling of embryonic and adult stem cells", Science 298, 597-600; 18 Oct 2002.

[9] Dao MA et al., "Reversibility of CD34 expression on human hematopoietic stem cells that retain the capacity for secondary reconstitution", Blood 101, 112-118; 1 Jan 2003; Handgretinger R et al., "Biology and plasticity of CD133+ hematopoietic stem cells", Annals of the New York Academy of Sciences 996, 141-151; 2003; Huss R et al., "Evidence of peripheral blood-derived, plastic-adherent CD34−/low hematopoietic stem cell clones with mesenchymal stem cell characteristics", Stem Cells 18, 252-260; 2000; Huss R. "Isolation of primary and immortalized CD34− hematopoietic and mesenchymal stem cells from various sources", Stem Cells 18, 1-9; 2000.

[10] Lodie TA et al; "Systematic analysis of reportedly distinct populations of multipotent bone marrow-derived stem cells reveals a lack of distinction"; Tissue Engineering 8, 739-751; 2002.

[11] Tagami M et al., "Genetic and ultrastructural demonstration of strong reversibility in human mesenchymal stem cell", Cell and Tissue Research 312, 31-40; Apr 2003.

[12] Theise ND and Krause DS; "Suggestions for a New Paradigm of Cell Differentiative Potential", Blood Cells, Molecules, and Diseases 27, 625–631; 2001; Thiese ND and Krause DS, "Toward a new paradigm of cell plasticity", Leukemia 16, 542-548; 2002.

[13] Blau HM et al., "The evolving concept of a stem cell: entity or function?", Cell 105, 829-841; 29 June 2001.

[14] Tsai RYL and McKay RDG, "Cell Contact Regulates Fate Choice by Cortical Stem Cells." Journal of Neuroscience 20, 3725-3735; May 2000.

[15] Wagers AJ et al.; "Little evidence for developmental plasticity of adult hematopoietic stem cells"; Science 297, 2256-2259; 27 Sept 2002.

[16] Krause DS et al.; "Multi-Organ, Multi-Lineage Engraftment by a Single Bone Marrow-Derived Stem Cell"; Cell 105, 369-377; 4 May 2001.

[17] Castro RF et al., "Failure of bone marrow cells to transdifferentiate into neural cells in vivo", Science 297, 1299; 23 Aug 2002.

[18] Brazelton, TR et al.; "From marrow to brain: expression of neuronal phenotypes in adult mice"; Science 290, 1775-1779; 1 Dec 2000.

[19] Moore BE and Quesenberry PJ, "The adult hemopoietic stem cell plasticity debate: idols vs new paradigms", Leukemia 17, 1205-1210; 2003.

[20] Alison MR et al.; "Plastic adult stem cells: will they graduate from the school of hard knocks?", Journal of Cell Science 116, 599-603; 2003.

[21] Tada M et al., "Nuclear reprogramming of somatic cells by in vitro hybridization with ES cells", Current Biology 11, 1553-1558; 2 Oct 2001; Terada N et al., "Bone marrow cells adopt the phenotype of other cells by spontaneous cell fusion", Nature 416, 542-545; 4 Apr 2002; Ying Q-L et al., "Changing potency by spontaneous fusion", Nature 545-548; 4 Apr 2002; Tada M et al., "Embryonic germ cells induce epigenetic reprogramming of somatic nucleus in hybrid cells", EMBO Journal 16, 6510-6520; 1997.

[22] See, for example: Sorieul S and Ephrussi B, "Karyological demonstration of hybridization of mammalian cells in vitro" Nature 190, 653–654; 1961; Littlefield JW, "Selection of hybrids from matings of fibroblasts in vitro and their presumed

recombinants", *Science* 145, 709-710; 14 Aug 1964; Weiss MC and Green H, "Human-mouse hybrid cell lines containing partial complements of human chromosomes and functioning human genes", *Proceedings of the National Academy of Sciences USA* 58, 1104-1111; Sept 1967; Ladda RL and Estensen RD, "Introduction of a heterologous nucleus into enucleated cytoplasms of cultured mouse L-cells", *Proceedings of the National Academy of Sciences USA* 67, 1528-1533; Nov 1970; Köhler G Milstein C, "Continuous culture of fused cells secreting antibody of predefined specificity" *Nature* 256, 495-497, 1975.

[23] Wang X *et al.*; "Cell fusion is the principal source of bone-marrow-derived hepatocytes"; *Nature* 422, 897-901; 24 April 2003; Vassilopoulos G *et al.*; "Transplanted bone marrow regenerates liver by cell fusion"; *Nature* 422, 901-904; 24 April 2003.

[24] Spees JL *et al.*, "Differentiation, cell fusion, and nuclear fusion during *ex vivo* repair of epithelium by human adult stem cells from bone marrow stroma", *Proceedings of the National Academy of Sciences USA* 100, 2397-2402; 4 Mar 2003.

[25] Blau HM, "Stem-cell fusion: A twist of fate", *Nature* 419, 437; 3 Oct 2002.

[26] Collas P and Håkelien A-M, "Reprogramming Somatic Cells for Therapeutic Applications"; *Journal of Regenerative Medicine* 4, 7-13; February 2003; Håkelien AM *et al*; "Reprogramming fibroblasts to express T-cell functions using cell extracts"; *Nature Biotechnology* 20, 460-466; May 2002; Håkelien AM and Collas P, "Novel approaches to transdifferentiation", *Cloning and Stem Cells* 4, 379-387; 2002.

[27] Tran SD *et al.*; "Differentiation of human bone marrow-derived cells into buccal epithelial cells in vivo: a molecular analytical study"; *Lancet* 361, 1084-1088; 29 March 2003.

[28] Newsome PN *et al.*, "Human cord blood-derived cells can differentiate into hepatocytes in the mouse liver with no evidence of cellular fusion", *Gastroenterology* 124, 1891-1900; June 2003.

[29] Mazurier F *et al*; "Rapid myeloerythroid repopulation after intrafemoral transplantation of NOD-SCID mice reveals a new class of human stem cells"; *Nature Medicine* 9, 959-963; July 2003.

[30] Bianco P *et al.*, "Bone marrow stromal stem cells: nature, biology, and potential applications", *Stem Cells* 19, 180-192; 2001.

[31] Sanchez-Ramos J *et al.*, "Adult bone marrow stromal cells differentiate into neural cells *in vitro*", *Experimental Neurology* 164, 247-256; 2000.

[32] Woodbury D *et al*; "Adult rat and human bone marrow stromal cells differentiate intoneurons" *Journal of Neuroscience Research* 61, 364-370; 2000.

[33] Pittenger MF *et al.*, "Multilineage potential of adult human mesenchymal stem cells", *Science* 284, 143-147; 2 Apr 1999.

[34] Gronthos S *et al.*, "Molecular and cellular characterisation of highly purified stromal stem cells derived from human bone marrow", *Journal of Cell Science* 116, 1827-1835; 2003.

[35] Woodbury D et al., "Adult bone marrow stromal stem cells express germline, ectodermal, endodermal, and mesodermal genes prior to neurogenesis", *Journal of Neuroscience Research* 96, 908-917; 2002.

[36] Koc ON et al., Allogeneic mesenchymal stem cell infusion for treatment of metachromatic leukodystrophy (MLD) and Hurler syndrome (MPS-IH), *Bone Marrow Transplant* 215-222; Aug 2002.

[37] Mezey, E et al.; "Turning blood into brain: Cells bearing neuronal antigens generated in vivo from bone marrow"; *Science* 290, 1779-1782; 1 Dec 2000.

[38] Kabos P et al.; "Generation of neural progenitor cells from whole adult bone marrow"; *Experimental Neurology* 178, 288-293; December 2002.

[39] Körbling MK et al.; "Hepatocytes and epithelial cells of donor origin in recipients of peripheral-blood stem cells"; *New England Journal of Medicine* 346, 738-746; 7 March 2002.

[40] Mezey E et al.; "Transplanted bone marrow generates new neurons in human brains"; *Proceedings of the National Academy of Sciences USA* 100, 1364-1369; 4 Feb 2003.

[41] Priller J et al; "Neogenesis of cerebellar Purkinje neurons from gene-marked bone marrow cells in vivo"; *Journal of Cell Biology* 155, 733-738; 26 Nov 2001.

[42] Weimann JM et al., "Contribution of transplanted bone marrow cells to Purkinje neurons in human adult brains", *Proceedings of the National Academy of Sciences USA* 100, 2088-2093; 18 Feb 2003.

[43] Hess DC et al., "Bone marrow as a source of endothelial cells and NeuN-expressing cells after stroke", *Stroke* 33, 1362-1368; May 2002.

[44] Chen J et al.; "Therapeutic benefit of intravenous administration of bone marrow stromal cells after cerebral ischemia in rats"; *Stroke* 32, 1005-1011; April 2001.

[45] Li Y et al.; "Human marrow stromal cell therapy for stroke in rat"; *Neurology* 59, 514-523; August 2002.

[46] Heissig B et al, "Recruitment of stem and progenitor cells from the bone marrow niche requires MMP-9 mediated release of kit-ligand" *Cell* 109, 625-637; 31 May 2002.

[47] Otani A et al; "Bone marrow derived stem cells target retinal astrocytes and can promotes or inhibit retinal angiogenesis"; *Nature Medicine* 8, 1004-1010; Sept 2002.

[48] Tomita M et al; "Bone marrow derived stem cells can differentiate into retinal cells in injured rat retina"; *Stem Cells* 20, 279-283; 2002.

[49] Sasaki M et al., "Transplantation of an acutely isolated bone marrow fraction repairs demyelinated adult rat spinal cord axons," *Glia* 35, 26-34; July 2001.

[50] Hofstetter CP et al., "Marrow stromal cells form guiding strands in the injured spinal cord and promote recovery", *Proceedings of the National Academy of Sciences USA* 99, 2199-2204; 19 February 2002.

[51] Sekiya I et al., "*In vitro* cartilage formation by human adult stem cells from bone marrow stroma defines the sequence of cellular and molecular events during chondrogenesis", *Proceedings of the National Academy of Sciences USA* 99, 4397-4402; 2 Apr 2002.

[52] Ferrari G et al; "Muscle regeneration by bone marrow-derived myogenic progenitors"; *Science* 279, 1528-1530; Mar 6, 1998.

[53] LaBarge MA and Blau HM, "Biological progression from adult bone marrow to multinucleate muscle fiber in response to injury"; *Cell* 111, 589-601; 15 November 2002

[54] Gussoni E. et al, "Long term persistence of donor nuclei in a Duchenne Muscular dystrophy patient receiving bone marrow transplantation" *Journal of Clinical Investigation* 110, 807-814; Sept 2002.

[55] Ito T et al., "Bone marrow is a reservoir of repopulating mesangial cells during glomerular remodeling", *Journal of the American Society of Nephrology* 12, 2625-2635; 2001.

[56] Masuya M et al.; "Hematopoietic origin of glomerular mesangial cells"; *Blood* 101, 2215-2218; 15 March 2003.

[57] Kale S et al., "Bone marrow stem cells contribute to repair of the ischemically injured renal tubule", *Journal of Clinical Investigation* 112, 42-49; July 2003.

[58] Lin F et al; "Hematopoietic stem cells contribute to the regeneration of renal tubules after renal ischemia-reperfusion injury in mice"; *Journal of the American Society of Nephrology* 14, 1188-1199; 2003.

[59] Poulsom R et al; "Bone marrow stem cells contribute to healing of the kidney"; *Journal of the American Society of Nephrology* 14, S48-S54; 2003.

[60] Miyazaki M et al.; "Improved conditions to induce hepatocytes from rat bone marrow cells in culture"; *Biochemical and Biophysical Research Reports* 298, 24-30; October 2002.

[61] Fiegel HC et al., "Liver-specific gene expression in cultured human hematopoietic stem cells", *Stem Cells* 21, 98-104; 2003.

[62] Lagasse et al., "Purified hematopoietic stem cells can differentiate into hepatocytes in vivo", *Nature Medicine* 6, 1229-1234; November 2000.

[63] Theise, N et al.; "Derivation of hepatocytes from bone marrow cells in mice after radiation-induced myeloablation"; *Hepatology* 31, 235-240; Jan. 2000.

[64] Theise, N et al.; "Liver from bone marrow in humans"; *Hepatology* 32, 11-16; July 2000; Alison, M et al.; "Cell differentiation: hepatocytes from non-hepatic adult stem cells"; *Nature* 406, 257; 20 July 2000.

[65] Wang X et al., "Kinetics of liver repopulation after bone marrow transplantation", *American Journal of Pathology* 161, 565-574; Aug 2002.

[66] Wulf GG et al., "Cells of the hepatic side population contribute to liver regeneration and can be replenished by bone marrow stem cells", *Haematologica* 88, 368-378; 2003.

[67] Ianus A et al.; In vivo derivation of glucose competent pancreatic endocrine cells from bone marrow without evidence of cell fusion; Journal of Clinical Investigation 111, 843-850; March 2003.

[68] Hess D et al., "Bone marrow-derived stem cells initiate pancreatic regeneration", Nature Biotechnology 21, 763-770; July 2003.

[69] Bittira B et al., "In vitro preprogramming of marrow stromal cells for myocardial regeneration", Annals of Thoracic Surgery 74, 1154-1160; 2002.

[70] Orlic D et al.; "Bone marrow cells regenerate infarcted myocardium"; Nature 410, 701-705; April 5, 2001.

[71] Jackson KA et al.; "Regeneration of ischemic cardiac muscle and vascular endothelium by adult stem cells"; Journal of Clinical Investigation 107, 1395-1402; June 2001.

[72] Orlic D et al., "Mobilized bone marrow cells repair the infarcted heart, improving function and survival"; Proceedings of the National Academy of Sciences USA 98, 10344-10349, 28 August 2001.

[73] Toma C et al.; "Human mesenchymal stem cells differentiate to a cardiomyocyte phenotype in the adult murine heart"; Circulation, 105, 93-98; 1/8 January 2002.

[74] Kocher AA et al.; "Neovascularization of ischemic myocardium by human bone-marrow-derived angioblasts prevents cardiomyocyte apoptosis, reduces remodeling and improves cardiac function"; Nature Medicine 7, 430-436; April 2001.

[75] Edelberg, JM et al; "Young adult bone marrow derived endothelial precursor cells restore aging impaired cardiac angiogenic function"; Circulation Research 90, e89-e93; 2002.

[76] Saito T et al; "Xenotransplant cardiac chimera: immune tolerance of adult stem cells"; Annals of Thoracic Surgery 74, 19-24; 2002.

[77] Laflamme MA et al; "Evidence for cardiomyocyte repopulation by extracardiac progenitors in transplanted human hearts", Circulation Research 90, 634-640; 5 Apr 2002.

[78] Deb A et al., "Bone marrow-derived cardiomyocytes are present in adult human heart", Circulation 107, 1207-1209; 11 March 2003.

[79] Perin EC et al., "Transendocardial, Autologous Bone Marrow Cell Transplantation for Severe, Chronic Ischemic Heart Failure", Circulation 107, r75-r83; 2003.

[80] Stamm C et al.; "Autologous bone-marrow stem-cell transplantation for myocardial regeneration"; The Lancet 361, 45-46; 4 Jan 2003.

[81] Tse H-F et al.; "Angiogenesis in ischaemic myocardium by intramyocardial autologous bone marrow mononuclear cell implantation"; The Lancet 361, 47-49; 4 January 2003.

[82] Strauer BE et al.; "Repair of infarcted myocardium by autologous intracoronary mononuclear bone marrow cell transplantation in humans"; Circulation 106, 1913-1918; 8 October 2002.

[83] Tateishi-Yuyama E *et al.*; "Therapeutic angiogenesis for patients with limb ischaemia by autologous transplantation of bone-marrow cells: a pilot study and a randomised controlled trial"; *Lancet* 360, 427-435; 10 August 2002.

[84] Kotton DN *et al.*; "Bone marrow-derived cells as progenitors of lung alveolar epithelium"; *Development* 128, 5181-5188; December 2001.

[85] Ortiz LA *et al.*, "Mesenchymal stem cell engraftment in lung is enhanced in response to bleomycin exposure and ameliorates its fibrotic effects", *Proceedings of the National Academy of Sciences USA* 100: 8407-8411; 8 July 2003.

[86] Okamoto R *et al*; "Damaged epithelia regenerated by bone marrow-derived cells in the human gastrointestinal tract"; *Nature Medicine* 8, 1011-1017; Sept 2002.

[87] Badiavas EV, "Participation of Bone Marrow Derived Cells in Cutaneous Wound Healing", *Journal Of Cellular Physiology* 196, 245–250; 2003.

[88] Kessinger A, Sharp JG, "The whys and hows of hematopoietic progenitor and stem cell mobilization", *Bone Marrow Transplant* 31, 319-329; Mar 2003.

[89] Willing AE *et al.*, "Mobilized peripheral blood cells administered intravenously produce functional recovery in stroke", *Cell Transplantation* 12, 449-454; 2003.

[90] Borlongan CV and Hess DC, "G-CSF-mobilized human peripheral blood for transplantation therapy in stroke", *Cell Transplantation* 12, 447-448; 2003.

[91] Zhao Y *et al.*; "A human peripheral blood monocyte-derived subset acts as pluripotent stem cells"; *Proceedings of the National Academy of Sciences USA* 100, 2426-2431; 4 March 2003.

[92] Abuljadayel IS, "Induction of stem cell-like plasticity in mononuclear cells derived from unmobilised adult human peripheral blood", *Current Medical Research and Opinion* 19, *Fast*Track PREPRINT: 355–375; 18 June 2003 (doi: 10.1185/030079903125001901).

[93] Pagano SF *et al*; "Isolation and Characterization of Neural Stem Cells from the Adult Human Olfactory Bulb"; *Stem Cells* 18, 295-300; 2000.

[94] Shihabuddin S *et al.*; "Adult spinal cord stem cells generate neurons after transplantation in the adult dentate gyrus"; *Journal of Neuroscience* 20, 8727-8735; December 2000.

[95] Palmer TD *et al.*, "Progenitor cells from human brain after death", *Nature* 411, 42-43; 3 May 2001.

[96] Bjornson CRR *et al.*, "Turning brain into blood: a hematopoietic fate adopted by adult neural stem cells in vivo", *Science* 283, 534-537; 22 Jan 1999.

[97] Galli R *et al*; "Neural stem cells: An overview"; *Circulation Research* 92, 598-602; Feb 2003.

[98] Clarke DL *et al.*; "Generalized potential of adult neural stem cells"; *Science* 288, 1660-1663, 2 June 2000.

[99] Galli R et al; "Skeletal myogenic potential of human and mouse neural stem cells"; *Nature Neuroscience* 3, 986-991; Oct 2000.

[100] Rietze RL et al. "Purification of a pluripotent neural stem cell from the adult mouse brain", *Nature* 412, 736-739; 16 Aug 2001.

[101] Englund U et al; "Grafted neural stem cells develop into functional pyramidal neutrons and integrate into host cortical circuitry"; *Proceedings of the National Academy of Sciences USA* 99, 17089-17094; 24 Dec 2002.

[102] Arvidsson A et al.; "Neuronal replacement from endogenous precursors in the adult brain after stroke"; *Nature Medicine* 8, 963-970; Sept 2002.

[103] Riess P et al.; "Transplanted neural stem cells survive, differentiate, and improve neurological motor function after experimental traumatic brain injury"; *Neurosurgery* 51, 1043-1052; Oct 2002.

[104] Chang MY et al., "Neurons and astrocytes secrete factors that cause stem cells to differentiate into neurons and astrocytes, respectively", *Molecular Cellular Neuroscience* 23, 414-426; July 2003.

[105] Hori J et al., "Neural progenitor cells lack immunogenicity and resist destruction as allografts", *Stem Cells* 21, 405-416; 2003; Klassen H et al., "The immunological properties of adult hippocampal progenitor cells", *Vision Research* 43, 947-956; Apr 2003.

[106] Pluchino S et al.; "Injection of adult neurospheres induces recovery in a chronic model of multiple sclerosis"; *Nature* 422, 688-694; 17 April 2003.

[107] Liker MA et al.; "Human neural stem cell transplantation in the MPTP-lesioned mouse"; *Brain Research* 971, 168-177; May 2003.

[108] Kim TE et al; "Sonic hedgehog and FGF8 collaborate to induce dopaminergic phenotypes in the Nurr1-overexpressing neural stem cells"; *Biochemical and Biophysical Research Communications* 305, 1040-1048; 2003.

[109] Ourednik J et al.; "Neural stem cells display an inherent mechanism for rescuing dysfunctional neurons"; *Nature Biotechnology* 20, 1103-1110; Nov 2002.

[110] Åkerud P et al.; "Persephin-overexpressing neural stem cells regulate the function of nigral dopaminergic neurons and prevent their degeneration in a model of Parkinson's disease"; *Molecular and Cellular Neuroscience* 21, 205-222; Nov 2002.

[111] Fallon J et al.; "In vivo induction of massive proliferation,directed migration, and differentiation of neural cells in the adult mammalian brain," *Proceedings of the National Academy of Sciences USA* 97, 14686-14691; 19 December 2000.

[112] Gill SS et al., "Direct brain infusion of glial cell line-derived neurotrophic factor in Parkinson disease", *Nature Medicine* 9, 589-595; May 2003.

[113] Lévesque M and Neuman T, "Autologous transplantation of adult human neural stem cells and differentiated dopaminergic neurons for Parkinson disease: 1-year postoperative clinical and functional metabolic result", American Association of Neurological Surgeons annual meeting, Abstract #702; 8 April 2002.

[114] Barnett et al.; "Identification of a human olfactory ensheathing cell that can effect transplant-mediated remyelination of demyelinated CNS axons," Brain 123, 1581-1588, Aug 2000.

[115] Ramón-Cueto A et al., "Functional recovery of paraplegic rats and motor axon regeneration in their spinal cords by olfactory ensheathing glia," Neuron 25, 425-435; February 2000.

[116] Ramón-Cueto A et al., "Long-distance axonal regeneration in the transected adult rat spinal cord is promoted by olfactory ensheathing glia transplants", The Journal of Neuroscience 18, 3803–3815; 15 May 1998.

[117] Ramer MS et al.; "Functional regeneration of sensory axons into the adult spinal cord," Nature 403, 312-316; Jan 20, 2000.

[118] Jiang S et al; "Enteric glia promote regeneration of transected dorsal root axons inot spinal cord of adult rats"; Experimental Neurology 181, 79-83; 2003.

[119] Menet V et al., "Axonal plasticity and functional recovery after spinal cord injury in mice deficient in both glial fibrillary acidic protein and vimentin genes", Proceedings of the National Academy of Sciences USA 100, 8999-9004; 22 July 2003.

[120] Iacovitti L et al., "Differentiation of human dopamine neurons from an embryonic carcinomal stem cell line", Brain Research 912, 99-104; 2001.

[121] Garbuzova-Davis S et al; "Positive effect of transplantation of hNT neurons (Ntera 2/D1 cell-line) in a model of familial amyotrophic sclerosis"; Experimental Neurology 174, 169-180; Apr 2002; Garbuzova-Davis S et al; "Intraspinal implantation of hNT neurons into SOD1 mice with apparent motor deficit"; Amyotrophic Lateral Sclerosis and Other Motor Neuron Disorders 2, 175-180; Dec 2001.

[122] Kondziolka MD et al; "Transplantation of cultured human neuronal cells for patients with stroke"; Neurology 55, 565-569; 2000; Meltzer CC et al., "Serial [18F]Fluorodeoxyglucose Positron Emission Tomography after Human Neuronal Implantation for Stroke", Neurosurgery 49, 586-592; 2001; Nelson PT et al., "Clonal human (hNT) neuron grafts for stroke therapy", American Journal of Pathology 160, 1201-1206; Apr 2002.

[123] Asakura A et al., "Myogenic specification of side population cells in skeletal muscle", Journal of Cell Biology 159, 123–134; 14 Oct 2002.

[124] Lee JY et al., Clonal isolation of muscle-derived cells capable of enhancing muscle regeneration and bone healing", Journal of Cell Biology 150, 1085-1099; 4 Sept 2000.

[125] Torrente Y et al.; "Intraarterial injection of muscle-derived CD34+Sca-1+ stem cells restores dystrophin in mdx mice"; Journal of Cell Biology 152, 335-348; January 22, 2001.

[126] Qu-Petersen Z et al.; "Identification of a novel population of muscle stem cells in mice: potential for muscle regeneration"; Journal of Cell Biology 157, 851-864; 27 May 2002.

[127] Polesskaya A et al., "Wnt signaling induces the myogenic specification of resident CD45+ adult stem cells during muscle regeneration", Cell 113, 841-852; 27 June 2003.

[128] Lee JY et al., "The effects of periurethral muscle-derived stem cell injection on leak point pressure in a rat model of stress urinary incontinence", *International Urogynecology Journal of Pelvic Floor Dysfunction* 14, 31-37; Feb 2003.

[129] Atkins, B et al.; "Intracardiac Transplantation of Skeletal Myoblasts Yields Two Populations of Striated Cells In Situ"; *Annals of Thoracic Surgery* 67, 124-129; 1999
[130] Iijima Y et al., "Beating is necessary for transdifferentiation of skeletal muscle-derived cells into cardiomyocytes", *FASEB Journal* 17, 1361-1363; July 2003 (full text Epub 8 May 2003, doi:10.1096/fj.02-1048fje).

[131] Menasché P et al. "Myoblast transplantation for heart failure." *Lancet* 357, 279-280; 27 January 2001.

[132] Horb ME et al., "Experimental conversion of liver to pancreas", *Current Biology*, 13, 105–115; 21 Jan 2003.

[133] Yang L et al.; "*In vitro* trans-differentiation of adult hepatic stem cells into pancreatic endocrine hormone-producing cells"; *Proceedings of the National Academy of Sciences USA,* 99, 8078-8083; 11 June 2002.

[134] Malouf, NN et al.; "Adult-derived stem cells from the liver become myocytes in the heart in vivo"; *American Journal of Pathology* 158, 1929-1935; June 2001.

[135] Wang, X et al; "Liver repopulation and correction of metabolic liver disease by transplanted adult mouse pancreatic cells" *American Journal of Pathology* 158, 571-579; Feb 2001.

[136] Shapiro AMJ et al., "Islet transplantation in seven patients with type 1 diabetes mellitus using a glucocorticoid-free immunosuppressive regimen", *New England Journal of Medicine* 343, 230-238; 27 July 2000; Ryan EA et al., "Clinical outcomes and insulin secretion after islet transplantation with the Edmonton protocol", *Diabetes* 50, 710-719; Apr 2001.

[137] Ramiya VK et al.; "Reversal of insulin-dependent diabetes using islets generated in vitro from pancreatic stem cells," *Nature Medicine* 6, 278-282, March 2000.

[138] Bonner-Weir S et al.; "In vitro cultivation of human islets from expanded ductal tissue"; *Proceedings of the National Academy of Sciences USA* 97, 7999-8004; 5 July 2000; Gmyr V et al., "Adult human cytokeratin 19-positive cells reexpress insulin promoter factor 1 in vitro: Further evidence for pluripotent pancreatic stem cells in humans"; *Diabetes* 49, 1671-1680; Oct 2000; de la Tour D et al.; "Beta-cell differentiation from a human pancreatic cell line in vitro and in vivo"; *Molecular Endocrinology* 15, 476-483, Mar 2001; Abraham et al.; "Insulinotropic hormone glucagon-like peptide-1 differentiation of human pancreatic islet-derived progenitor cells into insulin-producing cells"; *Endocrinology* 143, 3152-3161; August 2002.

[139] Suzuki A et al.; "Glucagon-like peptide 1 (1-37) converts intestinal epithelial cells into insulin-producing cells"; *Proceedings of the National Academy of Sciences USA* 100, 5034-5039; 29 April 2003.

[140] Ferber S et al; "Pancreatic and duodenal homeobox gene 1 induces expression of insulin genes in liver and ameliorates streptozocin induced hyperglycemia"; *Nature Medicine* 6, 568-572; May 2000.

[141] Meller D et al., "Ex vivo preservation and expansion of human limbal epithelial stem cells on amniotic membrane cultures", British Journal of Ophthalmology 86, 463-471; Apr 2002.

[142] Tsai et al.; "Reconstruction of damaged corneas by transplantation of autologous limbal epithelial cells."; New England Journal of Medicine 343, 86-93, 2000; Schwab IR et al.; "Successful transplantation of bioengineered tissue replacements in patients with ocular surface disease"; Cornea 19, 421-426; July 2000; Tsubota K et al.; "Treatment of severe ocular-surface disorders with corneal epithelial stem-cell transplantation"; New England Journal of Medicine 340, 1697-1703; 3 June 1999; Henderson TR et al., "The long term outcome of limbal allografts: the search for surviving cells", British Journal of Ophthalmology 85, 604-609; May 2001.

[143] Seigel GM et al., "Human corneal stem cells display functional neuronal properties", Molecular Vision 9, 159-163; 30 Apr 2003.

[144] Chacko DM et al., "Transplantation of ocular stem cells: the role of injury in incorporation and differentiation of grafted cells in the retina", Vision Research 43, 937-946; Apr 2003.

[145] Dontu G et al., "In vitro propagation and transcriptional profiling of human mammary stem/progenitor cells", Genes and Development 17, 1253-1270; 15 May 2003.

[146] Alvi AJ et al., "Functional and molecular characterisation of mammary side population cells", Breast Cancer Research 5, R1-R8; 2003.

[147] Okumura K et al., "Salivary gland progenitor cells induced by duct ligation differentiate into hepatic and pancreatic lineages", Hepatology 38, 104-113; July 2003.

[148] Toma JG et al; "Isolation of multipotent adult stem cells from the dermis of mammalian skin"; Nature Cell Biology 3, 778-784; 3 Sept 2002; Oshima H et al., "Morphogenesis and renewal of hair follicles from adult multipotent stem cells. Cell 104, 233-245; 2001; Taylor G et al., "Involvement of follicular stem cells in forming not only the follicle but also the epidermis", Cell 102, 451-461; 2000.

[149] Lako M et al.; "Hair follicle dermal cells repopulate the mouse haematopoietic system"; Journal of Cell Science 115, 3967-3974; Sept 2002.

[150] Salingcarnboriboon R et al., "Establishment of tendon-derived cell lines exhibiting pluripotent mesenchymal stem cell-like property", Experimental Cell Research 287, 289-300; 15 July 2003.

[151] De Bari C et al., "Multipotent mesenchymal stem cells from adult human synovial membrane", Arthritis Rheumatism 44, 1928-1942; 2001.

[152] De Bari C et al.; "Skeletal muscle repair by adult human mesenchymal stem cells from synovial membrane"; Journal of Cell Biology 160, 909-918; 17 March 2003.

[153] Beltrami, AP et al.; "Evidence That Human Cardiac Myocytes Divide after Myocardial Infarction"; New England Journal of Medicine 344, 1750-1757; 7 June 2001.

[154] Robinson D et al., "Characteristics of cartilage biopsies used for autologous chondrocytes transplantation", Cell Transplant 10, 203-208; Mar-Apr 2001.

[155] Horwitz EM et al., "Transplantability and therapeutic effects of bone marrow-derived mesenchymal cells in children with osteogenesis imperfecta", *Nature Medicine* 5, 309–313; 1999.

[156] Horwitz EM et al., "Isolated allogeneic bone marrow-derived mesenchymal cells engraft and stimulate growth in children with osteogenesis imperfecta: Implications for cell therapy of bone", *Proceedings of the National Academy of Sciences USA* 99, 8932-8937; 25 June 2002.

[157] Horwitz EM et al., "Clinical responses to bone marrow transplantation in children with severe osteogenesis imperfecta", *Blood* 97, 1227-1231; 1 March 2001.

[158] Bennett AR et al., "Identification and characterization of thymic epithelial progenitor cells", *Immunity* 803-814; June 2002.

[159] Gill J et al; "Generation of a complete thymic microenvironment by MTS24+ thymic epithelial cells"; *Nature Immunology* 3, 635–642; 1 Jul 2002.

[160] Gronthos S et al., "Postnatal human dental pulp stem cells (DPSCs) in vitro and in vivo", *Proceedings of the National Academy of Sciences USA* 97, 13625-13630; 5 Dec 2000.

[161] Miura M et al; "SHED: Stem cells from human exfoliated deciduous teeth"; *Proceedings of the National Academy of Sciences USA* 100, 5807-5812; 13 May 2003.

[162] Zuk PA et al; "Multilineage cells from human adipose tissue: implications for cell-based therapies" *Tissue Engineering* 7, 211-228; 2001; Halvorsen Y-DC et al., "Extracellular matrix mineralization and osteoblast gene expression by human adipose tissue–derived stromal cells", *Tissue Engineering* 7, 729-741; 2001; Erickson GR et al., "Chondrogenic potential of adipose tissue-derived stromal cells *in vitro* and *in vivo*", *Biochemical and Biophysical Research Communications* 290, 763-769; 2002.

[163] Zuk PA et al., "Human adipose tissue is a source of multipotent stem cells", *Molecular Biology of the Cell* 13, 4279-4295; Dec 2002; Safford KM et al., "Neurogenic differentiation of murine and human adipose-derived stromal cells", *Biochemical and Biophysical Research Communications* 294, 371–379; 2002.

[164] See, for example: Kurtzberg J et al., "Placental blood as a source of hematopoietic stem cells for transplantation into unrelated recipients", *New England Journal of Medicine* 335, 157-166; 18 July 1996; Laughlin MJ et al., "Hematopoietic engraftment and survival in adult recipients of umbilical-cord blood from unrelated donors", *New England Journal of Medicine* 344, 1815-1822; 14 June 2001; Gore L et al.; "Successful cord blood transplantation for sickle cell anemia from a sibling who is human leukocyte antigen-identical: implications for comprehensive care", *Journal of Pediatriatic Hematology and Oncology* 22, 437-440; Sep-Oct 2000.

[165] Rocha V et al., "Graft-versus-host disease in children who have received a cordblood or bone marrow transplant from an HLA-identical sibling", *New England Journal of Medicine*, 342, 1846-1854; 22 June 2000.

[166] Rainsford E, Reen DJ, "Interleukin 10, produced in abundance by human newborn T cells, may be the regulator of increased tolerance associated with cord blood stem cell transplantation", *British Journal of Haematology* 116, 702-709; Mar 2002.

[167] Beerheide W *et al.*, "Downregulation of ß2-microglobulin in human cord blood somatic stem cells after transplantation into livers of SCID-mice: an escape mechanism of stem cells?", *Biochemical and Biophysical Research Communications* 294, 1052–1063; 2002.

[168] Broxmeyer HE *et al.*, "High-efficiency recovery of functional hematopoietic progenitor and stem cells from human cord blood cryopreserved for 15 years", *Proceedings of the National Academy of Sciences USA* 100, 645-650; 12 Jan 2003.

[169] Sanchez-Ramos J *et al.*, "Expression of neural markers in human umbilical cord blood", *Experimental Neurology* 171, 109-115; 2001.

[170] Chen J *et al.*, "Intravenous administration of human umbilical cord blood reduces behavioral deficits after stroke in rats", *Stroke* 32, 2682-2688; Nov 2001.

[171] Saporta S *et al.*, "Human umbilical cord blood stem cells infusion in spinal cord injury: engraftment and beneficial influence on behavior", *Journal of Hematotherapy & Stem Cell Research* 12, 271-278; 2003.

[172] Garbuzova-Davis S *et al.*, "Intravenous administration of human umbilical cord blood cells in a mouse model of amyotrophic lateral sclerosis: distribution, migration, and differentiation", *Journal of Hematotherapy & Stem Cell Research* 12, 255–270; 2003.

[173] Buzanska L *et al.*, "Neural stem cell line derived from human umbilical cord blood - morphological and functional properties", *Journal of Neurochemistry* 85, Suppl 2, 33; June 2003.

[174] Kakinuma S *et al*; "Human Umbilical Cord Blood as a Source of Transplantable Hepatic Progenitor Cells"; *Stem Cells* 21, 217-227; 2003; Ishikawa F *et al*; "Transplantied human cord blood cells give rise to hepatocytes in engrafted mice"; *Annals of the New York Academy of Sciences* 996, 174-185; 2003; Wang X *et al.*, "Albumin-expressing hepatocyte-like cells develop in the livers of immune-deficient mice that received transplants of highly purified human hematopoietic stem cells", *Blood* 101, 4201-4208; 15 May 2003.

[175] Mitchell K *et al*; "Matrix Cells from Wharton's Jelly Form Neurons and Glia"; *Stem Cells* 21, 51-60; 2003.

[176] Weiss ML *et al.*, "Transplantation of porcine umbilical cord matrix cells into the rat brain", *Experimental Neurology* Article in Press, Corrected Proof; available online 11 July 2003.

[177] Okawa H *et al.*, "Amniotic epithelial cells transform into neuron-like cells in the ischemic brain", *NeuroReport* 12, 4003-4007; 2001.

[178] Prusa A-R *et al.*, "Oct-4-expressing cells in human amniotic fluid: a new source for stem cell research?", *Human Reproduction* 18, 1489-1493; 2003.

[179] *Minasi MG et al., "*The meso-angioblast: a multipotent, self-renewing cell that originates from the dorsal aorta and differentiates into most mesodermal tissues", *Development* 129, 2773-2783; 2002.

[180] Sampaolesi M *et al.*, "Cell therapy of alpha-sarcoglycan null dystrophic mice through intra-arterial delivery of mesoangioblasts", *Science* Published online 10 July 2003 (doi: 10.1126/science.1082254).

[181] Jensen GS and Drapeau C, "The use of in situ bone marrow stem cells for the treatment of various degenerative diseases"; *Medical Hypotheses* 59, 422-428; 2002.

[182] Grove JE *et al*; "Marrow-Derived Cells as Vehicles for Delivery of Gene Therapy to Pulmonary Epithelium"; *Stem Cells*, 645-651; July 2002.

[183] Steptoe RJ *et al.*; "Transfer of hematopoietic stem cells encoding autoantigen prevents autoimmune diabetes"; *Journal of Clinical Investigation* 111, 1357-1363; May 2003.

[184] McKee JA *et al*; "Human arteries engineered *in vitro*"; *EMBO Reports* 4, 633-638; May 2003.

[185] Shi S *et al.*, "Bone formation by human postnatal bone marrow stromal cells is enhanced by telomerase expression", *Nature Biotechnology* 20, 587-591; June 2002; Simonsen JL *et al.*, "Telomerase expression extends the proliferative life-span and maintains the osteogenic potential of human bone marrow stromal cells", *Nature Biotechnology* 20, 592-596; June 2002.

[186] Cavazzana-Calvo M *et al.*, "Gene therapy of human severe combined immunodeficiency (SCID)-X1 disease", *Science* 288, 669–672; 2000.

[187] Hacein-Bey-Abina A *et al.*, "Sustained correction of X-linked severe combined immunodeficiency by ex vivo gene therapy", *New England Journal of Medicine* 346, 1185-1193; 18 Apr 2002.

[188] Aiuti A *et al.*, "Correction of ADA-SCID by stem cell gene therapy combined with nonmyeloablative conditioning", *Science* 296, 2410-2413; 28 June 2002.

[189] Benedetti S *et al.*, "Gene therapy of experimental brain tumors using neural progenitor cells", *Nature Medicine* 6, 447-450; Apr 2000.

[190] Jin, HK *et al*; "Intracerebral transplantation of mesenchymal stem cells into acid sphingomyelinase-deficient mice delays the onset of neurological abnormalities and extends their life span"; *Journal of Clinical Investigation* 109, 1183-1191; May 2002.

[191] Lee JY *et al.*, "Effect of bone morphogenetic protein-2-expressing muscle-derived cells on healing of critical-sized bone defects in mice", *The Journal of Bone and Joint Surgery* 83-A, 1032-1039; July 2001.

[192] Zeisberg M *et al.*, "BMP-7 counteracts TGF-ß1-induced epithelial-to-mesenchymal transition and reverses chronic renal injury", *Nature Medicine* 7, 964-968; July 2003.

Appendix L.

Stem Cells and Tissue Regeneration: Lessons from Recipients of Solid Organ Transplantation

SILVIU ITESCU, M.D.

Director of Transplantation Immunology, Departments of Medicine and Surgery, Columbia University, New York, NY

CONTENTS

1. OVERVIEW

The Major Histocompatibility Complex (MHC) is located on the short arm of chromosome 6 in humans and encodes the alloantigens known as Human Leukocyte Antigens (HLA), polymorphic cell surface molecules which enable the immune system to recognize both self and foreign antigens. The class II HLA molecules (HLA-DR, HLA-DP, and HLA-DQ) are usually found only on antigen-

347

presenting cells such as B lymphocytes, macrophages, and dendritic cells of lymphoid organs, and initiate the immune response to foreign proteins, including viruses, bacteria, and foreign HLA antigens on transplanted organs.

Following binding of foreign proteins, class II HLA on antigen-presenting cells activate CD4+ T cells, which in turn activate cytotoxic CD8+ T cells to recognize the same foreign antigen bound to HLA class I (HLA-A, HLA-B, and HLA-C, molecules found on the surface of all cells) and destroy the target. The actual recognition of foreign HLA transplantation antigens by T cells is referred to as allorecognition. Two distinct pathways of allorecognition have been described, direct and indirect. The direct pathway involves receptors on the host T cells that directly recognize intact HLA antigens on the cells of the transplanted organ. The indirect pathway requires an antigen-presenting cell that internalizes the foreign antigen and presents it via its own HLA class II molecule on the surface of an antigen-presenting cell to the CD4+ helper T cells.

Once recognition has taken place, an important cascade of events is initiated at the cellular level, culminating in intracellular release of ionized calcium from intracellular stores. The calcium binds with a regulatory protein called calmodulin, forming a complex that activates various phosphatases, particularly calcineurin. Calcineurin dephosphorylates an important cytoplasmic protein called nuclear factor of activated T cells (NFAT), resulting in its migration to the nucleus and induction of the production of various cytokines such as IL-2. These cytokines recruit other T cells to destroy the transplanted organ, ultimately resulting in rejection and loss of the graft.

Immunosuppressive regimens used to prevent allograft rejection are aimed at inhibiting the various arms of the immune response, typically require multiagent combinations, and need to be maintained for the duration of life. The currently used armamentarium confers significant side-effect risks, including infectious and neoplastic complications. Moreover, despite success at preventing early allograft rejection, long-term survival of transplanted organs remains difficult to achieve and novel methods to achieve long-term tolerance are being actively sought.

Stem cells obtained from embryonic or adult sources differ from other somatic cells in that they express very low levels of HLA molecules on their cell surfaces. This endows these cell types with the theoretical potential to escape the standard mechanisms of immune rejection discussed above. However, under conditions that enable cellular differentiation in vitro and in vivo each of these stem cell populations acquires high level expression of HLA molecules, suggesting that their long-term survival following transplantation in vivo may be limited by typical immune rejection phenomena. Recent experimental data, however, provide striking counterintuitive examples that stem cells from both embryonic and adult sources may evade the recipient's immune system and result in long-term engraftment in the absence of immunosuppression despite acquisition of surface HLA molecule expression. These observations may have significant impact on the emerging field of regenerative medicine.

2. IMMUNOBIOLOGY OF ORGAN TRANSPLANTATION

The Human Leukocyte Antigens (HLA)

Differences between individuals which enable immune recognition of non-self from self are principally due to the extreme polymorphism of genes in the Major Histocompatibility Complex (MHC) on chromosome 6 in man which encode the cell surface HLA molecules. These molecules are cell surface glycoproteins whose biologic function is to bind antigenic peptides (epitopes) derived from viruses, bacteria, or cancer cells, and present them to T cells for subsequent immune recognition. Each HLA gene includes a large number of alleles and the peptide binding specificity varies for each different HLA allele. The 1996 WHO HLA Nomenclature Committee report lists more than 500 different HLA class I and class II alleles.

Crystallographic x-ray studies have demonstrated that the hypervariable regions encoded by polymorphic regions in the alleles correspond to HLA binding pockets which engage specific "anchor" residues of peptide ligands. One HLA molecule will recognize a range of possible peptides, whereas another HLA molecule will recognize a different range of peptides. Consequently, no two individuals will have the same capability of stimulating an immune response, since they do not bind the same range of immunogenic peptides. It is estimated that >99% of all possible peptides derived from foreign antigens are ignored by any given HLA molecule. Since in the absence of HLA polymorphism a large number of immunogenic peptides would not be recognized, the extensive HLA polymorphism in the population reduces the chance that a given virus or bacterium would not be recognized by a sizable proportion of the population, reducing the likelihood of major epidemics or pandemics.

T Cell Recognition Of Antigen Presented By HLA Molecules

Since HLA molecules regulate peptide display to and activation of the immune system, considerable effort has been devoted to understanding the molecular basis of peptide-HLA interactions.

These issues are important for defining the biology of T cell antigen recognition and the properties of a protein that make it immunogenic or non-immunogenic. Specific antigen recognition by T cells is dependent on recognition by the T cell receptor of a three-dimensional complex on the surface of antigen-presenting cells (APC) comprised of the HLA molecule and its bound peptide. The peptides are produced by complex antigen processing machineries within the APC (i.e. proteolytic enzymes, peptide transporters and molecular chaperones) which generate a pre-selected peptide pool for association with the HLA molecules. The different types of T cells require different HLA molecules for antigen presentation, so-called "HLA restriction" phenomena. T cell receptors on CD8+ cytotoxic T cells (CTLs) bind peptides presented by HLA class I molecules, whereas CD4+ T helper cells (Th) recognize peptides bound to HLA class II molecules. Of the 8-13 amino acid residues of a bound peptide within a class I or II HLA molecule, only three to four amino acid side chains are accessible to the T cell receptor, and a similar number of amino acids are involved in binding to the HLA molecule.

Thymic Education Of T Cells

T cells mature in the thymus to appropriately respond to foreign pathogens without inadvertently attacking the host. Under the influence of various thymic resident cells and factors they elaborate, maturing T cells fall into two categories: those that are able to discriminate between self and non-self and can appropriately respond to foreign pathogens without inadvertently attacking the host, and those which are unable to appropriately discriminate between self and non-self. Dendritic cells have been implicated in the deletion, or inhibition, of T cells reactive to self-antigens, particularly in the thymus during T cell development or in peripheral lymphoid organs. The process of self/non-self discrimination by the maturing T cells is dependent on thymic dendritic cell (DC) presentation of self-antigens in the context of self-HLA molecules. When maturing thymic T cells are highly reactive with self-antigen/ HLA complexes, they are deleted so that potentially autoreactive T cells will not be released into the periphery. If a particular foreign antigen can be presented in such a way in the thymus as to fool the maturing T cells into believing that the antigen is part of self tissue, then T cells capable of reacting with this antigen will also be eliminated. Indeed, it has been demonstrated that when mouse thymic DC present transgenically introduced foreign antigens to developing T cells, the mature peripheral T cell repertoire of the mouse lacks T cells capable of reacting with the specific foreign antigen, i.e. it is tolerant to the foreign antigen. This has raised the possibility that injection of dendritic cells into an allogeneic recipient might induce tolerance to a subsequent allograft by causing deletion or inhibition of alloreactive T cells.

T Cell Recognition Of Alloantigens

Recognition of foreign, or allogeneic, HLA antigens by the recipient immune system is the major limitation to the survival of solid organ grafts. The central role of HLA molecules in allograft rejection is due to their role as restriction elements for T cell recognition of donor antigens and the extensive polymorphism displayed by the HLA molecules, which elicit host immune responses. Although progress has been made in the short-term survival of transplants, chronic immunologic rejection remains an impediment to long-term survival.

The primary cause of acute rejection of transplanted organs is so-called "direct" recognition of whole allogeneic HLA antigens by receptors on the surface of recipient T cells. The direct recognition pathway involves recognition by recipient T cells of donor HLA class I and class II molecules, resulting in the generation of cytotoxic and helper T lymphocytes which play a pivotal role in the rejection process. In contrast, chronic rejection of transplanted organs results from so-called "indirect recognition" of donor HLA peptides derived from the allogeneic HLA molecules shed by the donor tissue. These foreign HLA molecules are taken up and processed by recipient antigen presenting cells (APC), and peptide fragments of the allogeneic HLA molecules containing polymorphic amino acid residues are bound and presented by recipient's (self) HLA molecules to recipient (self) T cells. Although direct and indirect recognition of alloantigen generally leads to adverse graft outcome, tolerance induction may occur following exposure of the recipient to donor alloantigens prior to transplantation. Since this strategy is based on the nature and dose of the antigen as well as the route of administration, understanding how to control the balance between activation and unresponsiveness mediated by the direct and/or indirect recognition of alloantigen is a an area of active research which could lead to development of new therapies to prolong graft survival.

Indirect allorecognition has been implicated in recurrent rejection episodes in various transplantation models of cardiac, kidney and skin grafts. Determinants on donor HLA molecules can be divided into two main categories: (a) the dominant allodeterminants that are efficiently processed and presented to alloreactive T cells during allograft rejection; and (b) the cryptic allodeterminants that are potentially immunogenic but do not normally induce alloreactive responses, presumably due to incomplete processing and/or presentation. Indirect recognition of allo-HLA peptides is important for the initiation and spreading of the immune response to other epitopes within the allograft. So-called "spreading" of indirect T cell responses to other allo-HLA epitopes expressed by graft tissue is strongly predictive of recurring episodes of rejection. Tolerance induction to the dominant donor determinants represents potential effective strategy for blocking indirect alloresponses and ensuring long-term graft survival in animal models.

Tolerance Induction

Advances in surgical methods and current immunosuppressive therapies have led to significant improvement in short-term graft survival, however long-term survival rates remain poor. For example, whereas both kidney and heart allografts have one-year graft survival rates of 85 to 95 percent, only about 50% of transplanted hearts survive five years and only about 50% of kidney grafts survive ten years. Thus, despite being able to achieve short-term success, these relatively poor long-term graft survival rates demonstrate the limitations of the current clinical immunosuppressive regimens to enable long-term immune evasion by the graft. Consequently, a major goal of transplantation immunobiologists is to induce donor-specific tolerance, allowing the long-term survival of human allografts without the need of HLA-compatibility and without the continuous recipient immunosupression leading to the concomitant risks of infection, malignancy, and/or other specific drug side effects. This would theoretically improve long-term graft survival, reduce or eliminate the continuing need for expensive, toxic and non-specific immunosuppressive therapy and enhance the quality of life.

Insight into some of the mechanisms involved in tolerance induction has been gained from pre-clinical and clinical studies in numerous animal models and in patients, particularly those with liver allografts which typically do not induce a prominent immune response leading to rejection. One possible mechanism by which liver transplantation results in allograft tolerance tolerance may be that the donor or "passenger" lymphoid cells in the transplanted liver emigrate and take up residence in the recipient's immune organs, such as the thymus or lymph nodes. Donor lymphocytes at these sites might "re-educate" the recipient immune system so that the donor organ is not recognized as foreign. In an attempt to initiate a similar process in other organ recipients, transfusions of donor blood or bone marrow have been used to enhance solid organ graft survival in animal models and in clinical trials. These studies are currently ongoing in various organ systems.

Molecular understanding of the cellular immune response has led to new strategies to induce a state of permanent tolerance after transplantation. Several approaches have shown promise, including the use of tolerizing doses of class I HLA-molecules in various forms for the induction of specific unresponsiveness to alloantigens, and the use of synthetic peptides corresponding to HLA class II sequences. Other approaches include alteration in the balance of cytokines that direct the immune response away from the TH1 type of inflammatory response and graft rejection to the TH2 type of response that might lead to improved graft survival, and the use of agents to induce "co-stimulatory blockade" of T cell activation. This latter approach is based on the concept that blockade of a "second signal" to the T cell enables the signal provided to the T cell receptor by the HLA-peptide complex to induce antigen specific tolerance.

The experimental use of human dendritic cells as tolerogenic agents has been limited due to the low frequency of circulating dendritic cells in peripheral human blood, the limited accessibility to human lymphoid organs, and the terminal state of differentiation of circulating human dendritic cells making their further expansion ex vivo difficult. Dendritic cells are migratory cells of sparse, but widespread, distribution in both lymphoid and non-lymphoid tissues. Although the earliest precursors are ultimately of bone marrow origin, the precise lineage of dendritic cells is controversial and includes both myeloid-derived and lymphoid-derived populations. Recent work has revealed that an expanded population of mature human dendritic cells can be derived from non-proliferating precursors in vitro is by culturing bone-marrow derived cells with a combination of cytokines. This method of enrichment for human dendritic cells from a precursor population can result in the production of dendritic cells that are tolerogenic to foreign antigens. Whether such cells could be useful when co-administered with an allograft transplant remains to be determined. Nevertheless, it is clear that considerable progress has been made in the past few years using approaches to manipulate the immune response to enable routine donor-specific tolerance, and there is reason to be optimistic that with better understanding of molecular and cellular mechanisms this goal could be attained.

3. IMMUNOSUPPRESSIVE AGENTS COMMONLY USED IN ORGAN TRANSPLANT RECIPIENTS: BENEFITS AND ADVERSE OUTCOMES

Cyclosporine

Cyclosporine has been the single most important factor associated with improved outcomes after organ transplantation over the past two decades. CyA binds to a cytosolic cell protein, cyclophilin (CyP). The CyA-CyP complex then binds to calcineurin and subsequently blocks interleukin-2 (IL-2) transcription. The binding of IL-2 to the IL-2 receptors on the surface of T lymphocytes is a key stimulant in promoting lymphocyte proliferation, activation, and ultimately allograft rejection. A review of the first decade of experience with heart transplantation revealed a total of 379 cardiac allograft recipients worldwide; actuarial survival rates in this cohort of patients at 1 year and 5 years were 56% and 31% respectively; the main causes of death being acute rejection and the side effects of immunosuppression. With the introduction and widespread use of CyA over the next decade, survival rates dramatically improved to 85% and 75% at 1 and 5 years respectively. Similar results were obtained with other organ transplants, including kidney and lung.

The major adverse effects of CyA are nephrotoxicity, hypertension, neurotoxocity and hyperlipidemia; less common side effects include hirsuitism, gingival hyperplasia and liver dysfunction. CyA nephrotoxicity can manifest as either acute or chronic renal dysfunction. It is important to note that a number of drugs commonly used in transplant patients, such as aminoglycosides, amphotericin B and ketoconazole can potentiate the nephrotoxicity induced by CyA. More than half the patients receiving CyA will require treatment for hypertension within the first year following transplantation. Corticosteroids also potentiate the side effects of CyA such as hypertension, hyperlipidemia and hirsuitism. Frequent monitoring of the serum level is essential to minimize the adverse effects. One of the major limitations of the original oil-based CyA formulation (Sandimmune) is its variable and unpredictable bioavailability. In the mid-90s Neoral was introduced, a new microemulsion formula of CyA, which has greater bioavailability and more predictable pharmacokinetics than Sandimmune.

Tacrolimus

Tacrolimus (FK506) is a macrolide antibiotic that inhibits T-cell activation and proliferation and inhibits production of other cytokines. The product of *Streptomyces tsurubaensis* fermentation, FK 506 was first discovered in 1984 and first used in clinical studies in 1988 at the University of Pittsburgh. While the mechanism of action of tacrolimus is similar to that of CyA, and comparative clinical trials have suggested similar efficacy, it has been suggested that some groups of patients may benefit from tacrolimus rather than CyA as primary immunosuppressive therapy. Unlike CyA, hirsuitism and gingival hyperplasia occur infrequently with tacrolimus; thus, tacrolimus-based therapy may improve compliance and quality of life in female and pediatric transplant recipients. It should be noted that alopecia has been documented with tacrolimus, but is known to improve with dose reductions. The decreased incidence of hypertension and hyperlipidemia with tacrolimus makes it preferable to CyA in patients with difficult to treat hypertension or hyperlipidemia. A final indication for tacrolimus has been as a rescue immunosuppressant in cardiac transplant recipients on CyA with refractory rejection or intolerance to immunosuppression (severe side effects). Since tacrolimus is metabolized using the same cytochrome P450 enzyme system as CyA, drug interactions are essentially the same. Thus, drugs that induce this system may increase the metabolism of tacrolimus, thereby decreasing its blood levels. Conversely, drugs that inhibit the P450 system decrease the metabolism of tacrolimus, thereby increasing its blood levels. It is important to note that some studies have indicated a higher incidence of nephrotoxicity with tacrolimus as compared to CyA.

Azathioprine and Mycophenolate Mofetil (MMF)

Despite being available for more than 35 years, azathioprine is still a useful agent as an immunosuppressive agent. Following administration, azathioprine is converted into 6-mercaptopurine, with subsequent transformation to a series of intracellularly active metabolites. These inhibit both an early step in *de novo* purine synthesis and several steps in the purine salvage pathway. The net effect is depletion of cellular purine stores, thus inhibiting DNA and RNA synthesis, the impact of which is most marked on actively dividing lymphocytes responding to antigenic stimulation. In currently used immunosuppressive protocols, azathioprine is used as part of a triple therapy regimen along with CyA or tacrolimus and prednisone. Mycophenolate mofetil (MMF), which is rapidly hydrolyzed after ingestion to mycophenolic acid, is a selective, noncompetitive, reversible inhibitor of onosine monophosphate dehydrogenase, a key enzyme in the de novo synthesis of guanine nucleotides. Unlike other marrow-derived cells and parenchymal cells that use the hypoxanthine-guanine phosphoribosyl transferase (salvage) pathway, activated lymphocytes rely predominantly on the de novo pathway for purine synthesis. This functional selectivity allows lymphocyte proliferation to be specifically targeted with less anticipated effect on erythropoiesis and neutrophil production than is seen with azathioprine.

Early studies in human kidney and heart transplant recipients showed that MMF, when substituted for azathioprine in standard triple-therapy regimens, is well tolerated and more efficacious than azathioprine. In a large, double-blind, randomized multicenter study comparing MMF versus azathioprine (with CyA and prednisone) involving 650 patients, the MMF group was associated with significant reduction in mortality as well as a reduction in the requirement for rejection treatment. However, there was noted to be an increase in the incidence of opportunistic viral infections in the MMF group. The overall greater efficacy of MMF compared to azathioprine has resulted in MMF generally replacing azathioprine in triple immunosuppressive protocols together with steroids and cyclosporine in most solid organ recipients.

Corticosteroids

Steroids are routinely used in almost all immunosuppressive protocols after organ transplantation. The metabolic side effects of steroids are well known and lead to significant morbidity and mortality in the post-transplant period. Almost 90% of organ recipients continue to receive prednisone at 1-year post-transplant and 70% at three-years post-transplant. A recent review of over 1800 patients from a combined registry outlined the morbid complications that patients suffer within the first year after transplantation. Many of these complications are known side effects of prednisone, including hypertension (16%), diabetes mellitus (16%), hyperlipidemia (26%), bone disease (5%) and cataracts (2%). It is thereby obvious that avoidance of steroids may decrease morbidity and mortality after organ transplantation. Two general approaches are used to institute prednisone-free immunosuppression: early and late withdrawal.

Withdrawal of prednisone during the first month post-transplant has resulted in long-term success of steroid withdrawal in 50–80% of patients. In these studies, the use of antilymphocyte antibody induction therapy appears to increase the likelihood of steroid withdrawal. Several centers have reported their results with immunosuppressive regimens that did not include steroids in the early post-transplant period. Studies reporting high success rates of 80% have used specific enrolment criteria, such as excluding patients with recurrent acute rejections or those with female gender.

Review of numerous studies demonstrate that steroid free maintenance immunosuppression is possible in atleast 50% of patients, is as safe as triple drug therapy and may reduce some of the long-term complications of steroids. Owing to the fact that the majority of acute rejection episodes occur in the first three months post-transplant, steroid withdrawal is made after this time period, resulting in long-term success in about 80% of patients. Generally, there is no need for conventional induction agents when late withdrawal of steroids is done.

Anti-Lymphocyte Antibody Therapy

Despite the extensive use of induction therapy using anti-lymphocyte antibody in solid organ transplantation, their exact role is unclear. There is no doubt that routine use of these agents is unwarranted as the generalized immunosuppression induced by then increased the risk of infections and malignancy. Despite the lack of consistent data supporting the routine use of induction therapy with anti-lymphocyte antibody agents, there is a role in certain select situations. Specifically, patients with early post-operative renal or hepatic dysfunction may benefit especially by the avoidance of cyclosporine therapy while using these induction agents. Anti-lymphocyte antibody therapy can provide effective immunosuppression for atleast 10 to 14 days without CyA or tacrolimus therapy. It has also been suggested that patients with overwhelming postoperative bacterial infections or diabetics with severe postoperative hyperglycemia may benefit from the comparatively low doses of corticosteroids required during anti-lymphocyte induction therapy.

The two main types of induction agents have been either the polyclonal antilymphocyte or antithymocyte globulins and more recently the murine monoclonal antibody OKT3. While these agents have been shown to be effective in terminating acute allograft rejection and in treating refractory rejection, the results of comparative studies of outcomes with and without monoclonal induction therapy have varied, with most studies demonstrating an effect on rejection that is maintained only while antibody therapy is ongoing. Without repeated administration, these agents only delay the time to a first rejection episode without decreasing the overall frequency or severity of rejection. More importantly, their use has been associated with an increased risk of short-term (infections) and long-term (lympho-proliferative disorders) complications. A complication specific to OKT3 is the development of a "flu-like syndrome" characterized by fever, chills and mild hypotension, typically seen with the first dose.

Since antilymphocyte antibodies are produced in nonhuman species, their use is associated with the phenomenon of sensitization, leading to decreased effectiveness with repeated use as well as the possibilty of serum sickness. The development of sensitization has been linked with an increased risk of acute vascular rejection. While this association has not been reported by other centers using OKT3 prophylaxis, it is believed that the development of immune-complex disease, inadequate immunosuppression due to decreased OKT3 levels or that OKT3 sensitization may be a marker for patients at higher risk for humoral rejection may be responsible for this phenomenon.

Interleukin-2 Receptor Inhibition

A new class of drugs has been developed which targets the high affinity IL-2 receptor. This receptor is present on nearly all activated T cells but not on resting T cells. In vivo activation of the high-affinity IL-2 receptor by IL-2 promoted the clonal expansion of the activated T cell population. A variety of rodent monoclonal antibodies directed against the α chain of the receptor have been used in animals and humans to achieve selective immunosuppression by targeting only T-cell clones responding to the allograft. Chimerisation or humanisation of these monoclonal antibodies resulted in antibodies with a predominantly human framework that retained the antigen specificity of the original rodent monoclonal antibodies. A fully humanized anti-IL2R monoclonal antibody, daclizumab, and a chimeric anti-IL-2R monoclonal antibody, basiliximab, have undergone successful phase III trials demonstrating their efficacy in the immunoprophylaxis of patients undergoing renal and cardiac transplantation.

Both agents have immunomodulatory effects that are similar to those of other monoclonal antibody-based therapies (i.e., induction of clonal anergy rather than clonal deletion). The advantages of these agents include their lack of immunogenicity, long half-lives, ability to repeat dosing, and short-term safety profile. Daclizumab appears to be an effective adjuvant immunomodulating agent in cardiac allograft recipients. It has advantages over conventional induction therapy as it is more selective and can be used for prolonged and potentially repeated periods. Studies with larger cohorts are needed to further study the short-term and long-term survival benefits for patients following organ transplantation and should determine the optimal dosing schedules of these new agents.

4. STEM CELL TRANSPLANTATION AND IMMUNO-SUPPRESSION

Materno-Fetal Tolerance

As outlined above, when tissues from an HLA-disparate donor are transplanted into a recipient they are always recognized as foreign, and immunosuppression is required to prevent rejection. An important exception to this is observed in pregnant women who tolerate their unborn fetus despite the fact that it expresses a full set of non-maternal HLA antigens inherited from the father. The mechanisms by which embryonic tissue demonstrates immune privilege during prenatal development have not yet been fully elucidated, however it is evident that interactions between fetus and mother differ substantially from the events triggered by a classical allograft. Consequently, much work is being dedicated to the emerging field of materno-fetal immunobiology in order to enable the development of innovative strategies to induce tolerance and prevent allogeneic graft rejection.

When maternal T cells encounter the fetus they demonstrate adaptive tolerance. In part this may be due to the absence of expression of MHC class II antigens and low levels of expression of MHC class I antigens on fetal cells. However, this can only partly explain the state of prolonged maternal tolerance since induction of HLA class I and II molecules inevitably occurs as the fetus matures and differentiates, yet rejection still does not occur. Consequently, non-fetal aspects of the placental barrier must be of critical importance in maintaining prolonged tolerance to the fetus. An important mechanism may relate to upregulation of the human non-classical HLA class Ib antigen, designated HLA-G, by the syncytiotrophoblast. HLA-G molecules bind to inhibitory receptors on natural killer cells and subsequently protect against maternal rejection responses. The placenta produces high levels of the anti-inflammatory cytokine interleukin 10 which stimulates HLA-G synthesis while concomitantly downregulating MHC class I antigen production, thus contributing to the tolerance-inducing local environment. The trophoblast also produces high levels of the enzyme indoleamine 2,3-dioxygenase, which catabolizes tryptophan, an essential amino acid necessary for rapid T cell proliferation. Annexin II, found in isolated placental membranes in vitro is present in placental serum, exerts immunosuppressive properties, and additionally contributes to fetal allograft survival. Together, these features indicate that materno-fetal tolerance results from a combination of transiently reduced antigenicity of the fetus in combination with a complex tolerance-inducing milieu at the placental barrier.

Immunogenic Characteristics Of Embryonic And Adult Stem Cells

Murine and human embryonic stem (ES) cells do not express HLA class I and II antigens, and demonstrate reduced surface expression of co-stimulatory molecules important for T cell activation. Transplantation of murine ES cells demonstrates long-term graft survival despite the fact that these cells do acquire HLA class II antigen expression after in vivo differentiation. Since they are able to accomplish long-term engraftment without the need for immunosuppression, their inability to induce an immune response is not likely to be the result of escaping immune surveillance, but rather due to their ability to colonize the recipient thymus and induce intrathymic deletion of alloreactive recipient T cells.

Recently, a population of cells has been described in human adult bone marrow that has similar functional characteristics to embryonic stem cells in that they have high self-regenerating capability and capacity for differentiation into multiple cell types, including muscle, cartilage, fat, bone, and heart tissue. While such cells, termed adult mesenchymal stem cells (MSC), appear to have a more restricted self-renewal capacity and differentiation potential than ES cells, their functional characteristics may be sufficient for clinically meaningful tissue regeneration. A striking recent observation is that MSC can broadly inhibit T-cell proliferation and activation by various types of antigenic stimulation, including allogeneic stimuli. MSCs have been shown to inhibit both naive and memory T cell responses in a dose-dependent fashion and affect cell proliferation, cytotoxicity, and the number of interferon gamma (IFN-gamma)-producing T cells. MSCs appear to inhibit T cell activation through direct contact, and do not require other regulatory cellular populations. Similarly to ES cells, adult bone marrow-derived mesenchymal stem cells (MSCs) do not express HLA class II molecules, and only low levels of HLA class I molecules. Despite the fact that MSC can be induced to express surface HLA class II molecules by in vitro culture with cytokines such as interferon-gamma, their ability to inhibit T cell activation results in induction of T cell non-responsiveness to the MSC themselves, endowing them with potential survival advantages in the setting of transplantation.

Tolerogenic Effects Of Stem Cell Transplantation

Extending the approaches discussed above using donor-derived blood transfusions to induce a tolerogenic state to the subsequent organ, the most promising clinical strategy for tolerance induction at present is the use of donor-derived hematopoietic stem cells in conjunction with reduced myeloablative conditioning. The objective of this therapy is to achieve a state of so-called mixed chimerism, or the permanent co-existence of donor- and recipient-derived blood cells comprising all the different hematopoietic lineages in the same host. This approach has been tested in a variety of small and large animal settings and currently available data suggest that stable engraftment of donor bone marrow reliably renders the host tolerant to donor antigens and subsequently to any cellular or solid organ graft of the same donor.

The two underlying mechanisms by which creation of a mixed-chimeric host results in tolerance induction are (1) thymic deletion of potentially donor-specific alloreactive T cells, and (2) nonthymic peripheral mechanisms, such as blocking costimulatory T cell activation, which facilitate the process of donor bone-marrow or stem cell engraftment. However, despite the efficacy of an approach using fully HLA-mismatched stem cells in an allogeneic host to induce tolerance to a subsequent organ allograft, the host is placed at a high risk of substantial morbidity and mortality due to toxicity of the myeloablative conditioning regimen and potential for graft-versus-host disease, or immune-mediated attack of the host by the implanted allogeneic stem cells.

In an attempt to overcome these potential limiting toxicities, investigators have suggested the use of either adult bone marrow-derived mesenchymal stem cells or preimplantation-derived embryonic stem (ES) cells for induction of mixed chimerism. The theoretical advantages of these cell types is their low level of surface expression of HLA class I and II antigens, and reduced surface expression of co-stimulatory molecules important for T cell activation. Rat preimplantation stage derived embryonic-like stem cells have been shown to successfully engraft in the recipient bone marrow without the need for pre-conditioning therapies such as irradiation, cytotoxic drug regimens or T cell depletion. Long-term partial mixed chimerism by use of rat preimplantation stage derived embryonic-like stem cells did not trigger graft-versus-host reactions, in contrast to the high frequency of this complication in the clinical setting of allogeneic hematopoietic stem cell transplantation. Of most interest, the induced partial chimerism enabled the recipient animals to be tolerant to a subsequent heart allograft. Allograft acceptance required the presence of an intact thymus, and rat ES cells were present in the recipient thymus.

Similar results have been reported following transplantation of human adult bone marrow-derived mesenchymal stem cells (MSC) into fetal sheep early in gestation, before and after the expected development of immunologic competence. In this xenogeneic system, human MSC engrafted, differentiated in a site-specific manner, and persisted in multiple tissues for as long as 13 months after transplantation, including the thymus. Since MSCs do not present alloantigen and do not require MHC expression to exert their inhibitory effect on alloimmune reactivity, the possibility exists that they could theoretically be derived from a donor irrespective of their HLA type and used to inhibit T-cell responses to transplantation antigens of an unrelated third party. In initial human clinical studies, the use human adult bone marrow-derived mesenchymal stem cells has been shown to successfully enable engraftment of subsequently infused allogeneic bone marrow in transplant recipients, reduce the risk of graft-versus-host disease, and reduce the need for concomitantly administered immunosuppression. Whether similar results will be obtained when combining adult bone marrow-derived mesenchymal stem cells with solid organ allografts remains to be determined, and this is an area of active research for clinical transplant immunobiologists. Of broader relevance, if the results relating to long-term engraftment and survival of adult bone marrow-derived MSC are confirmed and extended in human clinical studies, they will have broad implications for the field of tissue and organ regeneration.

SELECTED REFERENCES

1. Billingham RE, Brent L, Medawar PB (1953) Actively acquired tolerance of foreign cells. Nature 172:603-606.

2. Auchincloss H Jr (2001) In search of the elusive holy grail: the mechanisms and prospects for achieving clinical transplantation tolerance. Am J Transplant 1:6-12.

3. Wekerle T, Kurtz J, Ito H, Ronquillo JV, Dong V, Zhao G, Shaffer J, Sayegh MH, Sykes M (2000) Allogeneic bone marrow transplantation with co-stimulatory blockade induces macrochimerism and tolerance without cytoreductive host treatment. Nat Med 6:464-469.

4. Watkins DI (1995) The evolution of major histocompatibility class I genes in primates. Crit Rev Immunol 15:1-29.

5. Salter-Cid L, Flajnik MF (1995) Evolution and developmental regulation of the major histocompatibility complex. Crit Rev Immunol 15:31-75.

6. Fowlkes BJ, Ramsdell F (1993) T-cell tolerance. Curr Opin Immunol 5:873-879.

7. Marrack P, Kappler J (1990) T cell tolerance. Semin Immunol 2:45-49.

8. Van Parijs L, Abbas AK (1998) Homeostasis and self-tolerance in the immune system: turning lymphocytes off. Science 280:243-249.

9. Davies JD, Martin G, Phillips J, Marshall SE, Cobbold SP, Waldmann H (1996) T cell regulation in adult transplantation tolerance. J Immunol 157:529-533.

10. Thomson AW, Lu L (1999) Dendritic cells are regulators of immune reactivity: implications for transplantation. Transplantation 68:1-8.

11. Brent L (1997) Fetally and neonatally induced immunologic tolerance. In: Brent L (ed) A history of transplantation immunology. Academic, San Diego.

12. Matzinger P (1994) Tolerance, danger, and the extended family. Annu Rev Immunol 12:991-1045.

13. Janeway CAJ (1992) The immune system evolved to discriminate infectious nonself from noninfectious self. Immunol Today 13:11-16.

14. Mellor AL, Munn DH (2001) Extinguishing maternal immune responses during pregnancy: implications for immunosuppression. Semin Immunol 13:213-218.

15. Fändrich F, Lin X, Chai GX, Schulze M, Ganten D, Bader M, Holle J, Huang DS, Parwaresch R, Zavazava N, Binas B (2002) Preimplantation-stage stem cells induce allogeneic graft tolerance without supplementary host conditioning. Nat Med 8:171-178.

16. Liechty KW, et al (2000). Human mesenchymal stem cells engraft and demonstrate site-specific differentiation after in utero transplantation in sheep. Nat Med 6(11):1282-6.

17. Krampera M, et al (2002). Bone marrow mesenchymal stem cells inhibit the response of naive and memory antigen-specific T cells to their cognate peptide. Blood, 101(9):3722-3729.

18. Le Blanc K, Tammik L, Sundberg B, Haynesworth SE, Ringden O (2003). Mesenchymal stem cells inhibit and stimulate mixed lymphocyte cultures and mitogenic responses independently of the major histocompatibility complex.Scand J Immunol. 57(1):11-20.

Appendix M.

Potential Use of Cellular Therapy
for Patients With Heart Disease

SILVIU ITESCU, M.D.

*Departments of Medicine and Surgery, Columbia University,
New York, NY*

Summary

Congestive heart failure remains a major public health problem, and is frequently the end result of cardiomyocyte apoptosis and fibrous replacement after myocardial infarction, a process referred to as left ventricular remodelling. Cardiomyocytes undergo terminal differentiation soon after birth, and are generally considered to irreversibly withdraw from the cell cycle. In response to ischemic insult, adult cardiomyocytes undergo cellular hypertrophy, nuclear ploidy, and a high degree of apoptosis. A small number of human cardiomyocytes retain the capacity to proliferate and regenerate in response to ischemic injury, however whether these cells are derived from a resident pool of cardiomyocyte stem cells or from a renewable source of circulating bone marrow-derived stem cells that home to the damaged myocardium is at present not known. Replacement and regeneration of functional cardiac muscle after an ischemic insult to the heart could be achieved by either stimulating proliferation of endogenous mature cardiomyocytes or resident cardiac stem cells, or by implanting exogenous donor-derived or allogeneic cells such as fetal or embryonic cardiomyocyte precursors, bone marrow-derived mesenchymal stem cells, or skeletal myoblasts. The newly formed cardiomyocytes must integrate precisely into the existing myocardial wall in order to augment synchronized contractility and avoid potentially life-threatening alterations in the electrical conduction of the heart. A major impediment to survival of the implanted cells is altered immunogenicity by prolonged *ex vivo* culture conditions. In addition, concurrent myocardial revascularization is required to ensure viability of the repaired region and prevent further scar tissue formation. Human adult bone marrow contains endothelial precursors which resemble embryonic

365

angioblasts and can be used to induce infarct bed neovascularization after experimental myocardial infarction. This results in protection of cardiomyocytes against apoptosis, induction of cardiomyocyte proliferation and regeneration, long-term salvage and survival of viable myocardium, prevention of left ventricular remodelling and sustained improvement in cardiac function. It is reasonable to anticipate that cell therapy strategies for ischemic heart disease will need to incorporate (1) a renewable source of proliferating, functional cardiomyocytes, and (2) angioblasts to generate a network of capillaries and larger size blood vessels for supply of oxygen and nutrients to both the chronically ischemic endogenous myocardium and to the newly-implanted cardiomyocytes.

Introduction

Congestive heart failure remains a major public health problem, with recent estimates indicating that end-stage heart failure with two-year mortality rates of 70-80% affects over 60,000 patients in the US each year [1]. In Western societies heart failure is primarily the consequence of previous myocardial infarction [2]. As new modalities have emerged which have enabled significant reduction in early mortality from acute myocardial infarction, affecting over 1 million new patients in the US annually, there has been a paradoxical increase in the incidence of post-infarction heart failure among the survivors. Current therapy of heart failure is limited to the treatment of already established disease and is predominantly pharmacological in nature, aiming primarily to inhibit the neurohormonal axis that results in excessive cardiac activation through angiotensin- or norepinephrine-dependent pathways. For patients with end-stage heart failure treatment options are extremely limited, with less than 3000 being offered cardiac transplants annually due to the severely limited supply of donor organs [3,4], and implantable left ventricular assist devices (LVADs) being expensive, not proven for long-term use, and associated with significant complications [5-7]. Clearly, development of approaches that prevent heart failure after myocardial infarction would be preferable to those that simply ameliorate or treat already established disease.

Heart Failure After Myocardial Infarction Results From Progressive Ventricular Remodelling. Heart failure after myocardial infarction occurs as a result of a process termed myocardial remodelling. This process is characterized by myocyte apoptosis, cardiomyocyte replacement by fibrous tissue deposition in the ventricular wall [8-10], progressive expansion of the initial infarct area and dilation of the left ventricular lumen [11,12]. Another integral component of the remodelling process appears to be the development of neoangiogenesis within the myocardial infarct scar [13,14], a process requiring activation of latent collagenase and other proteinases [15]. Under normal circumstances, the contribution of neoangiogenesis to the infarct bed capillary network is insufficient to keep pace with the tissue growth required for contractile compensation and is unable to support the greater demands of the hypertrophied, but viable, myocardium. The relative lack of oxygen and nutrients to the hypertrophied myocytes may be an important etiological factor in the death of otherwise viable myocardium, resulting in progressive infarct extension and fibrous replacement. Since late reperfusion of the infarct vascular bed in both humans and experimental animal models significantly benefits ventricular remodelling and survival [16-18], we have postulated that methods to successfully augment vascular bed neovascularization might improve cardiac function by preventing loss of hypertrophied, but otherwise viable, cardiac myocytes.

Inability Of Damaged Myocardium To Undergo Repair Due To Cell Cycle Arrest Of Adult Cardiomyocytes. Cardiomyocytes undergo terminal differentiation soon after birth, and are thought by most investigators to irreversibly withdraw from the cell cycle. Analysis of cardiac myocyte growth during early mammalian development indicates that cardiac myocyte DNA synthesis occurs primarily in utero, with proliferating cells decreasing from 33% at mid-gestation to 2% at birth [19]. While ventricular karyokinesis and cytokinesis are coupled during fetal growth, resulting in increases in mononucleated cardiac myocytes, karyokinesis occurs in the absence of cytokinesis for a transient period during the post-natal period, resulting in binucleation of ventricular myocytes without an overall increase in cell number. A similar dissociation between karyokinesis and cytokinesis characterizes the primary adult mammalian cardiac response to ischemia, resulting in myocyte hypertrophy and increase in nuclear ploidy rather than myocyte hyperplasia [20,21]. Moreover, in parallel with an inability to progress through cell cycle, ischemic adult cardiomyocytes undergo a high degree of apoptosis.

When cells proliferate, the mitotic cycle progression is tightly regulated by an intricate network of positive and negative signals. Progress from one phase of the cell cycle to the next is controlled by the transduction of mitogenic signals to cyclically expressed proteins known as cyclins and subsequent activation or inactivation of several members of a conserved family of serine/threonine protein kinases known as the cyclin-dependent kinases (cdks) [22]. Growth arrest observed with such diverse processes as DNA damage, terminal differentiation, and replicative senescence is due to negative regulation of cell cycle progression by two functionally distinct families of Cdk inhibitors, the Ink4 and Cip/Kip families [19]. The cell cycle inhibitory activity of p21Cip1/WAF1 is intimately correlated with its nuclear localization and participation in quaternary complexes of cell cycle regulators by binding to G1 cyclin-CDK through its N-terminal domain and to proliferating cell nuclear antigen (PCNA) through its C-terminal domain [23-26]. The latter interaction blocks the ability of PCNA to activate DNA polymerase, the principal replicative DNA polymerase [27]. For a growth-arrested cell to subsequently enter an apoptotic pathway requires signals provided by specific apoptotic stimuli in concert with cell-cycle regulators. For example, caspase-mediated cleavage of p21, together with upregulation of cyclin A–associated cdk2 activity, have been shown to be critical steps for induction of cellular apoptosis by either deprivation of growth factors [28] or hypoxia of cardiomyocytes [29].

Throughout life, a mixture of young and old cells is present in the normal myocardium. Although most myocytes seem to be terminally differentiated, there is a fraction of younger myocytes (15–20%) that retains the capacity to replicate [30]. Moreover, recent observations have suggested that some human ventricular cardiomyocytes also have the capacity to proliferate and regenerate in response to ischemic injury [31,32]. The dividing myocytes can be identified on the basis of immunohistochemical staining of proliferating nuclear structures such as Ki67 and cell surface expression of specific surface markers, including c-kit (CD117). Whether these cells are derived from a resident pool of cardiomyocyte stem cells or are derived from a renewable source of circulating bone marrow-derived stem cells that home to the damaged myocardium remains to be determined. More importantly, the signals required for homing, in situ expansion and differentiation of these cells are, at present, unknown. Gaining an understanding of these issues would open the possibility of manipulating the biology of endogenous cardiomyocytes in order to augment the healing process after myocardial ischemia.

Strategies For The Use Of Cellular Therapy To Improve Myocardial Function. Replacement and regeneration of functional cardiac muscle after an ischemic insult to the heart could be achieved by either stimulating proliferation of endogenous mature cardiomyocytes or resident cardiac stem cells, or by implanting exogenous donor-derived or allogeneic cardiomyocytes. The newly formed cardiomyocytes must integrate precisely into the existing myocardial wall in order to augment contractile function of the residual myocardium in a synchronized manner and avoid alterations in the electrical conduction and syncytial contraction of the heart, potentially resulting in life-threatening consequences. In addition, whatever the source of the cells used, it is likely that concurrent myocardial revascularization must also occur in order to ensure viability of the repaired region and prevent further scar tissue formation. The following section discusses various methods of using cellular therapies to replace damaged myocardium or re-initiate mitosis in mature endogenous cardiomyocytes, including transplanted bone marrow-derived cardiomyocyte or endothelial precursors, fetal cardiomyocytes, and skeletal myoblasts.

Potential Role For Bone Marrow-Derived Or Embryonic Cardiomyocyte Lineage Stem Cells In Myocardial Repair/ Regeneration. Over the past several years, a number of studies have suggested that stem cells can be used to generate cardiomyocytes *ex vivo* for potential use in a range of cardiovascular diseases [33-37]. Multipotent bone marrow-derived mesenchymal stem cells have been identified in adult murine and human bone marrow functionally by their ability to differentiate to lineages of diverse mesenchymal tissues, including bone, cartilage, fat, tendon, and both skeletal and cardiac muscle [36], and phenotypically by their expression of specific surface markers and lack of hematopoietic lineage markers such as CD34 or CD45 [35]. It is well established that murine embryonic stem (ES) cells can give rise to cardiomyocytes in vitro and in vivo [38,39]. Recently, Kehat et al. were able to demonstrate that human embryonic stem cells can also differentiate *in vitro* into cells with characteristics of cardiomyocytes [37]. However, there are striking differences in the human and murine stem cell models, and this needs to be taken into account when extrapolating results of mouse experiments to the human condition. For example, human ES cells have a very low efficiency of differentiation to cardiomyocytes compared with murine ES cells, and a considerably slower time course (a median of 11 days vs 2 days). Whether these differences reflect true variations between species, or differences in the experimental protocols, remains to be determined.

Irrespective whether the cardiomyocyte lineage stem cell precursors are obtained from adult bone marrow or embryonic sources, the newly generated cardiomyocytes appear to resemble normal cardiomyocytes in terms of phenotypic properties, such as expression of actinin, desmin and troponin I, and function, including positive and negative chronotropic regulation of contractility by pharmacological agents and production of vasoactive factors such as atrial and brain natriuretic peptides. However, *in vivo* evidence for functional cardiac improvement following transplantation of adult bone marrow-derived or ES-derived cardiomyocytes has been exceedingly difficult to show to date. In part this may be because the signals required for cardiomyocyte differentiation and functional regulation are complex and poorly understood. For example, phenotypic and functional differentiation of mesenchymal stem cells to cardiomyocyte lineage cells *in vitro* requires culture with exogenously added 5-azacytidine [33,34]. Alternatively, the poor functional data obtained to date may reflect immune-mediated rejection of cells which have been modified during the *ex vivo* culture process or poor viability due to the lack of a sufficient vascular supply to the engrafted cells (see below).

Potential Role For Autologous Skeletal Myoblasts In Myocardial Repair. An alternative approach to replacing damaged myocardium involves the use of autologous skeletal myoblasts [40]. The procedure involves harvesting a patient's skeletal muscle cells, expanding the cells in a laboratory, and re-injecting the cells into the patient's heart. Perceived advantages of the approach include ease of access to the cellular source, the fact that immunosuppression is not needed, and the lack of ethical dilemmas associated with the use of allogeneic or embryonic cells. It has also been argued that using relatively ischemia-resistant skeletal myoblasts rather than cardiomyocytes might enable higher levels of cell engraftment and survival in infarcted regions of the heart, where cardiomyocytes would probably perish [41].

Successful engraftment of autologous skeletal myoblasts into injured myocardium has been reported in multiple animal models of cardiac injury. These studies have demonstrated survival and engraftment of myoblasts into infarcted or necrotic hearts [40], differentiation of the myoblasts into striated cells within the damaged myocardium [40], and improved myocardial functional performance [40,42,43]. Other studies have shown that the survival of transplanted myoblasts can be improved by heat shock pretreatment [44], and have confirmed that the benefits of skeletal myoblast transfer are additive with those of conventional therapies, such as angiotensin-converting enzyme inhibition [45]. More recently, the procedure has been reported anecdotally to result in improved myocardial function in humans [46]. On the basis of these preliminary results, clinical trials have begun both in Europe and in the United States. In addition to demonstrating functional improvement in large, prospective series, questions that remain to be addressed include whether the skeletal myoblasts can make meaningful electromechanical connections to the surrounding endogenous cardiomyocytes through gap junctions, whether the cellular mass will contract in concert, and whether electrical impulses will be transmitted to the myoblast tissue without inducing significant tachyarrhythmias.

Poor Survival Of Cells Transplanted Into Damaged Myocardium After *Ex Vivo* Culture. A major limitation to successful cellular therapy in animal models of myocardial damage has been the inability of the introduced donor cells to survive in their host environment, whether such transplants have been congenic (analogous to the autologous scenario in humans) or allogeneic. It has become clear that a major impediment to survival of the implanted cells is the alteration of their immunogenic character by prolonged *ex vivo* culture conditions. For example, whereas myocardial implantation of skeletal muscle in the absence of tissue culture does not induce any adverse immune response and results in grafts showing excellent survival for up to a year, injection of cultured isolated (congenic) myoblasts results in a massive and rapid necrosis of donor myoblasts, with over 90% dead within the first hour after injection [47-50]. This rapid myoblast death appears to be mediated by host natural killer (NK) cells [50] which respond to immunogenic antigens on the transplanted myoblasts altered by exposure to tissue culture conditions [48]. It seems likely that a similar mechanism of host NK cell-mediated rejection will apply also to transplanted ES-derived, cultured cardiomyocytes [51] since massive death of injected donor cells is recognized as a major problem with transplanted cardiomyocytes, especially in the inflammatory conditions that follow infarction [52]. In this regard, the report that cultured mesenchymal stem cells obtained from adult human bone marrow are not rejected on transfer to other species is intriguing [36], needs confirmation in humans, and requires detailed investigation into possible tolerogenic mechanisms.

Concomitant Induction Of Vascular Structures Augments Survival and Function Of Cardiomyocyte Precursors. An additional explanation for the poor survival of transplanted cardiomyocytes or skeletal myoblasts may be that viability and prolonged function of transplanted cells requires an augmented vascular supply. Whereas many transplanted cardiomyocytes die by apoptosis, cultured cardiomyocytes that incorporate more vascular structures in vivo demonstrate significantly greater survival[52]. Moreover, in situations where transplanted cardiomyocyte precursors contained an admixture of cells also giving rise to vascular structures, survival and function of the newly formed cardiomyocytes has been significantly augmented.

In one study, direct injection of whole rat bone marrow into a cryo-damaged heart resulted in neovascularization, cardiomyocyte regeneration and functional improvement [34]. More recently, systemic delivery of highly purified bone marrow-derived hematopoietic stem cells in lethally irradiated mice contributed to the formation of both endothelial cells and long-lived cardiomyocytes in ischemic hearts [53]. Most strikingly, significant improvement in cardiac function of mice who had previously undergone LAD ligation was demonstrated after direct myocardial injection of syngeneic bone marrow-derived stem cells, defined on the basis of c-kit (CD117) expression [54]. This population of cells contains a mixture of cellular elements in addition to cardiomyocyte precursors, including CD117-positive endothelial progenitors (see below). These cells were found to proliferate and differentiate into myocytes, smooth muscle cells and endothelial cells, resulting in the partial regeneration of the destroyed myocardium and prevention of ventricular scarring. Together, these findings raise the intriguing possibility that for long-term in vivo viability and functional integrity of stem cell-derived cardiomyocytes it may be necessary to induce neovascularization by co-administration of endothelial cell progenitors (see below).

Endothelial Precursors And Formation Of Vascular Structures During Embryogenesis. In order to develop successful methods for inducing neovascularization of the adult heart, one needs to understand the process of definitive vascular network formation during embryogenesis. In the pre-natal period, hemangioblasts derived from the human ventral aorta give rise to cellular elements involved in both vasculogenesis, or formation of the primitive capillary network, and hematopoiesis [55,56]. In addition to hematopoietic lineage markers, embryonic hemangioblasts are characterized by expression of the vascular endothelial cell growth factor receptor-2, VEGFR-2, and have high proliferative potential with blast colony formation in response to VEGF [57-60]. Under the regulatory influence of various transcriptional and differentiation factors, embryonic hemangioblasts mature, migrate and differentiate to become endothelial lining cells and create the primitive vasculogenic network. The differentiation of embryonic hemangioblasts to pluripotent stem cells and to endothelial precursors appears to be related to co-expression of the GATA-2 transcription factor, since GATA-2 knockout embryonic stem cells have a complete block in definitive hematopoiesis and seeding of the fetal liver and bone marrow[61]. Moreover, the earliest precursor of both hematopoietic and endothelial cell lineage to have diverged from embryonic ventral endothelium has been shown to express VEGF receptors as well as

GATA-2 and alpha4-integrins [62].   Subsequent to capillary tube formation, the newly-created vasculogenic vessels undergo sprouting, tapering, remodelling, and regression under the direction of VEGF, angiopoietins, and other factors, a process termed angiogenesis. The final component required for definitive vascular network formation to sustain embryonic organogenesis is influx of mesenchymal lineage cells to form the vascular supporting structures such as smooth muscle cells and pericytes.

Characterization Of Endothelial Progenitors, Or Angioblasts, In Human Adult Bone Marrow. In studies using various animal models of peripheral ischemia a number of groups have shown the potential of adult bone marrow-derived elements to induce neovascularization of ischemic tissues [63-69]. In the most successful of these [63], bone marrow-derived cells injected directly into the thighs of rats who had undergone ligation of the left femoral artery and vein induced neovascularization and augmented blood flow in the ischemic limb as documented by laser doppler and immunohistochemical analyses. Although the nature of the bone marrow-derived endothelial progenitors was not precisely identified in these studies, the cumulative reports indicated that this site may be an important source of endothelial progenitors which could be useful for augmenting collateral vessel growth in ischemic tissues, a process termed therapeutic angiogenesis.

In more recent studies, our group has identified such endothelial progenitors in human adult bone marrow [70]. By employing both *in vitro* and *in vivo* experimental models we have sought to precisely identify the surface characteristics and biological properties of these bone marrow-derived endothelial progenitor cells. Following G-CSF treatment, mobilized mononuclear cells were harvested and CD34+ cells were separated using anti-CD34 mAb coupled to magnetic beads. 90-95% of CD34+ cells co-expressed the hematopoietic lineage marker CD45, 60-80% co-expressed the stem cell factor receptor CD117, and <1% co-expressed the monocyte/macrophage lineage marker CD14. By quadruple parameter analysis, the VEGFR-2 positive cells within the CD34+CD117[bright] subset displayed phenotypic characteristics of endothelial progenitors, including co-expression of Tie-2, as well as AC133, but not markers of mature endothelium such as ecNOS, vWF, E-selectin, and ICAM. Sorting CD34+ cells on the basis of CD117 bright or dim expression demonstrated that GATA-2 mRNA and protein levels were significaantly higher in the CD117[bright] population, indicating that human adult bone marrow contains cells with an angioblast-like phenotype.

Since the frequency of circulating endothelial cell precursors in animal models has been shown to be increased by either VEGF [71] or regional ischemia [63-66], phenotypically-defined angioblasts were examined for proliferative responses to VEGF and to factors in ischemic serum. CD117^{bright}GATA-2^{hi} angioblasts demonstrated significantly higher proliferative responses relative to CD117^{dim} GATA-2^{lo} bone marrow-derived cells from the same donor following culture for 96 hours with either VEGF or ischemic serum. The expanded angioblast population consisted of large blast cells, defined by forward scatter, which continued to express immature markers, including GATA-2 and CD117^{bright} but not markers of mature endothelial cells, including eNOS or E-selectin, indicating blast proliferation without differentiation under these culture conditions. However, culture on fibronectin with endothelial growth medium resulted in outgrowth of monolayers with endothelial morphology and functional and phenotypic features characteristic of endothelial cells, including uniform uptake of acetylated LDL, and co-expression of CD34, factor VIII, and eNOS. Thus, G-CSF treatment of adult humans mobilizes into the peripheral circulation a bone-marrow derived population with phenotypic and functional characteristics of embryonic angioblasts, as defined by specific surface phenotype, high proliferative responses to VEGF and cytokines in ischemic serum, and ability differentiate into endothelial cells by culture in medium enriched with endothelial growth factors.

Human Angioblasts Induce Neovascularization Of The Myocardial Infarct Zone. Intravenous injection of freshly-obtained human angioblasts into athymic nude rats who had undergone ligation of the left anterior descending (LAD) coronary artery resulted in infarct zone infiltration within 48 hours. Few human cells were detected in unaffected areas of hearts with regional infarcts or in myocardium of sham-operated rats. Histologic examination at two weeks post-infarction revealed that injection of human angioblasts was accompanied by a significant increase in infarct zone microvascularity, cellularity, and numbers of capillaries, and by reduction in matrix deposition and fibrosis in comparison to controls. Neovascularization was significantly increased within both the infarct zone and in the peri-infarct rim in rats receiving angioblasts compared with controls receiving saline or other cells which infiltrated the heart (e.g. CD34- cells or saphenous vein endothelial cells, SVEC). The neovascularization induced by human angioblasts was due to both an increase in capillaries of human origin as well as of rat origin, as defined by monoclonal antibodies with specificity for human or rat CD31 endothelial markers. Capillaries of human origin, defined by co-expression of DiI fluorescence and human CD31, but not rat CD31, accounted for 20-25% of the total myocardial capillary vasculature, and was located exclusively within the central infarct zone of collagen deposition. In contrast, capillaries of rat origin, as determined by expression of rat, but not human, CD31, demonstrated a distinctively different pattern of localization, being absent within the central zone of collagen deposition and abundant both at the peri-infarct rim between the region of collagen deposition and myocytes, and between myocytes.

Human Angioblasts Protect Hypertrophied Endogenous Cardiomyocytes Against Apoptosis. By concomitantly staining rat tissues for the myocyte-specific marker desmin and performing DNA end-labeling using the TUNEL technique, temporal examinations demonstrated that the infarct zone neovascularization induced by injection of human angioblasts prevented an eccentrically-extending pro-apoptotic process evident in saline controls. Thus, at two weeks post-infarction, myocardial tissue of LAD-ligated rats who received saline demonstrated 6-fold higher numbers of apoptotic myocytes compared with that from rats receiving intravenous injections of human angioblasts. Moreover, these myocytes had distorted appearance and irregular shape. In contrast, myocytes from LAD-ligated rats who received human angioblasts had regular, oval shape, and were significantly larger than myocytes from control rats.

Human Angioblasts Induce Sustained Regeneration/ Proliferation Of Endogenous Cardiomyocytes. In addition to protection of hypertrophied myocytes against apoptosis, human angioblast-dependent neovascularization resulted in a striking induction of regeneration/proliferation of endogenous rat cardiomyocytes at the peri-infarct rim [72]. At two weeks after LAD ligation, rats receiving human angioblasts demonstrated numerous "fingers" of cardiomyocytes of rat origin, as determined by expression of rat MHC class I molecules, extending from the peri-infarct region into the infarct zone. The islands of cardiomyocytes at the peri-infarct rim in animals receiving human angioblasts contained a high frequency of rat myocytes with DNA activity, as determined by dual staining with mAbs reactive against cardiomyocyte-specific troponin I and rat Ki67. In contrast, in animals receiving saline there was a high frequency of cells with fibroblast morphology and reactivity with rat Ki67, but not troponin I, within the infarct zone. The number of cardiomyocytes progressing through cell cycle at the peri-infarct region of rats receiving human angioblasts was 40-fold higher than that at sites distal to the infarct, 20-fold higher than that found in non-infarcted hearts, and 5-fold higher than that at the peri-infarct rim of animals receiving saline [72].

Neovascularization Of Acutely Ischemic Myocardium By Human Angioblasts Prevents Ventricular Remodelling And Causes Sustained Improvement In Cardiac Function. By 15 weeks post-infarction, rats receiving human angioblasts demonstrated markedly smaller scar sizes together with increased mass of viable myocardium within the anterior free wall. Whereas collagen deposition and scar formation extended almost through the entire left ventricular wall thickness in controls, with aneurysmal dilatation and typical EKG abnormalities, the infarct scar extended only to 20-50% of the left ventricular wall thickness in rats receiving CD34+ cells. Moreover, pathological collagen deposition in the non-infarct zone was markedly reduced in rats receiving CD34+ cells. At 15 weeks, the mean proportion of scar/normal left ventricular myocardium was 13% in rats receiving CD34+ cells compared with 36-45% for each of the other groups studied (saline, CD34-, SVEC).

Remarkably, by two weeks after injection of human angioblasts, and in a parallel time-frame with the observed neo-vascularization, left ventricular ejection fraction (LVEF) recovered by a mean of 22%. This effect was long-lived, with LVEF recovering by a mean of 34% at the end of follow-up, 15 weeks post injection. Neither CD34- cells nor SVEC demonstrated similar effects. At 15 weeks post-infarction, mean cardiac index in rats injected with human angioblasts was only reduced by 26% relative to normal rats, whereas for each of the other groups it was reduced by 48-59%. Together, these results indicate that the neovascularization, reduction in peri-infarct myocyte apoptosis and increase in cardiomyocyte regeneration/proliferation observed at two weeks prevented myocardial replacement with fibrous tissue and caused sustained improvement in myocardial function.

Potential Use Of Angioblasts In Combination With Cardiomyocyte Progenitors For Repair and Regeneration of Ischemic Myocardium. While increasing capillary density through angioblast-dependent neovascularization is a promising approach for preventing apoptotic death and inducing regeneration of endogenous cardiomyocytes following acute myocardial infarction, the role of angioblast therapy for the treatment of congestive heart failure following chronic ischemia is at present unknown. Nevertheless, it is reasonable to anticipate that cellular therapies for congestive heart failure due to ischemic cardiomyopathy will need to address two interdependent processes: (1) a renewable source of proliferating, functional cardiomyocytes, and (2) development of a network of capillaries and larger size blood vessels for supply of oxygen and nutrients to both the chronically ischemic, endogenous myocardium and to the newly-implanted cardiomyocytes. To achieve these endpoints it is likely that co-administration of angioblasts and mesenchymal stem cells will be needed in order to develop regenerating cardiomyocytes, vascular structures, and supporting cells such as pericytes and smooth muscle cells. Future studies will need to address the timing, relative concentrations, source and route of delivery of each of these cellular populations in animal models of acute and chronic myocardial ischemia.

In addition to synergistic cellular therapies, it is likely that optimal regimens for the treatment of acute and chronically ischemic hearts will require a combined approach employing additional pharmacological strategies. For example, augmentation in myocardial function might be achieved by combining infusion of human angioblasts and cardiomyocyte progenitors together with beta blockade, ACE inhibition or AT_1-receptor blockade to reduce angiotensin II-dependent cardiac fibroblast proliferation and collagen secretion [73-76]. Understanding the potential of defined lineages of stem cells or undifferentiated progenitors, and their interactions with pharmacological interventions, will lead to better and more focussed clinical trial designs using each cell type independently or in combination, depending on which particular clinical indication is being targeted.

REFERENCES

1. Effects of enalapril on mortality in severe congestive heart failure. Results of the Cooperative North Scandinavian Enalapril Survival Study (CONSENSUS) Trial Study Group. N Engl J Med 1987; 316(23):1429-1435.

2. Mahon NG, O'Roke C, Codd MB, et al. Hospital mortality of acute myocardial infarction in the thrombolytic era. Heart 1999;81:478-82.

3. Hognes JR. In The Artificial Heart: Prototypes, Policies and Patients. Washingtion, DC: National Academy Press; 1991:1-312.

4. Annual Report of the US Scientific Registry for Organ Transplantation and the Organ Procurement and Transplantation Network - 1990. Washingtion, DC: US Department of Health and Human Services: 1990.

5. Frazier OH, Rose EA, Macmanus Q, et al. Multicenter clinical evaluation of the Heartmate 1000 IP left ventricular assist device. Ann Thorac Surg 1992; 102:578-587.

6. McCarthy PM, Rose EA, Macmanus Q, et al. Clinical experience with the Novacor ventricular assist system. J Thorac Cardiovasc Surg. 1991; 102:578-587.

7. Oz MC, Argenziano M, Catanese KA, et al. Bridge experience with long-term implantable left ventricular assist devices. Are they an alternative to transplantation? Circulation. 1997; 95:1844-1852.

8. Colucci WS. Molecular and cellular mechanisms of myocardial failure. Am J Cardiol 1997;80:15L-25L.

9. Ravichandran LV and Puvanakrishnan R. In vivo labeling studies on the biosynthesis and degradation of collagen in experimental myocardial myocardial infarction. Biochem Intl 1991;24:405-414.

10. Agocha A, Lee H-W, Eghali-Webb M. Hypoxia regulates basal and induced DNA synthesis and collagen type I production in human cardiac fibroblasts: effects of TGF-beta, thyroid hormone, angiotensis II and basic fibroblast growth factor. J Mol Cell Cardiol 1997;29:2233-2244H.

11. Pfeffer JM, Pfeffer MA, Fletcher PJ, Braunwald E. Progressive ventricular remodeling in rat with myocardial infarction. Am J Physiol 1991; 260:H1406-14.

12. White HD, Norris RM, Brown MA, Brandt PWT, Whitlock RML, Wild CJ. Left ventricular end systolic volume as the major determinant of survival after recovery from myocardial infarction. Circulation 1987; 76:44-51.

13. Nelissen-Vrancken H, Debets J, Snoeckx L, Daemen M, Smits J. Time-related normalization of maximal coronary flow in isolated perfused hearts of rats with myocardial infarction. Circulation 1996;93:349-355.

14. Kalkman EAJ, Bilgin YM, van Haren P, van Suylen R-J, Saxena PR, Schoemaker RG.Determinants of coronary reserve in rats subjected to coronary artery ligation or aortic banding. Cardiovasc Res 1996.

15. Heymans S, Luutun A, Nuyens D, et al. Inhibition of plasminogen activators or matrix metalloproteinases prevents cardiac rupture but impairs therapeutic angiogenesis and causes cardiac failure. Nat Med 1999;5:1135-1142.

16. Hochman JS and Choo H. Limitation of myocardial infarct expansion by reperfusion independent of myocardial salvage. Circulation 1987;75:299-306.

17. White HD, Cross DB, Elliot JM, et al. Long-term prognostic importance of patency of the infarct-related coronary artery after thrombolytic therapy for myocardial infarction. Circulation 1994;89:61-67.

18. Nidorf SM, Siu SC, Galambos G, Weyman AE, Picard MH. Benefit of late coronary reperfusion on ventricular morphology and function after myocardial infarction. J Am Coll Cardiol 1992;20:307-313.

19. MacLellan WR and Schneider MD. Genetic dissection of cardiac growth control pathways Annu. Rev. Physiol. 2000. 62:289-320.

20. Soonpaa MH, Field LJ. Assessment of cardiomyocyte DNA synthesis in normal and injured adult mouse hearts. *Am J Physiol* **272**, H220-6 (1997).

21. Kellerman S, Moore JA, Zierhut W, Zimmer HG, Campbell J, Gerdes AM. Nuclear DNA content and nucleation patterns in rat cardiac myocytes from different models of cardiac hypertrophy. *J Mol Cell Cardiol* **24**, 497-505 (1992).

22. Hill MF, Singal PK. Right and left myocardial antioxidant responses during heart failure subsequent to myocardial infarction. Circulation 1997 96:2414-20.

23. Li, Y., Jenkins, C.W. , Nichols, M.A. and Xiong, Y. (1994) Cell cycle expression and p53 regulation of the cyclin-dependent kinase inhibitor p21. Oncogene, 9, 2261-2268

24. Steinman, R.A. , Hoffman, B. , Iro, A. , Guillouf, C. , Liebermann, D.A. and El-Houseini, M.E. (1994) Induction of p21 (WAF1/CIP1) during differentiation. Oncogene, 9, 3389-3396.

25. Halevy, O. , Novitch, B.G. , Spicer, D.B. , Skapek, S.X. , Rhee, J. , Hannon, G.J. , Beach, D. and Lassar, A.B. (1995) Correlation of terminal cell cycle arrest of skeletal muscle with induction of p21 by MyoD. Science, 267, 1018-1021.

26. Andres, V. and Walsh, K. (1996) Myogenin expression, cell cycle withdrawal and phenotypic differentiation are temporally separable events that precedes cell fusion upon myogenesis. J. Cell Biol., 132, 657-666.

27. Tsurimoto, T. PCNA Binding Proteins. Frontiers in Bioscience, 4:849-858, 1999.

28. Levkau B, Koyama H, Raines EW, Clurman BE, Herren B, Orth K, Roberts JM, Ross R. Cleavage of p21cip1/waf1 and p27 kip1 mediates apoptosis in endothelial cells through activation of cdk2: role of a caspase cascade. Mol Cell. 1998;1:553–563.

29. Adachi S, et al. Cyclin A/cdk2 activation is involved in hypoxia-induced apoptosis in cardiomyocytes Circ Res. 88:408, 2001.

30. Anversa P, and Nadal-Ginard B. Myocyte renewal and ventricular remodelling Nature 415, 240 - 243, 2002.

31. Kajstura J, Leri A, Finato N, di Loreto N, Beltramo CA, Anversa P. Myocyte proliferation in end-stage cardiac failure in humans. Proc Natl Acad Sci USA 95, 8801-8805 (1998).

32. Beltrami AP, et al. Evidence that human cardiac myocytes divide after myocardial infarction. *N Engl J Med* 344, 1750-7 (2001).

33. Makino S, Fukuda K, Miyoshi S, et al. Cardiomyocytes can be generated from marrow stromal cells in vitro. J Clin Invest 1999, 103:697-705.

34. Tomita S, Li R-K, Weisel RD, et al. Autologous transplantation of bone marrow cells improves damaged heart function. Circulation 1999; 100:II-247.

35. Pittenger MF, Mackay AM, Beck SC, et al. Multilineage potential of adult human mesenchymal stem cells. Science 1999, 284:143-147.

36. Liechty KW, MacKenzie TC, Shaaban AF, et al. Human mesenchymal stem cells engraft and demonstrate site-specific differentiation after in utero transplantation in sheep. Nat Med 2000, 6:1282-6.

37. Kehat I, Kenyagin-Karsenti D, Snir M, Segev H, Amit M, Gepstein A, Livne E, Binah O, Itskovitz-Eldor J, Gepstein L. Human embryonic stem cells can differentiate into myocytes with structural and functional properties of cardiomyocytes. J Clin Invest. 2001; 108: 407-14.

38. Klug MG, Soonpaa MH, Koh GY, Field LJ (1996) Genetically selected cardiomyocytes from differentiating embryonic stem cells form stable intracardiac grafts. J Clin Invest 98:216-224.

39. Hescheler J, Fleischmann BK, Wartenberg M, Bloch W, Kolossov E, Ji G, Addicks K, Sauer H (1999) Establishment of ionic channels and signalling cascades in the embryonic stem cell-derived primitive endoderm and cardiovascular system. Cells Tissues Organs 165:153-164.

40. DA Taylor, BZ Atkins, P Hungspreugs, TR Jones, MC Reedy, KA Hutcheson, DD Glower, WE. Regenerating functional myocardium: improved performance after skeletal myoblast transplantation. Nature Medicine 1998, 4: 929-933.

41. L Field: Future therapy for cardiovascular disease. In Proceedings of the NHLBI Workshop Cell Transplantation: Future Therapy for Cardiovascular Disease? Columbia, MD; 1998.

42. CE Murry, RW Wiseman, SM Schwartz, SD Hauschka: Skeletal myoblast transplantation for repair of myocardial necrosis. J Clin Invest 1996, 98: 2512-2523.

43. BZ Atkins, MT Hueman, JM Meuchel, MJ Cottman, KA Hutcheson, DA Taylor: Myogenic cell transplantation improves in vivo regional performance in infarcted rabbit myocardium. J Heart Lung Transpl 1999, 18: 1173-1180.

44. Suzuki K, Smolenski RT, Jayakumar J, Murtuza B, Brand NJ, Yacoub MH (2000) Heat shock treatment enhances graft cell survival in skeletal myoblast transplantation to the heart. Circulation 102:III216-221.

45. Pouzet B, Ghostine S, Vilquin J-T, Garcin I, Scorsin M, Hagege AA, Duboc D, Schwartz K, Menasche P (2001) Is skeletal myoblast transplantation clinically relevant in the era of angiotensin-converting enzyme inhibitors? Circulation 104:I223-228.

46. Menasche P, Hagege AA, Scorsin M, Pouzet B, Desnos M, Duboc D, Schwartz K, Vilquin J-T, Marolleau J-P (2001) Myoblast transplantation for heart failure. Lancet 347:279.

47. Beauchamp JR, Morgan JE, Pagel CN, Partridge TA (1999) Dynamics of myoblast transplantation reveal a discrete minority of precursors with stem cell-like properties as the myogenic source. J Cell Biol 144:1113-1122.

48. Smythe GM, Grounds MD (2000) Exposure to tissue culture conditions can adversely affect myoblast behaviour in vivo in whole muscle grafts: implications for myoblast transfer therapy. Cell Transplant 9:379-393.

49. Smythe GM, Hodgetts SI, Grounds MD (2001) Problems and solutions in myoblast transfer therapy. J Cell Mol Med 5:33-47.

50. Hodgetts SI, Beilharz MW, Scalzo T, Grounds MD (2000) Why do cultured transplanted myoblasts die in vivo? DNA quantification shows enhanced survival of donor male myoblasts in host mice depleted of CD4+ and CD8+ or NK1.1+ cells. Cell Transplant 9:489-502.

51. Maier S, Tertilt C, Chambron N, Gerauer K, Huser N, Heidecke C-D, Pfeffer K (2001) Inhibition of natural killer cells results in acceptance of cardiac allografts in CD28-/- mice. Nature Med 7:557-562.

52. Zhang M, Methot D, Poppa V, Fujio Y, Walsh K, Murry CE (2001) Cardiomyocyte grafting for cardiac repair: graft cell death and anti-death strategies. J Mol Cell Cardiol 33:907-921.

53. Jackson KA, Majka SM, Wang H, Pocius J, Hartley CJ, Majesky MW, Entman ML, Michael LH, Hirschi KK, Goodell MA (2001) Regeneration of ischemic cardiac muscle and vascular endothelium by adult stem cells. J Clin Invest 107:1395-1402.

54. Orlic D, Kajstura J, Chimenti S, et al. Bone marrow cells regenerate infarcted myocardium. Nature 2001, 410:701-705.

55. Tavian M, Coulombel L, Luton D, San Clemente H, Dieterlen-Lievre F, Peault B. Aorta-associated CD34 hematopoietic cells in the early human embryo. Blood 1996;87:67-72.

56. Jaffredo T, Gautier R, Eichmann A, Dieterlen-Lievre F. Intraaortic hemopoietic cells are derived from endothelial cells during ontogeny. Development 1998;125:4575-4583.

57. Kennedy M, Firpo M, Choi K, Wall C, Robertson S, Kabrun N, Keller G. A common precursor for primitive erythropoiesis and definitive haematopoiesis. Nature 1997;386:488-493.

58. Choi K, Kennedy M, Kazarov A, Papadimitriou, Keller G. A common precursor for hematopoietic and endothelial cells. Development 1998;125:725-732.

59. Elefanty AG, Robb L, Birner R, Begley CG. Hematopoietic-specific genes are not induced during in vitro differentiation of scl-null embryonic stem cells. Blood 1997;90:1435-1447.

60. Labastie M-C, Cortes F, Romeo P-H, Dulac C, Peault B. Molecular identity of hematopoietic precursor cells emerging in the human embryo. Blood 1998;92:3624-3635.

61. Tsai FY, Keller G, Kuo FC, Weiss M, Chen J, Rosenblatt M, Alt FA, Orkin SH. An early hematopoietic defect in mice lacking the transcription factor GATA-2. Nature 1994;371:221-225.

62. Ogawa M, Kizumoto M, Nishikawa S, Fujimoto T, Kodama H, Nishikawa SI. Blood 1999;93:1168-1177.

63. Asahara, T. et al. Isolation of putative progenitor cells for endothelial angiogenesis. Science 1997; 275:964-967.

64. Folkman, J. Therapeutic angiogenesis in ischemic limbs. Circulation 1998; 97:108-110.

65. Takahashi, T. et al. Ischemia- and cytokine-induced mobilization of bone marrow-derived endothelial progenitor cells for neovascularization. Nat. Med. 1999; 5:434-438.

66. Kalka, C. et al. Transplantation of ex vivo expanded endothelial progenitor cells for therapeutic neovascularization. Proc. Natl. Acad. Sci. USA 2000; 97:3422-3427.

67. Rafii S, Shapiro F, Rimarachin J, Nachman R, Ferris B, Weksler B, Moore AS, Asch AS. Isolation and characterization of human bone marrow microvascular endothelial cells: hematopoietic progenitor cell adhesion. Blood 1994;84:10-19.

68. Shi Q, Rafii S, Wu MH-D, et al. Evidence for circulating bone marrow-derived endothelial cells. Blood 1998;92:362-367.

69. Lin Y, Weisdorf DJ, Solovey A, Hebbel RP. Origins of circulating endothelial cells and endothelial outgrowth from blood. J Clin Invest 2000;105:71-77.

70. Kocher AA, Schuster MD, Szabolcs MJ, Takuma S, Burkhoff D, Wang J, Homma S, Edwards NM, Itescu S. Neovascularization of ischemic myocardium by human bone-marrow-derived angioblasts prevents cardiomyocyte apoptosis, reduces remodeling and improves cardiac function. Nature Med. 2001; 7: 430-6.

71. Asahara T, Takahashi T, Masuda H, Kalka C, Chen D, Iwaguro H, Inai Y, Silver M, Isner JM. VEGF contributes to postnatal neovascularization by mobilizing bone marrow-derived endothelial progenitor cells. EMBO J 1999; 18:3964-3972.

72. Kocher A, Schuster M, Szabolcs M, Itescu S. Cardiomyocyte regeneration after neovascularization of ischemic myocardium by human bone marrow-derived angioblasts. Nature Medicine 2002, in press.

73. McEwan PE, Gray GA, Sherry L, Webb DJ, Kenyon CJ. Differential effects of angiotensin II on cardiac cell proliferation and intramyocardial perivascular fibrosis in vivo. Circulation 1998;98:2765-2773.

74. Kawano H, Do YS, Kawano Y, Starnes V, Barr M, Law RE, Hsueh WA. Angiotensin II has multiple profibrotic effects in human cardiac fibroblasts. Circulation 2000;101:1130-1137.

75. Pfeffer MA, Braunwald E, Moye LA, et al. Effect of captopril on mortality and morbidity in patients with left ventricular dysfunction after myocardial infarction. Results of the survival and ventricular enlargement trial. The SAVE investigators. N Engl J Med 1992;327:669-677.

76. Pitt B, Segal R, Martinez FA, et al. Randomised trial of losartan versus captopril in patients over 65 with heart failure (Evaluation of Losartan in the Elderly Study, ELITE). Lancet 1997;349:747-752.

Appendix N.

The Biology of Nuclear Cloning and the Potential of Embryonic Stem Cells for Transplantation Therapy

RUDOLF JAENISCH, M.D.

Whitehead Institute, 9 Cambridge Center, Cambridge, MA

SUMMARY

An emerging consensus is that somatic cell nuclear transfer (SCNT) for the purpose of creating a child (also called "reproductive cloning") is not acceptable for both moral and scientific reasons. In contrast, SCNT with the goal of generating an embryonic stem cell line ("therapeutic cloning") remains a controversial issue. Although therapeutic cloning holds the promise of yielding new ways of treating a number of degenerative diseases, it is not acceptable to many because the derivation of an embryonic stem cell line from the cloned embryo (an essential step in this process) necessarily involves the loss of an embryo and hence the destruction of potential human life.

In this article, I will develop two main arguments that are based on the available scientific evidence. 1) In contrast to an embryo derived by *in vitro* fertilization (IVF), a cloned embryo has little if any potential to ever develop into a normal human being. This is because, by circumventing the normal processes of gametogenesis and fertilization, nuclear cloning prevents the proper reprogramming of the clone's genome, which is a prerequisite for development of an embryo to a normal individual. It is unlikely that these biological barriers to normal development can be solved in the foreseeable future. Therefore, from a biologist's point of view, the cloned human embryo, used for the derivation of an embryonic stem cell and the subsequent therapy of a needy patient, has *little if any potential* to create a normal human life. 2) Embryonic stem cells developed from a cloned embryo are functionally indistinguishable from those that have been generated from embryos derived by in vitro fertilization (IVF). Both types of embryonic stem cells have an *identical potential* to serve as a source for therapeutically useful cells.

385

It is crucial that the ongoing debate on the possible therapeutic application of SNCT is based on biological facts. The goal of this article is to provide such a basis and to contribute to a more rational discussion that is founded on scientific evidence rather than on misconceptions or misrepresentations of the available scientific data.

I. Introduction

It is important to distinguish between "reproductive cloning" and "nuclear transplantation therapy" (also referred to as "SCNT" or "therapeutic cloning"). In reproductive cloning a cloned embryo is generated by transfer of a somatic nucleus into an enucleated egg with the goal to create a cloned individual. In contrast, the purpose of nuclear transplantation therapy is to generate an embryonic stem cell line (referred to as "ntES cells") that is "tailored" to the needs of a patient who served as the nuclear donor. The ntES cells could be used as a source of functional cells that would be suitable for treating an underlying disease by transplantation.

There is now experience from cloning of seven different mammalian species that is relevant for three main questions of public interest: 1) Would a cloned human embryo be "normal"? 2) Could the problems currently seen with cloning be solved in the foreseeable future? 3) Would ES cells derived from a cloned human embryo be "normal" and useful for cell therapy? The arguments advanced in this article are strictly based on molecular and biological evidence that has been obtained largely in the mouse. I will not attempt to review the cloning literature but only refer to selected papers on cloned mice. The relevant literature on cloning of mammals can be found in recent reviews (Byrne and Gurdon, 2002; Gurdon, 1999; Hochedlinger and Jaenisch, 2002b; Oback and Wells, 2002; Rideout et al., 2001; Wilmut, 2001; Young et al., 1998).

II. Most cloned animals die or are born with abnormalities

The majority of cloned mammals derived by nuclear transfer (NT) die during gestation, and those that survive to birth frequently display "Large Offspring Syndrome", a neonatal phenotype characterized by respiratory and metabolic abnormalities and enlarged and dysfunctional placentas (Rideout et al., 2001; Young et al., 1998). In order for a donor nucleus to support development into a clone, it must be reprogrammed to a state compatible with embryonic development. The transferred nucleus must properly activate genes important for early embryonic development and also suppress differentiation-associated genes that had been transcribed in the original donor cell. Inadequate "reprogramming"[1] of the donor nucleus is most likely the principal reason for developmental failure of clones. Since few clones survive to birth, the question remains whether survivors are fully normal or merely the least affected animals carrying through to adulthood despite harboring subtle abnormalities that originate in faulty reprogramming but that are not severe enough to interfere with survival to birth or beyond.

III. Reprogramming of the genome during normal development and after nuclear transfer

The fundamental difference between nuclear cloning and normal fertilization is that the nucleus used in nuclear cloning comes from a somatic (body) cell that has not undergone the developmental events required to produce the egg and sperm. Nuclear cloning involves the transplantation of a somatic nucleus into the oocyte from which the nucleus has been removed. However, the genes in the somatic nucleus are not in the same state as those in the fertilized egg because nuclear transplantation short-cuts the complex process of egg and sperm maturation which involves extensive "reprogramming" of the genome, a process that shuts some genes off and leaves others on. Reprogramming during gametogenesis prepares the genome of the two mature gametes with the ability to activate faithfully the genetic program that ensures normal embryonic development when they combine at fertilization (Fig 1a). This reprogramming of the genome begins at gastrulation, when primordial germ cells (PGCs) are formed, and continues during differentiation into mature gametes resulting, in a radically different chromatin configuration of sperm and oocyte (Rideout et al., 2001).

Experiments have shown that uniparental embryos (embryos whose genomes are derived solely from either the maternal or paternal parent) do not develop normally. Uniparental embryos first seem normal; they direct cleavage (early development to the blastocyst stage) despite profound differences in their epigenetic organization (Reik et al., 2001). However, uniparental embryos fail soon after the implantation of the embryo into the wall of the uterus, indicating that both parental genomes are needed and functionally complement each other beginning at this later step of embryogenesis. Presumably, the different epigenetic organization of the two genomes is crucial for achieving normal development.

Moreover, it has been well established that the imbalance of imprinted [2] gene expression represents an important cause of embryonic failure.

In order for cloned embryos to complete development, genes normally expressed during embryogenesis but silent in the somatic donor cell, must be reactivated. This complex process of epigenetic[3] remodeling (i.e., the reconfiguration of the genome by turning on and turning off specific genes) that occurs during gametogenesis in normal development ensures that the genome of the zygote can faithfully activate early embryonic gene expression (Fig 1a). In a cloned embryo, reprogramming, which in normal gametogenesis requires months to years to complete, must occur in a cellular context radically different from gametogenesis and within the short interval (probably within hours) between transfer of the donor nucleus into the egg and the time when zygotic transcription becomes necessary for further development. Given these radically different conditions, one can envisage a spectrum of different outcomes to the reprogramming process ranging from (i) no reprogramming of the genome, resulting in immediate death of the NT embryo; through (ii) partial reprogramming, allowing initial survival of the clones, but resulting in an abnormal phenotype and/or lethality at various stages of development; to (iii) faithful reprogramming producing fully normal animals (Fig 1b). The phenotypes observed over the past five years in cloned embryos and newborns suggest that complete reprogramming is the exception, if it occurs at all.

IV. Development of clones depends on the differentiation-state of the donor nucleus

The majority of cloned embryos fail at an early step of embryonic development, soon after implantation in the wall of the uterus, an early step of embryonic development (Hochedlinger and Jaenisch, 2002b; Rideout et al., 2001). Those that live to birth often display common abnormalities irrespective of the donor cell type (Table 1). In addition to symptoms referred to as "Large Offspring Syndrome", neonate clones often suffer from respiratory distress and kidney, liver, heart or brain defects (Cibelli et al., 2002). However, the abnormalities characteristic of cloned animals are not inherited by their offspring (Tamashiro et al., 2002), indicating that epigenetic aberrations (i.e., failure of genome reprogramming) rather than genetic aberrations (changes in the sequences within the DNA) are the cause.

The efficiency of creating cloned animals is strongly influenced by the differentiation-state of the donor nucleus (Table 1). In the mouse, for example, only 1-3% of cloned blastocysts derived from somatic donor nuclei, e.g., those prepared from–fibroblasts or cumulus cells, will develop to adult cloned animals (Hochedlinger and Jaenisch, 2002b). In certain cases, such as those using terminally differentiated B or T cell donor nuclei, the efficiency of cloning is so low as to preclude the direct derivation of cloned animals. In stark contrast to these examples, cloning using donor nuclei prepared from embryonic stem (ES) cells is significantly more efficient (between 15 and 30 %, Table 1). This correlation with differentiation-state suggests that embryonic nuclei require less reprogramming of their genome, ostensibly because the genes essential for embryonic development are already active and need not be reprogrammed. In fact, the nucleus of an embryonic cell such as an ES cell may well have the same high efficiency to generate postnatal mice after nuclear transfer as the nucleus prepared from a recently fertilized egg (Table 1, compare Fig. 4). Nonetheless, most if not all mice that have been cloned from ES cell donor nuclei, in contrast to mice derived through natural fertilization from the zygote, are abnormal, indicating that the processes of gametogenesis (development of sperm and of egg) and fertilization endows the zygote nucleus with the ability to direct *normal* development. In summary, these data indicate that the potential of a nucleus to generate a normal embryo is lost progressively with development.

V. Adult cloned animals: how normal are they?

The observation that apparently healthy adult cloned animals have been produced in seven mammalian species (albeit at low efficiency) is being used by some as a justification for attempting to clone humans. In fact, even those that survive to adulthood, such as Dolly, may succumb relatively early in adulthood because of numerous health problems. Insights into the mechanisms responsible for clone failure before and after birth have come from molecular and biological analyses of mouse clones that have reached (i) the blastocyst stage, (ii) the perinatal period and (iii) adulthood.

(i) Most clones fall short of activating key embryonic genes and fail early.
As stated above in order for clones to develop, the genes that are normally expressed during embryogenesis, but are silent in the somatic donor cell, must be reactivated (Hochedlinger and Jaenisch,

2002b; Rideout et al., 2001). It is the failure to activate key "embryonic" genes that are required for early development that leads to the demise of most clones just after implantation. Recently, a set of about 70 key embryonic genes termed "Oct-4 like" genes have been identified that are active in early embryos but not in somatic donor cells. Importantly, the failure to faithfully activate this set of genes can be correlated with the frequent death of cloned animals during the immediate post-implantation period (Bortvin et al., 2003). These results define "faulty reprogramming" as the cause of early demise of cloned embryos as the failure to reactivate key embryonic genes that are silent in the donor cell.

(ii) Newborn clones misexpress hundreds of genes.
Clones that survive to birth suffer from serious problems, many of which appear to be due to an abnormal placenta. The most common phenotypes observed in animals cloned from either somatic or ES cell nuclei are fetal growth abnormalities such as increased placental and birth weight. This has suggested that surviving clones had accurately reprogrammed the "Oct-4 like" genes that are essential for the earliest stages of development, i.e. those immediately following implantation of the embryo into the uterus. The abnormal phenotype of those clones that do survive through these early stages and develop to birth indicates that other genes that are important for later stages of development but are not essential for early survival are not correctly reprogrammed. To assess the extent of abnormal expression of various genes in the cells of clones, global gene expression has been assessed by microarray analysis of RNA prepared from the placentas and livers of neonatal cloned mice, i.e., clones that survived development and were viable at birth; these clones had been derived by nuclear transfer (NT) of nuclei prepared either from cultured ES cells or from freshly isolated cumulus cells (somatic cells that surround the egg) (Humpherys et al., 2002). Direct comparison of gene expression profiles of over 10,000 genes (of the 30,000 or so in the mammalian genome) showed that for both classes of cloned neonatal mice, approximately 4% of the expressed genes in their placentas differed dramatically in expression levels from those in controls, and that the majority of abnormally expressed genes were common to both types of clones. When imprinted genes, a class of genes that express only one allele (either from maternal or paternal origin), were analyzed, between 30 and 50% were not correctly activated. These data represent strong molecular evidence that cloned animals, even those that survive to birth, suffer from serious gene expression abnormalities.

(iii) Cloned animals develop serious problems with age
The generation of adult and seemingly healthy adult cloned animals has been taken as evidence that normal cloned animals can be generated by nuclear transfer, albeit with low efficiency. Indeed, a routine physical and clinical laboratory examination of 24 cloned cows of 1 to 4 years of age failed to reveal major abnormalities (Lanza et al., 2001). Cloned mice of a corresponding age as that of the cloned cows (2 - 6 months in mice vs. 1 - 4 years in cows) also appear "normal" by superficial inspection. However, when cloned mice were aged, serious problems, not apparent at younger ages, became manifest. One study found that the great majority of cloned mice died significantly earlier than normal mice, succumbing with immune deficiency and serious pathological alterations in multiple organs (Ogonuki et al., 2002). Another study found that aged cloned mice became overweight with major metabolic disturbances (Tamashiro et al., 2002). Thus, serious abnormalities in cloned animals may often become manifest only when the animals age.

Firm evidence about aging and "normalcy" of cloned farm animals is incomplete or anecdotal because cloned animals of these species are still comparatively young (relative to their respective normal life span). For example, the premature death of Dolly (Giles and Knight, 2003) is entirely consistent with serious abnormalities in cloned sheep that become manifest only at later ages. Also, two of the analyzed cloned cows developed disease soon after the study on "healthy and normal cattle" (Lanza et al., 2001) had appeared: one animal developed an ovarian tumor and another one suffered brain seizures (J. Cibelli, pers. comm.). While it cannot be ruled out that these are "spontaneous" maladies unconnected with the cloning procedure, a more likely alternative is that these problems were direct consequences of the nuclear transfer procedure.

(iv) Are there any "normal" clones?
It is a key question in the public debate whether it is ever possible to produce a normal individual by nuclear cloning, even if only with low efficiency. The available evidence suggests that it may be difficult if not impossible to produce normal clones for the following reasons: 1) As summarized above, all analyzed clones at birth showed dysregulation of hundreds of genes. The development of clones to birth and beyond despite widespread epigenetic abnormalities suggests that mammalian development can tolerate dysregulation of many genes. 2) Some clones survive to adulthood by compensating for gene dysregulation. Though this "compensation" assures *survival*, it may not prevent maladies to

become manifest at later ages. Therefore, most if not all clones are expected to have at least subtle abnormalities that may not be so severe as to result in an obvious phenotype at birth but will cause serious problems later as seen in aged mice. Clones may just differ in the extent of abnormal gene expression: if the key "Oct-4 like" genes are not activated, clones die immediately after implantation. If those genes are activated, the clone may survive to birth and beyond.

As schematically shown in Fig. 2, the two stages when the majority of clones fail are immediately after implantation and at birth. These are two critical stages of development that may be particularly vulnerable to faulty gene expression. Once cloned newborns have progressed through the critical perinatal period, various compensatory mechanisms may counterbalance abnormal expression of other genes that are not essential for the subsequent postnatal survival. However, the stochastic occurrence of disease and other defects at later age in many or most adult clones implies that such compensatory mechanisms do not guarantee "normalcy" of cloned animals. Rather, the phenotypes of surviving cloned animals may be distributed over a wide spectrum from abnormalities causing sudden demise at later postnatal age or more subtle abnormalities allowing survival to advanced age (Fig. 2). These considerations illustrate the complexity of defining subtle gene expression defects and emphasize the need for more sophisticated test criteria such as environmental stress or behavior tests. However, the available evidence suggests that truly normal clones may be the exception.

It should be emphasized that "abnormality" or "normalcy" is defined here by molecular and biological criteria that distinguish cloned embryos or animals from control animals produced by sexual reproduction. The most informative data for the arguments presented above come from the mouse. There is, however, every reason to believe that these difficulties associated with producing mice and a variety of other mammalian embryos by nuclear transplantation will also afflict the process of human reproductive cloning (Jaenisch and Wilmut, 2001).

(v) Is it possible to overcome the problems inherent in reproductive cloning?
It is often argued that the "technical" problems in producing normal cloned mammals will be solved by scientific progress that will be made in the foreseeable future. The following considerations argue that this may not be so.

A principal biological barrier that prevents clones from being normal is the "epigenetic" difference (such as distinct patterns of DNA methylation[4]) between the chromosomes inherited from mother and from father, i.e. the difference between the "maternal" and the "paternal" genome of an individual. Such methylation of specific DNA sequences is known to be responsible for shutting down the expression of nearby genes. Parent-specific methylation marks are responsible for the expression of imprinted genes and cause only one copy of an imprinted gene, derived either from sperm or egg, to be active while the other allele is inactive (Ferguson-Smith and Surani, 2001). When sperm and oocyte genomes are combined at fertilization, the parent-specific marks established during oogenesis and spermatogenesis persist in the genome of the zygote (Fig 3A). Of interest for this discussion is that within hours after fertilization, most of the global methylation marks (with the exception of those on imprinted genes) are stripped from the sperm genome whereas the genome of the oocyte is resistant to this active demethylation process (Mayer et al., 2000). This is because the oocyte genome is in a different "oocyte-appropriate" epigenetic state than the sperm genome. The oocyte genome becomes only partially demethylated within the next few days by a passive demethylation process. The result of these post-fertilization changes is that the two parental genomes are epigenetically different (as defined by the patterns of DNA methylation) in the later stage embryo and remain so in the adult in imprinted as well as non-imprinted sequences.

In cloning, the epigenetic differences that are established during gametogenesis may be erased because both parental genomes of the somatic donor cell are introduced into the egg from the outside and are thus exposed equally to the demethylation activity present in the egg cytoplasm (Fig 3B). This predicts that imprinted genes should be particularly vulnerable to inappropriate methylation and associated dysregulation in cloned animals. The results summarized earlier are consistent with this prediction. For cloning to be made safe, the two parental genomes of a somatic donor cell would need to be physically separated and separately treated in an "oocyte-appropriate" and a "sperm-appropriate" way, respectively. At present, it seems that this is the only rational approach to guarantee the creation of the epigenetic differences that are normally established during gametogenesis. Such an approach is beyond our present abilities. These considerations imply that *serious biological barriers* exist that interfere with faithful reprogramming after nuclear transfer. It is a safe conclusion that these biological barriers represent a major stumbling block to efforts aimed at making nuclear cloning a safe reproductive procedure for the foreseeable future.

It has been argued that the problems in mammalian cloning are similar to those encountered with IVF 30 years ago: Thus, following this argument, the methods of culture and embryo manipulations just would need to be improved to develop reproductive cloning into a safe reproductive technology that is as acceptable as IVF. This argument appears to be fundamentally flawed. It is certainly correct that merely "technical" problems needed to be solved to make IVF efficient and safe. It is important to distinguish between the perfection of technical skills to imitate a biological event and the development of wholly new science to overcome the blocks to events that have severe biological restrictions. Nuclear cloning faces serious biological barriers that cannot be addressed by mere adjustments in experimental technique. Indeed, since the birth of Dolly *no* progress has been made in solving any of the underlying *biological issues* of faulty gene reprogramming and resulting defective development.

VI. Therapeutic applications of SCNT

(i) Reproductive cloning vs. therapeutic cloning
In spite of the biological and ethical barriers associated with reproductive cloning, nuclear transfer technology has significant therapeutic potential that is within our grasp. There is an enormous distinction between the goals and the end product of these two technologies. The purpose of reproductive cloning is to generate a cloned embryo that is then implanted in the uterus of a female to give rise to a cloned individual. In contrast, the purpose of nuclear transplantation therapy is to generate an embryonic stem cell line that is derived from a patient (referred to as "ntES cells") and can be used subsequently for tissue replacement.

Many scientists recognize the potential of NtES cells for organ transplantation (for recent review see (Hochedlinger and Jaenisch, 2003). This procedure is currently complicated by immune rejection due to immunological incompatibility. Thus, virtually all organ transplants undertaken at present involve the use of donor organs that are recognized as foreign by the immune systems of the recipient and thus are targeted for destruction by these immune systems. To treat this "host versus graft" disease, immunosuppressive drugs are routinely given to transplant recipients in order to suppress this organ rejection. Such immunosuppressive treatment has serious side effects including increased risks of infections and malignancies. In principle, ES cells can be created from a patient's nuclei using nuclear transfer. Because ntES cells will be genetically identical to the patient's cells, the risks of immune rejection and the requirement for immunosuppression are eliminated. Moreover, ES cells provide a renewable source of replacement tissue allowing for repeated therapy whenever needed. Finally, if ES cells are derived from a patient carrying a known genetic defect, the mutation in question can be corrected in the ntES cells using standard gene targeting methods before introducing these ES cells (or derived tissue-specific stem cells) back into the patient's body.

(ii) Combining nuclear cloning with gene and cell therapy

In a "proof of principle" experiment, nuclear cloning in combination with gene and cell therapy has been used to treat a mouse genetic disorder that has a human counterpart (Figure 4). To do so, the well-characterized *Rag2* mutant mouse was used as "patient" (Rideout et al., 2002). This mutation causes *severe combined immune deficiency (SCID)*, because the enzyme that catalyzes immune receptor rearrangements in lymphocytes is non-functional. Consequently, these mice are devoid of mature B and T cells, a disease resembling human *Omenn syndrome* (Rideout et al., 2002).

In a first step, somatic (fibroblast) donor cells were isolated from the tails of *Rag2*-deficient mice and their nuclei were injected into enucleated eggs. The resultant embryos were cultured to the blastocyst stage and isogenic ES cells were isolated. Subsequently, one of the mutant *Rag2* alleles was targeted by homologous recombination in ES cells to restore normal *Rag2* gene structure and function. In order to obtain somatic cells for treatment, these genetically repaired ES cells were differentiated into embryoid bodies and further into hematopoietic precursors by expressing *HoxB4*, a transcription factor that is responsible for programming the behavior of the hematopoietic stem cells, i.e., those cells that are able to generate the full range of red and white cells in the blood. Resulting hematopoietic precursors were transplanted into irradiated *Rag2*-deficient animals in order to treat the disease caused by their *Rag2* mutation. Initial attempts to engraft these cells were, however, unsuccessful because of an increased level of natural killer (NK) cells in the *Rag* mutant host. ES cell derived hematopoietic cells express low levels of the MHC antigens and thus are a preferred target for NK mediated destruction. Elimination of NK cells by antibody depletion or genetic ablation allowed the ntES cells to efficiently populate the myeloid and to a lesser degree the lymphoid lineages of these mice. Functional B and T cells that had undergone proper rearrangements of their immunoglobulin and T cell receptor alleles as well as serum immunoglobulins were detected in the transplanted mutants. Hence, important cellular components of the immune system were restored in mice that previously were unable to produce these cells.

This experiment demonstrated that embryonic stem (ES) cells derived by NT from somatic cells of a genetically afflicted individual can be combined with gene therapy to treat the underlying genetic disorder. Because Rag2 deficiency causes an increase in NK activity and necessitated the elimination of NK cells prior to transplantation in the above-described experiments, some have concluded that "The experiment failed to show success with therapeutic cloning" (Coalition and Ethics, 2003) and that "This indicates that the only successful therapy using cloned embryos would be through 'reproductive' cloning, to produce born clones who can serve as tissue donors for patients" (Prentice, 2002). This is a troubling misinterpretation of the data. (i) It has been shown that ES cell-derived hematopoietic cells can successfully engraft and rescue lethally irradiated mice indicating that increased NK activity is a peculiarity of Rag2-deficiency (Kyba et al., 2002). Therefore, it would seem that for most diseases, no anti-NK treatment would be required to assure engraftment of ES cell-derived somatic cells. (ii) It is correct that treatment of a human patient with *Omenn syndrome,* which is equivalent to Rag2 deficiency, by SCNT may also require anti-NK treatment to *transiently* reduce NK activity. This would allow the transplanted cells to engraft as in the mouse experiment. Once these cells are successfully engrafted, there is every reason to believe that such anti-NK treatment would no longer be necessary.

In conclusion, the mouse experiment indicates that, unlike the situation with reproductive cloning, no *biological* barriers exist that in principle prevent the use of SCNT to treat human diseases. The *technical* issues in using SCNT and human stem cells for therapeutic purposes need, however, to be solved, but there are no indications at present that these represent formidable problems that will resist relatively rapid solution.

VII. Faulty reprogramming after nuclear transfer: does it interfere with the therapeutic potential of ES cells derived by SCNT?

As summarized above, most if not all cloned animals are abnormal because of faulty reprogramming after nuclear transfer. Does this epigenetic dysregulation affect the potential of ntES cells to generate functional somatic cells that can be used for cell therapy? To address this question, I will first compare the *in vivo* development of embryos with the *in vitro* process of ES cell derivation from explanted embryos. This will be followed by discussing the epigenetic state of the ES cell genome. Finally, I will contrast the phenotype of cloned mice derived from ES cell donor nuclei with that of chimeric mice generated by injection of ES cells into blastocysts.

(i) The phenotype of an embryo is determined by its donor nucleus
As mentioned repeatedly above, embryos can be derived from the fertilized egg or from a somatic nucleus by SCNT. The potential of the resulting blastocyst, when implanted into the womb, to develop into a fetus and a postnatal animal depends strictly on the nature of the donor nucleus (Fig 5): (i) When derived from the zygote, most embryos develop to birth and generate a normal animal; (ii) Similarly, most blastocysts cloned from an embryonic stem cell donor nucleus develop to birth but, in contrast to the normally fertilized embryo, the great majority of the cloned animals will be abnormal ("Large offspring syndrome") (Eggan et al., 2001; Humpherys et al., 2001); (iii) The great majority of cloned blastocysts derived from somatic donor nuclei such as fibroblasts or cumulus cells will die soon after implantation and only a few clones will survive to birth and these too will be abnormal, suffering once again from the Large offspring syndrome (Wakayama and Yanagimachi, 2001); (iv) Finally, the likelihood of cloned blastocysts derived from another type of somatic donor nuclei - those present in terminally differentiated lymphoid cells - to generate a cloned animal is extremely low and has not been achieved except by using a two step procedure involving the intermediate generation of embryonic stem cells (Hochedlinger and Jaenisch, 2002a). These observations suggest that a blastocyst retains an "epigenetic memory" of its donor nucleus. This memory determines its potential for fetal development: while a fertilized embryo develops normally, any embryo derived by SCNT will be abnormal though the efficiency of a given clone to develop to birth is strongly influenced by the differentiation state of the donor cell (see Table 1). In other words, the cloned embryo after implantation into the womb will be abnormal because the cloned blastocyst retained an epigenetic memory of its donor nucleus and this causes faulty fetal development. This epigenetic memory is erased when a blastocyst, either derived by nuclear cloning or from the fertilized egg, is explanted into tissue culture and grown into an embryonic stem cell. Erasure of the epigenetic memory has major consequences for the "normalcy" of embryonic stem cells.

(ii) The derivation of embryonic stem cells is a highly selective process that erases the "epigenetic memory" of the donor nucleus
Embryonic stem cells, regardless of whether they have been generated from a fertilized egg or by SCNT, are derived from the

cells of a blastocyst that have been explanted and propagated in tissue culture. Of the blastocyst cells that are explanted in this way, those that derive from the portion of the blastocyst termed the inner cell mass (ICM) initially express "key" embryonic genes such as Oct-4. However, soon after explantation, most ICM cells extinguish Oct-4 expression and cease proliferating (Buehr et al., 2003). Only one or a few of the ICM-derived cells will eventually re-express Oct-4 and these few Oct-4-positive cells are those that resume rapid proliferation, yielding the cell populations that we designate as "embryonic stem" cells. These cells represent a cell population that has no equivalent in the normal embryo and may be considered a tissue culture artifact, though a useful one (Fig. 6).

The important point for this discussion is that the propagation of blastocyst cells *in vitro* results in a rare population of surviving cells that have erased the "epigenetic memory" of the donor nucleus. This process results ultimately in ES cells that have, regardless of donor nuclear origin, an identical developmental potential. In other words, ES cells derived from embryos produced by normal fertilization and those produced from cloned embryos are functionally indistinguishable (Hochedlinger and Jaenisch, 2002b; Rideout et al., 2002; Wakayama et al., 2001). Because the ES cells that derive from normally fertilized embryos are able to participate in the generation of all normal embryonic tissues, we can conclude that the ES cells derived from cloned embryos have a similar potential to generate the full range of normal tissues.

(iii) ES cells, epigenetic instability and therapeutic potential
Epigenetic instability appears to be a consistent characteristic of ES cells. This was shown when individual ES cells were analyzed for expression of imprinted genes: even cells in a recently subcloned ES cell line differed strongly in the expression of genes such as H19 or Igf2. The variable expression was correlated with the DNA methylation status of the genes, which switched from an unmethylated to a methylated state between sister cells (Humpherys et al., 2001). This was a surprising result in view of the known potential of ES cells to generate terminally differentiated cells that function normally after transplantation into an animal. Possible explanations include (i) that epigenetic instability in ES cells is a consequence of propagation of cells in tissue culture or (ii) that epigenetic instability is a prerequisite for cells to be pluripotent, i.e., this instability may be a manifestation of a plasticity in the gene expression program that is required to enable the ES cells to generate a wide variety of differentiated cell lineages.

Whatever the explanation for the observed epigenetic instability of ES cells may be, it supports the view that the process of generating ES cells erases all epigenetic memory of the donor nucleus and, as a consequence of the selection process, generates epigenetic instability in the selected cells. In other words, epigenetic instability appears to be an intrinsic characteristic of ES cells regardless of whether derived by SCNT or from a fertilized egg. This is consistent with the conclusion that both types of ES cells have an equivalent potency to generate functional cells in culture and, in the longer term, fully normal differentiated tissues upon implantation of these cells *in vivo*.

(iv) ES cells form normal chimeras but abnormal nuclear clones

As outlined above, faulty reprogramming leads to abnormal phenotypes of cloned mice derived from ES cell donor nuclei. Why is faulty reprogramming and epigenetic instability a problem for reproductive cloning but not for therapeutic applications? The main reason for this seeming paradox is that, in contrast to reproductive cloning, the therapeutic application of NT does not require the formation of a fetus. Therapeutic applications involve the ability of cloned ES cells to form a single tissue or organ, not to recapitulate all of fetal development. For example, normal fetal development requires faithful expression of the imprinted genes. As outlined above, nuclear cloning causes between 30% and 50% of imprinted genes to be dysregulated consistent with the notion that disturbed imprinting is a major contributing factor to clone failure. As most imprinted genes have no known function in the postnatal animal, the dysregulation of imprinting would not be expected to impede functionality of *in vitro* differentiated ES cells because this process does not require the formation of a fetus. Therefore, the functionality of mature cells derived in culture from ES cells would not depend on the faithful reprogramming of the imprinted genes. Dysregulation of some imprinted genes such as *Igf2* are known, however, to cause disease in the adult. Thus, it will be important to test whether dysregulation of such genes has adverse effects on the function of somatic cells derived from ES cells.

When injected into a blastocyst, ES cells form normal chimeras. It appears that the presence of surrounding "normal" cells, i.e. cells that are derived from a fertilized embryo, prevents an abnormal phenotype of the chimera such as the "Large Offspring Syndrome" that is typical for cloned animals. Any therapeutic application creates, of course, a chimeric tissue where cells derived from ntES cells are introduced into a diseased adult individual and interact with surrounding "normal" host cells. Therefore, no phenotypic abnormalities, such as those seen in cloned animals, would be expected in patients transplanted with cells derived from ntES cells.

VIII. SCNT for cell therapy: destruction of potential human life?

A key concern raised against the application of the nuclear transplantation technology for tissue therapy in humans is the argument that the procedure involves the destruction of potential human life. From a biological point of view, life begins with fertilization when the two gametes are combined to generate a new embryo that has a unique combination of genes and has a high potential to develop into a normal baby when implanted into the womb. A critical question for the public debate on SCNT is this one: is the cloned embryo equivalent to the fertilized embryo?

In cloning, the genetic contribution is derived from one individual and not from two. Obviously, the cloned embryo is the product of laboratory-assisted technology, not the product of a natural event. From a biological point of view, nuclear cloning does not constitute the creation of new life, rather the propagation of existing life because no meiosis, genetic exchange and conception are involved. Perhaps more important is, however, the overwhelming evidence obtained from the cloning of seven different mammalian species. As summarized above, the small fraction of cloned animals that survive beyond birth, even if they appear "normal" upon superficial inspection, are likely not so. The important conclusion is that a cloned human embryo would have little if any potential to develop into a normal human being. With other words, the cloned human embryo lacks essential attributes that characterize the beginning of *normal* human life.

Taking into account the potency of fertilized and cloned embryos, the following scenarios regarding their possible fates can be envisaged (Fig. 7). Fertilized embryos that are "left over" from IVF have three potential fates: disposal, generation of normal embryonic stem cells or generation of a normal baby when implanted into the womb. Similarly, the cloned embryo has three potential fates: it can be destroyed or could be used to generate a normal ntES cell line that has the same potential for therapy as an ES cell derived from a fertilized embryo. In contrast to the fertilized embryo, the cloned embryo has little if any potential to ever generate a normal baby. An embryonic stem cell line derived by nuclear may, however, help sustain existing life when used as a source for cell therapy that is "tailored" to the need of the patient who served as its nuclear donor.

If SCNT were accepted as a valid therapeutic option, a major concern of its implementation as medical procedure would be the problem of how to obtain sufficient numbers of human eggs that could be used as recipients. Commercial interests may pressure women into an unwanted role as egg donors. The recent demonstration that embryonic stem cells can be coaxed into a differentiation pathway that yields oocyte-like cells (Hubner et al., 2003) may offer a solution to this dilemma. If indeed functional oocytes could be generated from a generic human ES cell line, sufficient eggs could be generated in culture and serve as recipients for nuclear transfer without the need of a human egg donor. It seems that technical issues, not fundamental biological barriers, need to be overcome so that transplantation therapy can be carried out without the use of human oocytes.

Fig 1: Reprogramming in normal development and nuclear cloning.

    a. The genome of primordial germ cells (PGCs) is hypomethylated ("reset", white boxes). Reprogramming and establishment of parent specific epigenetic marks occurs over the course of gametogenesis so that the genome of sperm and egg are competent to express the genes that need to be activated in early embryonic (box with wavy lines) and later (hatched box) development. During cleavage and early postimplantation development "embryonic" genes, such as Oct 3/4, become activated (black box) and are repressed at later stages (stippled boxes) when tissue specific genes (hatched boxes) are activated in adult tissues (A, B, C). Epigenetic reprogramming of imprinted and non-imprinted genes occurs during gametogenesis in contrast to X inactivation and the readjustment of telomere length which take place postzygotically.

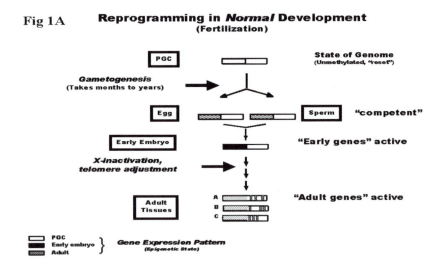

Fig 1A   **Reprogramming in *Normal* Development**
(Fertilization)

    b. Reprogramming of a somatic nucleus following nuclear transfer may result in (i) no activation of "embryonic" genes and early lethality, (ii) faulty activation of embryonic genes and an abnormal phenotype, or (iii) in faithful activation of "embryonic" and "adult" genes and normal development of the clone. The latter outcome is the exception if it occurs at all.

Fig 1B

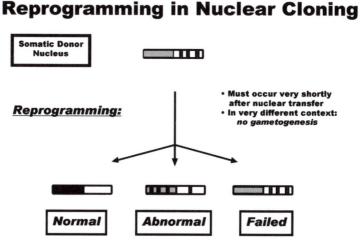

Reprogramming in Nuclear Cloning

Development of Clones

Fig. 2: The phenotypes are distributed over a wide range of abnormalities. Most clones fail at two defined developmental stages, implantation and birth. More subtle gene expression abnormalities result in disease and death at later ages.

Fig 2 **Degree of Abnormalities in Clones: A Continuum without Defined Stages**

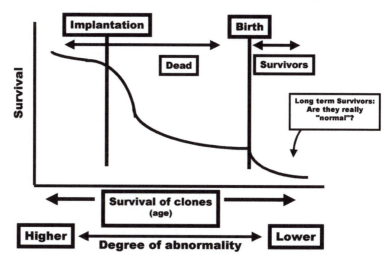

Fig 3: Parental epigenetic differences in normal and cloned animals

A: The genomes of oocyte and sperm are differentially methylated during gametogenesis and are different in the zygote when combined at fertilization. Immediately after fertilization the paternal genome (derived from the sperm) is actively demethylated whereas the maternal genome is only partially demethylated during the next few days of cleavage. This is because the oocyte genome is in a different chromatin configuration and is resistant to the active demethylation process imposed on the sperm genome by the egg cytoplasm. Thus, the methylation of two parental genomes is different at the end of cleavage and in the adult. Methylated sequences are depicted as filled lollipops and unmethylated sequences as empty lollipops.

Fig. 3A

Gametogenesis and Fertilization Establishes Epigenetic Differences in Parental Genomes

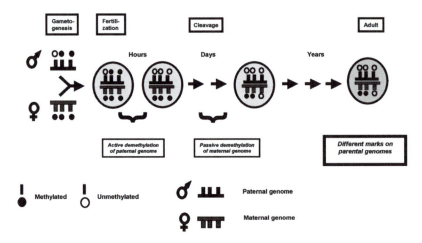

3B: In cloning a somatic nucleus is transferred into the enucleated egg and *both* parental genomes are exposed to the active demethylating activity of the egg cytoplasm. Therefore, the parent specific epigenetic differences are equalized.

Fig. 3B

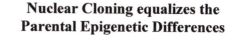

Nuclear Cloning equalizes the Parental Epigenetic Differences

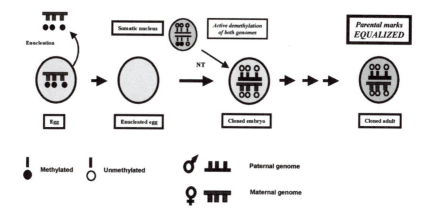

Fig. 4: Scheme for therapeutic cloning combined with gene and cell therapy.

A piece of tail from a mouse homozygous for the recombination activating gene 2 (Rag2) mutation was removed and cultured. After fibroblast-like cells grew out, they were used as donors for nuclear transfer by direct injection into enucleated MII oocytes using a Piezoelectric driven micromanipulator. Embryonic stem (ES) cells isolated from the NT-derived blastocysts were genetically repaired by homologous recombination. After repair, the ntES cells were differentiated *in vitro* into embryoid bodies (EBs), infected with the HoxB4iGFP retrovirus, expanded, and injected into the tail vein of irradiated, Rag2-deficient mice (after (Rideout et al., 2002)).

Fig 4 Correction of a Genetic Defect by Combination of Therapeutic Cloning and Gene Therapy

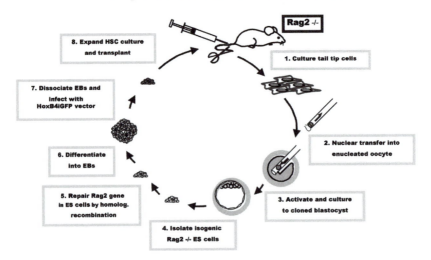

Fig. 5: Blastocysts retain epigenetic memory of donor nucleus

Blastocysts can be derived from the fertilized egg or by nuclear transfer. After implantation development of the embryo strictly depends on the donor nucleus: Blastocysts derived from a fertilized egg will develop with high efficiency to *normal* animals; blastocysts derived by NT from an ES cell donor will develop with high efficiency to *abnormal* animals; blastocysts derived by NT from a fibroblast or cumulus cell donor will develop with low efficiency to *abnormal* animals; blastocysts derived by NT from B or T donor cells will not develop to newborns by direct transfer into the womb (only by a 2 step procedure, compare (Hochedlinger and Jaenisch, 2002a).

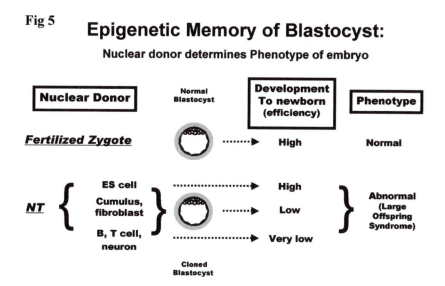

Fig 5
Epigenetic Memory of Blastocyst:
Nuclear donor determines Phenotype of embryo

Fig. 6: The establishment of ES cells from blastocysts erases epigenetic memory of donor nucleus

Most cells of the inner cell mass turn off Oct-4 like genes and die after explantation of blastocysts into tissue culture. Only one or a few cells turn on the Oct-4 like genes and proliferate. The surviving cells will give rise to ES cells. During this highly selective outgrowth of the surviving cells all epigenetic memory of the donor nucleus is erased. Therefore, regardless of donor nucleus (fertilized egg or somatic nucleus in cloned embryos), all ES cells have an equivalent potency to generate functional differentiated cells.

Fig 6

In vitro Selection of ES cells:

Erasure of Epigenetic Memory

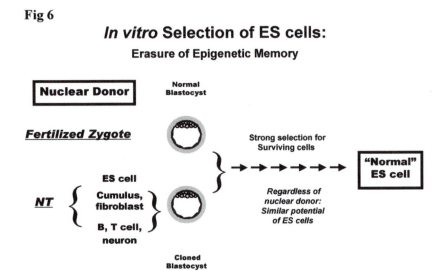

Fig. 7: Normal and cloned embryos have three possible fates

Embryos derived by IVF ("left over embryos") have three fates: they can be disposed, create *normal* babies if implanted or can generate ES cells if explanted into tissue culture. Cloned embryos have also three fates: they can be disposed, can generate *abnormal* babies if any when implanted or can generate ES cells when explanted. The ES cells derived from an IVF embryo or a cloned embryo are indistinguishable (same potency, see figure 6)

Fig 7

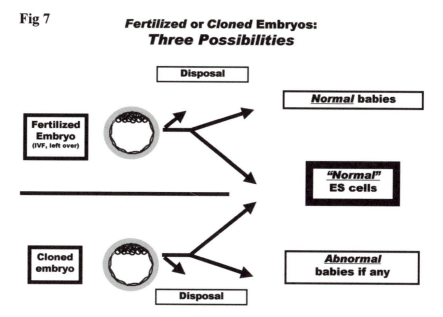

Fertilized or *Cloned* **Embryos:**
Three Possibilities

Table 1

Donor nucleus		Mice (% of blastocysts)	Phenotype	Ref.
Fertilized zygote		30 - 50 %	Normal	
Nuclear transfer from	ES cell	15 - 30 %	Most if not all clones are abnormal	1
	Cumulus cell, fibroblast	1 - 3 %		2
	B, T cell	< 1/3000		3

Development of normal embryos and embryos cloned from ES cell and somatic donor nuclei. Note that normal and ES cell derived blastocysts have a similar potency to develop to term if calculated from the fraction of transplanted blastocysts.

1: (Eggan et al., 2001; Eggan et al., 2002; Rideout et al., 2000); 2 (Wakayama et al., 1998; Wakayama and Yanagimachi, 1999); 3 (Hochedlinger and Jaenisch, 2002a).

Acknowledgements

I thank my colleagues Bob Weinberg, Gerry Fink, George Daley and Andy Chess for critical and constructive comments on this manuscript.

ENDNOTES

[1] Reprogramming: The genome of a somatic cell is in an epigenetic state that is appropriate for the respective tissue and assures the expression of the tissue specific genes (in mammary gland cells, for example, those genes important for mammary gland function such as milk production). In cloning, the somatic nucleus must activate those genes that are needed for embryonic development but which are silent in the donor cell in order for the cloned embryo to survive. The egg cytoplasm contains "reprogramming factors" that can convert the epigenetic state (see endnote

3) characteristic of the somatic donor nucleus to one that is appropriate for an embryonic cell. This process is very inefficient leading to inappropriate expression of many genes and causes most clones to fail early.

[2] Imprinted genes: For most genes, both copies, the one inherited from father and the one inherited from mother, are expressed. In contrast, only one of the two copies of an imprinted gene, either the maternal one or the paternal one, is active. The two copies are distinguished by methylation marks (see endnote 4) that are imposed on imprinted genes either during oogenesis (maternally imprinted genes) or during spermatogenesis (paternally imprinted genes). Thus, the two copies of imprinted genes are epigenetically different in the zygote and remain so in all somatic cells. These epigenetic marks distinguish the two copies and cause only one copy to be expressed whereas the other copy remains silent. It is estimated that between 100 and 200 genes (of the total of 30,000 genes) are imprinted. Disturbances of normal imprinted gene expression lead to growth abnormalities during fetal life and can be the cause of major diseases such as Beckwith-Wiedeman or Prader-Willi syndrome.

[3] Epigenetic changes: Cells of a multicellular organism are genetically identical but express, depending on the particular cell type, different sets of genes ("tissue specific genes"). These differences in gene expression arise during development and must be retained through subsequent cell divisions. Stable alterations of this kind are said to be "epigenetic", as they are heritable in the short term (during cell divisions) but do not involve mutations of the DNA itself.

[4] DNA methylation: Reversible modification of DNA (methylation of the base cytosine) that affects the "readability" of genes: usually, methylated genes are silent and unmethylated genes are expressed. DNA methylation represents an important determinant of the "epigenetic state" of genes and affects the state of the chromatin: methylated regions of the genome are in a "silent" state and unmethylated regions are in an "open" configuration that causes genes to be active.

References

Bortvin, A., Eggan, K., Skaletsky, H., Akutsu, H., Berry, D. L., Yanagimachi, R., Page, D. C., and Jaenisch, R. (2003). Incomplete reactivation of Oct4-related genes in mouse embryos cloned from somatic nuclei. Development *130*, 1673-1680.

Buehr, M., Nichols, J., Stenhouse, F., Mountford, P., Greenhalgh, C. J., Kantachuvesiri, S., Brooker, G., Mullins, J., and Smith, A. G. (2003). Rapid loss of oct-4 and pluripotency in cultured rodent blastocysts and derivative cell lines. Biol Reprod *68*, 222-229.

Byrne, J. A., and Gurdon, J. B. (2002). Commentary on human cloning. Differentiation *69*, 154-157.

Cibelli, J. B., Campbell, K. H., Seidel, G. E., West, M. D., and Lanza, R. P. (2002). The health profile of cloned animals. Nat Biotechnol *20*, 13-14.

Coalition, and Ethics, A. f. R. (2003). Do no harm - Reality check: proof of "therapeutic cloning"? wwwstemcellresearchorg/pr/pr_2003-03-10htm.

Eggan, K., Akutsu, H., Loring, J., Jackson-Grusby, L., Klemm, M., Rideout, W. M., 3rd, Yanagimachi, R., and Jaenisch, R. (2001). Hybrid vigor, fetal overgrowth, and viability of mice derived by nuclear cloning and tetraploid embryo complementation. Proc Natl Acad Sci U S A *98*, 6209-6214.

Eggan, K., Rode, A., Jentsch, I., Samuel, C., Hennek, T., Tintrup, H., Zevnik, B., Erwin, J., Loring, J., Jackson-Grusby, L., *et al.* (2002). Male and female mice derived from the same embryonic stem cell clone by tetraploid embryo complementation. Nat Biotechnol *20*, 455-459.

Ferguson-Smith, A. C., and Surani, M. A. (2001). Imprinting and the epigenetic asymmetry between parental genomes. Science *293*, 1086-1089.

Giles, J., and Knight, J. (2003). Dolly's death leaves researchers woolly on clone ageing issue. Nature *421*, 776.

Gurdon, J. B. (1999). Genetic reprogramming following nuclear transplantation in Amphibia. Semin Cell Dev Biol *10*, p239-243.

Hochedlinger, K., and Jaenisch, R. (2002a). Monoclonal mice generated by nuclear transfer from mature B and T donor cells. Nature *415*, 1035-1038.

Hochedlinger, K., and Jaenisch, R. (2002b). Nuclear transplantation: Lessons from frogs and mice. Curr Opin Cell Biol *14*, 741-748.

Hochedlinger, K., and Jaenisch, R. (2003). Nuclear transplantation, embryonic stem cells, and the potential for cell therapy. New England Journal of Medicine *349, in press* (July 17, 2003).

Hubner, K., Fuhrmann, G., Christenson, L. K., Kehler, J., Reinbold, R., De La Fuente, R., Wood, J., Strauss, I. J., Boiani, M., and Scholer, H. R. (2003). Derivation of Oocytes from Mouse Embryonic Stem Cells. Science.

Humpherys, D., Eggan, K., Akutsu, H., Friedman, A., Hochedlinger, K., Yanagimachi, R., Lander, E., Golub, T. R., and Jaenisch, R. (2002). Abnormal gene expression in cloned mice derived from ES cell and cumulus cell nuclei. Proc Natl Acad Sci U S A *99*, 12889-12894.

Humpherys, D., Eggan, K., Akutsu, H., Hochedlinger, K., Rideout, W., Biniszkiewicz, D., Yanagimachi, R., and Jaenisch, R. (2001). Epigenetic instability in ES cells and cloned mice. Science *293*, 95-97.

Jaenisch, R., and Wilmut, I. (2001). Developmental biology. Don't clone humans! Science *291*, 2552.

Kyba, M., Perlingeiro, R. C., and Daley, G. Q. (2002). HoxB4 confers definitive lymphoid-myeloid engraftment potential on embryonic stem cell and yolk sac hematopoietic progenitors. Cell *109*, 29-37.

Lanza, R. P., Cibelli, J. B., Faber, D., Sweeney, R. W., Henderson, B., Nevala, W., West, M. D., and Wettstein, P. J. (2001). Cloned cattle can be healthy and normal. Science *294*, 1893-1894.

Mayer, W., Niveleau, A., Walter, J., Fundele, R., and Haaf, T. (2000). Demethylation of the zygotic paternal genome. Nature *403*, 501-502.

Oback, B., and Wells, D. (2002). Donor cells for cloning-many are called but few are chosen. Cloning Stem Cells *4*, 147-168.

Ogonuki, N., Inoue, K., Yamamoto, Y., Noguchi, Y., Tanemura, K., Suzuki, O., Nakayama, H., Doi, K., Ohtomo, Y., Satoh, M., *et al.* (2002). Early death of mice cloned from somatic cells. Nat Genet *30*, 253-254.

Prentice, D. (2002). Why the "Successful" Mouse "Therapeutic" Cloning Really Didn't work. wwwcloninginformationorg/info/unsuccessful_mouse_therapyhtm.

Reik, W., Dean, W., and Walter, J. (2001). Epigenetic reprogramming in mammalian development. Science *293*, 1089-1093.

Rideout, W. M., 3rd, Hochedlinger, K., Kyba, M., Daley, G. Q., and Jaenisch, R. (2002). Correction of a genetic defect by nuclear transplantation and combined cell and gene therapy. Cell *109*, 17-27.

Rideout, W. M., Eggan, K., and Jaenisch, R. (2001). Nuclear cloning and epigenetic reprogramming of the genome. Science *293*, 1093-1098.

Rideout, W. M., Wakayama, T., Wutz, A., Eggan, K., Jackson-Grusby, L., Dausman, J., Yanagimachi, R., and Jaenisch, R. (2000). Generation of mice from wild-type and targeted ES cells by nuclear cloning. Nat Genet *24*, 109-110.

Tamashiro, K. L., Wakayama, T., Akutsu, H., Yamazaki, Y., Lachey, J. L., Wortman, M. D., Seeley, R. J., D'Alessio, D. A., Woods, S. C., Yanagimachi, R., and Sakai, R. R. (2002). Cloned mice have an obese phenotype not transmitted to their offspring. Nat Med *8*, 262-267.

Wakayama, T., Tabar, V., Rodriguez, I., Perry, A. C., Studer, L., and Mombaerts, P. (2001). Differentiation of embryonic stem cell lines generated from adult somatic cells by nuclear transfer. Science *292*, 740-743.

Wakayama, T., Whittingham, D. G., and Yanagimachi, R. (1998). Production of normal offspring from mouse oocytes injected with spermatozoa cryopreserved with or without cryoprotection. J Reprod Fertil *112*, 11-17.

Wakayama, T., and Yanagimachi, R. (1999). Cloning of male mice from adult tail-tip cells. Nat Genet *22*, 127-128.

Wakayama, T., and Yanagimachi, R. (2001). Mouse cloning with nucleus donor cells of different age and type. Mol Reprod Dev *58*, 376-383.

Wilmut, I. (2001). How safe is cloning? Cloning *3*, 39-40.

Young, L. E., Sinclair, K. D., and Wilmut, I. (1998). Large offspring syndrome in cattle and sheep. Rev Reprod *3*, p155-163.